WATER

Resource Augmentation, Management and Policies

WATER

Resource Augmentation, Management and Policies

H. Sarvothaman

ASIATECH PUBLISHERS, INC.
New Delhi

ASIATECH PUBLISHERS INC.
New Delhi

ISBN 81-87680-13-X

Published by N.K. Muraleedharan for Asiatech Publishers Inc.
E-20, Arya Nagar, CGHS, Plot No. 91, I.P. Extension, Patparganj, Delhi 110092.
Ph.: +91-11-22724601 Fax: +91-11-22724601/22727254 • e-mail: asiatech@bol.net.in

Typeset by Printplus, F/47, Madhu Vihar, Patparganj, Delhi-110092
Printed at Chaman Enterprises, 1603, Pataudi House, New Delhi 110 002.

This book is dedicated to those people of my country, who genuinely cannot afford to buy the water to drink.

Anna University
Chennai 600025 India

Phone +91-44-22351445
Fax +91-44-22350397
E-mail vc@annauniv.edu

E. BALAGURUSAMY
ME (Hons) Ph D RE FIETE C Eng
Vice Chancellor

FOREWORD

The richest possession that a country can boast of in the twenty-first century is its replenishable water resources. A sizable population of India has attained this awareness in the beginning of this century, if not a few years earlier, on account of the extreme dry weather during prolonged summers and weak monsoons in that period. A few other countries that are not fortunate enough to have adequate replenishable source of water have been gradually resorting to the mechanical means to augment their water wealth. These countries reap the dual benefit of generating adequate water for their needs and demonstrating their technological capabilities. The economic growth achieved by such countries by augmenting the water resources is so stupendous that other countries attempt to emulate them. Now it is India's turn to do so.

In the Indian scenario, a sharp divide exists between those States, which have abundant water resources and those that do not possess adequate water to meet the demands. The attempts to transfer the surplus water from a State to the water-deficient States are fraught with the geographic and socio-political constraints. Weak monsoons and truant rainfall make their own contribution to the constraints. Despite all these constraints, we need to tide over the water crises year after year and provide water to our people and their activities such as housing, agriculture and industries, besides bacteria-free drinking water.

Water resources are national wealth and every citizen has an equal right on it. The efforts to empower this right to its citizens and the management of water resources are already proving to be major issues to the governments. The water supply authorities and numerous other organizations set up both by the Central and the State Governments to look after the equitable water distribution to various segments of the human activities such as irrigation and industries are grappled with the problem of water inadequacy. The deficit mounts year after year on account of consistent overdraft of water vis-à-vis insufficient replenishment; the gap between the two widens in the event of the passage of each year of deficit rainfall. The Indian experience in terms of the management of water resources demonstrates that water-equity is difficult for various reasons, the main reason being insufficient annual accretion to the water resources due to the vagaries of the Nature. Should we learn from the experience, our prudence lies in augmentation of our water resources using technological means.

Treatment of hard water to potable levels, wastewater treatment to recycle the municipal and industrial wastewater, and desalting of seawater are proven technologies in augmenting the water resources. The recyclable treated wastewater is an ideal substitute in those human activities in which the quality of water is not critical. Wastewater treatment plants ensure not only addition of usable water for supporting human activities, but also flow of clean influent water into the inland water bodies, thus preserving the sanctity our rivers and lakes. Besides, few desalination plants on the east coast of India would ease the stress on the availability of water in the coastal cities that are plagued by water scarcity and ingression of seawater in the coastal aquifers. This book logically discusses these implications and offers a food for thought to all those concerned in the water resource management in the country.

The author has extensively traveled all over the country to collect the ground truth on the water resources and gather original inputs for this book. The management practices in respect of water resources as discussed in this book are bound to make tremendous impact, though it would need initial investments. The country, of course, is not starved of funds for crucial matters such as water resource augmentation. There is no dearth of talents either in India to invent, innovate and install the facilities for water resource augmentation techniques such as the municipal wastewater plants, filtration plants and membrane-based reverse-osmosis plants to desalinate the seawater. What we need are committed entrepreneurs, civil and mechanical engineers and a large labour force evenly distributed in the whole country to carry out the mission. Above all, it is the will of those who matter that matters most. It is hoped this book will give that spark that is necessary to ignite the thoughts into action in respect of the augmentation and management of the country's water resources.

(E Balagurusamy)

PREFACE

A nation's water resource is the richest wealth it can be proud of, apart from its heritage, language and culture. The distribution of this wealth in different parts of the Planet Earth is skewed and unequal. Some terrain segments are entirely deprived of water for various reasons. The paucity of the water wealth has affected roughly half of the world population one way or other. This population devotes considerable time and efforts daily to carry water to their homes from long distances. As if it is not enough, the very same water that they bring home after so much of labour often inflicts a host of diseases such as diarrhoea, gastroenteritis, worm infestation and fluorosis to them and their offspring, because they collect this water from a source that is neither protected nor treated. Sometimes that water gets infected during storage. There is no precise current estimate on how many people world over perish daily due to the non-availability of protected water and water-linked infections, but a 1992 statistics say that as much as 25,000 children alone die every day due to water-borne diseases. This rate of mortality is growing with each passing year, and reports of water-related mortality are coming in from those countries where they were not reported earlier. Such statistics and bare facts are indeed depressing, and are the root cause for this book.

I had no idea of water scarcity in my early phase of life. I realized that the water scarcity was a reality as late as around 1970 through the print medium, as there were reports that several districts in Tamil Nadu (such as Ramnathapuram and Tirunelveli) and Rajasthan's desert districts faced water crisis. Between the years1976 and 1985, I had the privilege of ferrying across the mighty Brahmaputra River several times at different places in the post-monsoon period. I was awe-struck by the vastness of the Brahmaputra River near Guwahati, Tezpur and Dibrugarh. I wishfully thought that if there were one such river in every State of India, there would not be water scarcity in the whole country, and that its water would immensely benefit many of our cities and towns. In December 2002, in a scientific meet held at Hyderabad on the theme of Remote Sensing, a young remote-sensing scientist based at Guwahati had told me that he was keen to learn remote-sensing techniques in sustainable development of water resource, as he was trying to find solutions to augment water resources in and around Guwahati. He said that several areas in Guwahati face water scarcity at regular frequency in recent times. The information about the scarcity of water in Guwahati, which is situated right at the left bank of mighty Brahmaputra River, stunned me.

In the early-1990s, the World Bank functionaries have observed that if the Indian government did not formulate a sensible Water Policy and also take concrete steps to conserve and develop water resources, India might have to face serious problems in economic slow down and downward trends in the agriculture and industry sectors. Ironically, after a decade, this statement has relevance not only to India, but to the whole world as well. There are instances in many parts of the world, including India, that despite availability of sufficient water and also there is scope to generate potable and/or recyclable water by mechanical means, major population in these countries faces extreme water deficiency. With prudent management of the available water resources and transfer of technologies to the society-related causes and to find practical solutions to augment water resources, such hardships to the masses of the population can be effectively mitigated. This precisely is the central theme of this book.

Dr. A.P.J. Abdul Kalam, the Honourable President of India mentions in his book 'INDIA 2020 A Vision for the New Millennium' that India *'should concentrate on development of key areas, namely agriculture production, food processing, materials and also on the emerging niche areas like computer software, biotechnologies...'*. These should currently be the thrust sectors in the country's activity, which would bring major transformation in Indian economy as also give an impetus to the economy to compete at global levels. Success in two prime sectors identified by Dr. Abdul Kalam, namely Agricultural Production and Food Processing, depends primarily on the availability of large quantities of good quality water. Whereas agriculture in most of the lands is rainfed, the water received from the monsoon source will have to be allocated for irrigation, in which case, the monsoon source water will be scarce for use in other human activities and industry-oriented purposes. The sustainable alternative to augment the water resources is to bank on technologies and start making small but continuous investments annually, both in terms of finances and human resources, in the technological means to water resource augmentation. This exercise should go on at least till 2025, even while we attempt prudent management of the annually replenishable water resources.

If the summer of 2002 in India was bad in terms of availability of water for daily needs of the people and for agriculture, the summer that succeeded was even worse. Those in the governments, the technologists, entrepreneurs, engineers, policy-makers, agriculturists and general public hoped during those two summers that the monsoon rains would bring plenty of rains and prosperity. Unfortunately, not much groundwork has even been initiated in the country to employ water augmentation technologies in those two difficult years, despite deficient rainfall in many terrain segments of the country. An Austrian mining consultant incredulously asked me in 1988: *"Water scarcity in India? In spite of long sea coasts?"* During the summer of 2003, these questions kept haunting me repeatedly, and gave me the spark to write this book.

The processes and technologies that are mentioned in this book are not new. Then, why is this book? Volumes of details are entrenched in literature, giving even the minute information on various technologies, their comparative advantages, plant dimension and so on. Such information is in bits and pieces, and probably, is not available in one single bunch. This book attempts achieve that objective by offering overviews of those technologies. At the same time, this book does not answer all aspects on the technologies highlighted. It is recommended that the readers may gain additional insights into the highlighted processes by directly referring to the quoted references.

Our educational institutions train few hundred thousands of technologists and engineers each year with the hope that they would make innovations in various technologies. Currently, the thrust is on the technologies related to Water Resource Augmentation. Through this book, I draw the attention of civil and mechanical engineers to invent, to innovate the designs to suit Indian conditions and affordability, and to install various water technology plants mentioned in this book. I urge all those in the governments to draw sustainable policies, both in terms of augmentation and management of water wealth. The entrepreneurs and investors too can benefit if they recognize water-augmentation technologies and plants as potential new system of wealth creation, both in terms of water resource, and the wealth that can be ploughed back into the country's economy, through the newly generated water resources. I emphasize that every citizen of a country has something to offer to the management of water resources, because management originates at an individual or grass-root level.

June, 2004 **H. Sarvothaman**

ACKNOWLEDGEMENT

I offer this book as a tribute to my mother Saraswathi, who merged into the Eternity with the Lord, a few days before it took final shape.

Mr. N. K. Muraleedharan of Asiatech Publishers, Inc. has been the main source of inspiration and sustenance to me from beginning to the end of writing this book. He had reviewed several chapters and offered suggestions for the approach and improvement in the contents of this book. Completion of this work was possible because the members of my family created congenial atmosphere for writing. Everything was nice at home, and that gave me focus. I had come across some fine persons in the Hyderabad Metropolitan Water Supply & Sewerage Board while pursuing the wastewater treatment study. They are: Mr. M.G. Gopal, I.A.S., Chairman & Managing Director, Mr. P. S. Suryanarayana, General Manager and Mr. Gopal Hazra, Foreman of the Board's Wastewater Treatment Plant. Discussions with Mr. M. Subrahmanyam, Managing Director of Nikhileshwari Consultants and In-Charge of the Board's Wastewater Treatment Plant's operations, about the Hussain Sagar Plant and wastewater treatment technology in general were enriching. Mrs. Ishrath and Prof. Sirish Chandra of the Osmania University, Hyderabad had provided their service with a smile in regeneration of the figures from rough drafts with the aid of computers. Dr. E. Balagurusamy, Vice-Chancellor, Anna University, Chennai, and Dr. B. Sushila Balagurusamy, who have distinguished themselves as writers, and my wife Dr. S. Kasthuri have induced me into writing this book by offering their positive strokes. I thank Mr. Aslam Parvez and his team of Printplus for the drawing and designing.

I had sought Lord Shiva's Blessings all through the writing.

CONTENTS

Chapter 1

This chapter highlights:

- Distribution of water on the surface of the Earth.
- Evolution of water supply; and, water resource management through history.
- Scenario of drinking water in India.
- A few approaches to pragmatic water policy.

INTRODUCTION

Planet Earth is the only place in the Solar System known to have enormous quantities of water, and that trait makes the Earth singularly unique. The existence of a *'sphere of life'* known as 'biosphere' or 'ecosphere' on the surface of the Earth is solely attributed to the presence of water. The forests, the meadows, the wild trees, the prairies, the seaweeds, the algae, the wide variety of fauna ranging from protozoa to whales, planktons to fish, corals to nautilus, and human beings to wild life, are parts of the Earth's biosphere. Oceans cover an area of approximately 361 million square-kilometre or 70.78% of the 510 million square-kilometre total area of the Earth. On the other hand, the landmasses of the five continents account for the rest of the 149 million square-kilometre or 29.22% of the Earth's total area. Biosphere covers both ocean and landmass alike. Humanity, on a global scale, is an extremely tiny constituent of the biosphere, as it survives only in the continental part of the biosphere.

Unlike various other natural resources of the Earth that are not renewable or permanently spent when used (like coal and minerals), water is a renewable resource since it is replenished on the Earth's surface by Hydrologic Cycle. The total quantity of water on the Earth is huge and its volume is *roughly* estimated as a whopping one-and-a-half billion cubic kilometers (Stern, 1989). Water holds the key to life, sustains life and, therefore, water and life are intimately intertwined in the universe. Unfortunately for the predominant constituents of biosphere, which depend on freshwater for their survival, 95% of this water is stocked as saltwater in oceans and seas. Only 5% of the water resource of the Earth is freshwater. The 149 million square-kilometre area or 29.22% of landmass constituting five continents cumulatively possesses this 5% freshwater stock. To add to the misery of living beings in the continental biosphere, about 4 out of the 5 percent freshwater stock is estimated as frozen freshwater located at the polar region and high-altitude areas. Effectively therefore only a meager about 1% of the total water resources of the Earth is available as liquid freshwater to support life on the continental biosphere.

Distribution of water on the Earth

It would be even more interesting to visualize the approximate volume and pattern of distribution of this 1% freshwater on the surface of the continents. There is a continuous passage of water to the tune of an estimated 520,000 cubic-kilometres in the form of vapours originating from oceans and continental landmasses (Stern, 1989). Nearly 108,000 cubic-kilometre volume of water comes back as direct precipitation of rainfall on the land surfaces and the rest 412,000 cubic-kilometres of water vapours are precipitated on the surface of oceans and seas. An estimated 71,000 cubic-kilometres out of the 108,000 cubic-kilometre volume of water received by the landmasses undergoes evaporation and transpiration, and transferred back into the atmosphere. It is the reminder of the approximately 37,000 cubic-kilometre volume of freshwater that fills the rivers and lakes, replenishes the groundwater by percolation and supports the plant and animal life on the continents. A considerable part of this freshwater gets drained into the oceans and seas by surface runoff, and nearly 0.001% is confined in the river system. About 0.002% of this freshwater saturates the soil cover and supports the agricultural needs of humanity, forests and the entire eco-system.

Incidentally, the 37,000 cubic-kilometre volume of freshwater is not evenly distributed on the landmasses of the Earth; its distribution is highly skewed. The distribution of freshwater in different parts of the world depends solely on the rainfall, which in turn is linked to the climate, geographic location, terrain condition and the wind system that navigates the atmospheric water vapours. The current world population of over 6.1 billion scattered all over the continental biosphere has to be contented with whatever share of water it gets in the part of the world or the country in which it lives. Not only that, this meager water resource has to be shared by the human population with the other living creatures in the eco-system.

Earth's biophysical environment is a complex system driven by the solar energy as well as the Earth's surfacial structures created by its own internal dynamics. Human beings, like any other species of the animal kingdom with which they share the Planet Earth, are intricately linked to this complex system. This system provides basic essentials—*air, water* and *food*—for survival of all. No wonder then that most of the intelligent ancient civilizations of the world flourished along the banks of perennial rivers, to take full advantage of the copious water available round the year, irrespective of the seasonal changes. There were ancient Indian civilizations on the banks of Indus, Ganges and the now extinct Saraswati Rivers. Euphrates River together with its tributary Tigris River was the cradle of some of the ancient civilizations in Middle East. Nile River had supported the ancient Egyptian civilization. There existed an ancient civilization at the banks of Yellow River in China. The ruins of Indus Valley Civilization (at Harappa and Mohanjadaro) and of Babylon, Sumer and Ur situated on the Euphrates River banks are testimony to the pivotal role played by the rivers in the lifestyle of those ancient civilizations.

Water resource management through ages

The needs of the people who belonged to the ancient civilizations were just basic—*air, water* and *food*—provided by the said complex system of the Earth. Those civilizations lived in perfect harmony with the Nature, worshipped it and depended solely on it for their basic needs. The Nature on its part had taken care of them and their needs for several generations. The significant aspect of the ancient civilizations is that they were *not* controlled by money, as money was not the main tool of social control and the people of those civilizations were not driven by the monetary system. History allows us to infer that the subjects of these ancient civilizations had to grow foodgrain by themselves for their consumption, near the riverbanks using the river water, apart from making their shelter and weaving cloths for themselves.

The '*grey matter*' in the human brains has a unique capability of constantly innovating ways and means for speedy acquisition of comforts and also to improve the quality of his or her life. Using this capability and also aided by the impulses from the '*selfish genes*' to succeed in survival, a fittest descendents among the progeny of the ancient civilization, had invented the idea of trade or business and came out of the routine life of growing foodgrain. Gradually, as the civilization became aware of the trade practices, the traders had resorted to bartering their trade items for foodgrain. The items traded in those times were gold, copper, gemstones, salt, animal hides, cotton and cloth. This segment of the society, under compulsion from their trade practices, had to live away from rivers, and therefore, the water and foodgrain had to be transported by them to the places where they lived and conducted their trade practice. That perhaps was the first time in human history when water was transported from rivers to the places farther from rivers, which a small part of the world's population adopted as their habitats.

Apart from the population belonging to the ancient civilizations at the riverbanks, there were several ancient races of aboriginal people in India, Africa and Australia, living away from major perennial rivers. In India, some of the notable races among them lived in areas far South (south of Narmada River) currently constituting areas in Tamil Nadu, Kerala, Andhra Pradesh, Karnataka, Madhya Pradesh and Maharashtra. Though it cannot be said with certainty that these aboriginals had actually invented the practice of collecting, using and preserving the rainwater for their needs by impounding it in tanks, there are evidences for this in several parts of India. Besides, the indulgent life pattern of some segments of the early and medieval civilizations under compulsion from trade practices had led to large groups of traders settling down to live in the then unconventional locations farther from rivers. These traders had to look for alternative sources of water for their survival, and consequently, commercialisation of lands laden with water resource had begun. Gradually, many of these traders had acquired vast stretches of land and became powerful landlords, satraps and kings who wielded limitless authority over the local aboriginal races.

The ancient Indian subcontinent was a crucible in which several new ethnic groups were generated by constant intermingling between the militant races that had arrived through the Khyber Pass, the existing civilizations and the aboriginal races. India has often been described as an 'ethnological museum' in which countless races of mankind converged, propagated and coexisted in groups ranging from savages to learned philosophers (Smith, 1958). The variety of races, ethnic groups, languages and customs is largely the cause of innumerable kingdoms and/or political sub-divisions characterizing Indian history from ancient times. Megasthenes had reported 118 kingdoms in the Indian subcontinent in the 4th Century B.C, but the actual number might have been much more. In the later centuries, several satraps had set up more and more kingdoms in different parts of the subcontinent that had three distinct sub-divisions namely, the Indo-Gangetic plains, the Deccan and the far South. Keen students of early and medieval history of India would realize that these satraps or kings had acquired vast stretches of lands by imposing their hegemony or by commerce or by mutiny. Taking cue from the aboriginals, the practice of collecting runoff rainwater by impounding it in small tanks and preserving it for a later use was patronized by the traders-turned-satraps and –kings, with a view to develop the areas under their influence and generate funds by agricultural means to meet the expenses towards ruling their territory. Some of those kings or satraps had taken masses of aboriginal and local population as slaves and forced them into developmental work such as construction of forts, stables, and also water reservoirs and check-dams for impounding runoff rainwater. Over centuries starting from ancient period through medieval times, in Deccan and far South regions of India, innumerable reservoirs, bunded tanks and lakes were constructed this way.

The ancient practice of rainwater collection, preservation in reservoirs and its use for irrigation and other day-to-day human activities are human beings' first form of sustainable water resource management in areas other than perennial river basins. Further, this age-old practice of India had sown seeds for what is known in modern terminology as 'watershed management' or 'watershed development'. This practice seemed to have originated in Deccan and South India, but its precise antiquity is *perhaps* not properly chronicled in Indian history, as it was a routine way of life in India from time immemorial and handed down to the generations next and the subsequent. Whereas this inexpensive and eco-friendly practice is still in vogue in several States of modern India, the intensity of its practice has certainly diminished, and it has resulted in substantial reduction in the stock of water resource. The current awareness in India of its utility and pitched pleas by the government authorities for the revival of this practice are welcome signs towards sustainable water resource management and positive step towards formulation of a meaningful Water Policy, provided the painstakingly formulated policy is implemented with seriousness and sincerity, and those in authority who implement the Water Policy exhibit unquestionable character ethics.

The concept and use of groundwater from open wells can be traced to more than the 4000-year-old Vedic period in India. In other countries, as Todd (1959) observed, "*utilization of groundwater greatly preceded understanding of its origin, occurrence and movement*". In the 'age of agriculture' or First Wave of civilization as Toffler (1990) puts it, all the countries of the world attempted to boost their economy by means of agriculture. More and more areas were brought under cultivation to increase the agricultural output. The use of groundwater for irrigation was resorted to in most of the countries. Thus, globally, overexploitation of two types of water resource—soil-water and groundwater—had resulted.

In India also, overexploitation of soil water, surface runoff and groundwater had taken place as a result of the pressure of providing food to the steadily growing population, Green Revolution and the dependence on agricultural profession by the majority of our population. In the last half a century, exploitation of the groundwater stock had far exceeded replenishment, and consequently the stock of groundwater had depleted substantially all over the country. As a remedy, a concept known as '*sustainable agriculture*', whereby inputs to agriculture such as water and soil nutrients are optimally used, is developed in current times. Sustainable agriculture is the only way to relieve the stress created by the agriculture segment on the groundwater stock, and therefore, salient aspects of sustainable agriculture are discussed in this book.

At the dawn of the Industrial Era or the Second Wave of modern civilization, there was frenzied movement of rural population to the industrial belts all over the world, and India was no exception. Industries of a wide variety had sprung up everywhere and it was a welcome sign for a growing economy as well as attempts to meet the aspirations and fight unemployment amidst the growing educated population of a young republic. Further, there was a necessity to keep pace with the development in all spheres and compete with global industrial development. More and more farmlands were converted into industrial belts and more and more people started living in colonies to observe punctuality in attendance to work as well as boost industrial production. There were new coal and gold mines, steel industries, oil refineries, automobile manufacturing and bodybuilding units, fertilizer and chemical plants, and industrial units that manufactured matchbox, petrochemicals and railway coaches. Gradually, India had started indigenously manufacturing a few import-substitute products too. Almost all those required by a developing country for self-sufficiency!

These positive aspects were not without negative fallouts. The owners of the industrial units, be it government or private entrepreneurs, felt the obligation to provide water to the colonies where the workers lived. As the industrial units were located at places far away from water bodies such as rivers, reservoirs and tanks, water had to be transported and supplied to the workers at or near their residence or workplace. The dawn and progress of the industrial age, thus, was the dawn of yet another (besides the ongoing industrial revolution) unnoticed revolution known in current times as the organized '*Water Supply*'. This indeed was a revolution because this was the dawn of the concept of organized provision of water to a section of population by transporting it from the water bodies through pipes of large diameter. Until then, the responsibility of fetching water from water bodies for use of the family entirely lay on individuals or family-heads, especially the womenfolk of the family.

A similar trend can be traced in the case of development and proliferation of the urban sprawls, which have become the hubs of bureaucratic, political, educational, cultural and a host of other human activities. Town planning departments, municipal corporations and municipalities were created at a certain stage of governance by the States, with specific purposes of regulating planned development and providing civic amenities in the urban centres. Consequently, the concerned governments assigned the responsibility of water supply to the subordinate bodies on which they exercised direct control. Further, the responsibility of an organized water supply was undertaken by the local governments since the Government of India had not only assumed the responsibility of providing *'protected water supply'* to its citizens through these local bodies, but also had been allocating funds in each Five Year Plan towards accomplishing that objective.

As an offshoot to the responsibility of providing the civic facility of water supply undertaken by the governments and their affiliated civic bodies such as municipal corporations and municipalities, the humanity had witnessed an innovative technique known as *'Piped Water Supply'*. This technique points to the process that begins from treatment and transportation of water from a reservoir through huge pipes or conduits, to supplying the treated water to the individual residences through taps connected to metal pipes of smaller diameter, which in turn, are linked to the main water supply pipes. The evolution of piped water supply in India was slow in the beginning, but had taken rapid strides subsequently. Today, professional plumbers of India engaged in providing pipelines to the individual residential buildings and linking them to the main water supply are some of the best in the world.

The 'piped water supply' to the individual residential units has its own merits and shortcomings. The citizens now get the water with so much ease for his or her daily needs just by opening a tap that they are not even aware of the difficulty faced by those providing such a facility, say the Water Supply Department. Hardly anyone realizes the arduous methods adopted by the civic departments in the treatment, upgradation of raw water to potable level, preservation of the water as 'protected water' till it is supplied to the citizens as drinking water, and the cost factor in the whole process of supplying this 'piped water'. The greatest disadvantage of the piped water supply undertaken by the civic authorities as a social responsibility is that this service is grossly abused and misused by a majority of the citizens, without realizing that they have a responsibility towards judicious use of the civic amenity provided to them. They squander away the valuable treated and protected water meant for direct oral consumption and for cooking, and misuse it for marginal human activities, such as washing their automobiles, watering the garden, flushing toilets and building construction. The citizens have also started realizing that the moment the urban planning or municipal authorities sanction their house plans after collecting the requisite fees, the responsibility of providing piped water supply to them is entirely that of the water supply authorities. Neither the people realize that they have a responsibility towards conserving the protected water, nor those in authority, including the water supply department, have attempted to highlight the processes and costs involved in upgrading the raw water by treatments to potable grade and the cost of 'protecting' it till it is supplied. No efforts were made till recently to educate the people about the easy and inexpensive technologies for conservation of the treated potable water by adopting practices such as 'rooftop rainwater harvesting', 'groundwater recharging' and 'wastewater recycling', as also the areas of human activity where recycled water can be employed without inviting health hazards.

Keen students of history of India, or for that matter any part of the world, would realize that the history is intimately linked to the geography, in the sense that the rivers had played important roles in the rise, progress and fall of civilizations, empires and kingdoms. Earth's own internal dynamism (known to produce 'tectonic' movements) coupled with the exterior agents have been largely responsible for altering the course of several rivers of the Himalayan, Sub-Himalayan and Peninsular regions of India. Temporal variations are too obvious in the course of rivers of all the said three regions when comparative studies on their active channels and flood-basins are made with the aid of the topographic maps produced in the early part of 20th Century, aerial photographs of early 1960s and the modern Indian Remote-Sensing Satellite (IRS) imagery of post-2000 period. If that is the case of river course changes in the last one hundred years due to the neo-tectonic movements and recent dynamics of the river water and sediments, the nature-related reasons for the disintegration of civilizations, empires and kingdoms need no further explanation.

It would be apt now to quote the historian Smith (1958) from the edited work of Percival Spear: 'Who can tell where the Indus flowed in the days of Alexander the Great? Yet books....trace his course minutely through the Panjab and Sind by the help of some modern map, and offer pretended identifications of sites upon the banks of rivers, which certainly were somewhere else twenty-two centuries ago. We know that they must have been somewhere else, *but where they were no man can tell* (emphasis is by the current author). So with the Vedic rivers, several of which bear the ancient names. The rivers of the Rishis were not the rivers of today. The descriptions prove that in the old days their character often differed completely from what it now is, and experience teaches that their courses must have been widely divergent. ......One large river system based on the extinct Hakra or Wahindah River, which once flowed down from the mountains of Bahawalpur has wholly disappeared, the final stages having been deferred until the 18th Century. Scores of mounds...and nameless towns bear testimony to the desolation wrought when the waters of life desert their channel'.

Whereas history does not often leave most of its footprints except a few evidences, the geography related to large rivers always leave the trails of their channels and flood basins, or what are now known as their paleo-channels, on the face of Earth's surface. Such paleo-channels stock lot of water within them, as they had in the past, allowed unimaginable proportions of fresh water to flow through them, and in that process, volumes of water had infiltrated and reached the groundwater beneath them. Remote-sensing scientists can trace these paleo-channels on modern satellite imagery and locate them on ground after examining empirical field data. In recent times, several such paleo-channels and other features associated with the lost rivers have been successfully located. The earlier opinion *'where they were no man can tell'* is now a myth, thanks to modern tools like satellite imagery. Illustrations are given in this book with the hope that *systematic* remote-sensing exercise covering the Deccan and far South regions of the country would facilitate in tracking down the paleo-channels to facilitate tapping the ancient water reserves. But, before a beginning is made to squander away the ancient waters, we have to explore possibilities of replenishing and augmenting the water resource by other means, so that the ancient waters can be conserved for the *'rainy days'*.

Drinking water scenario in India

Today, India is in a position to export milk and dairy products. Thanks to the White Revolution or the Operation: Flood, one litre of wholesome milk is available in a large part of the country at a price range of Rs.13 to Rs.15. This is in no way a mean achievement for our country, because from a stage of milk deficiency in the recent past, we have now marched into the club of milk exporting countries. Let us compare this unique achievement on the milk front, with the scenario of availability and cost of wholesome drinking water in the country. Popular brands of bottled 'mineral water' or 'pure water' are *on sale* with price tags ranging from Rs.12 to Rs.15 per litre. Other less known brands cost *not* less than Rs.10 per litre, but the methods employed by the firms in carrying out treatment and disinfection of water sold as 'drinking water' is questionable. The display of bottled water for consumption of the state executives and bureaucrats, who represent the governments that supply 'protected water' to its citizens, on the dais of public meetings addressed by them, narrates the comic side of the claims on the quality of the 'protected water', in a subtle but loud manner.

Indian hospitality of offering drinking water to the guests and visitors had been of a unique nature for centuries. Today, much to the disappointment, anguish and embarrassment of the hosts, the visitors get doubt on the quality of water offered for drinking. Still worse, the hospitality of offering and receiving the drinking water is fast disappearing in many parts of the country, especially among the urban masses, since drinking water has become a saleable commodity. Gone are the days when citizens could consume water directly from the rivers and public wells. It is rather distressing that in a country that has so many perennial rivers, in a country that receives a high average rainfall of 120 cm in a year, in a country that is surrounded on three sides by ocean and sea, in a country where a considerable people still straddle the poverty line or are living below it, in a country whose population has to consume about three litres of water for seven months in a year and at least one litre in other five months due to the compulsions of climatic factor, in a country where majority of its citizens still cling to the olive trees of culture, tradition and heritage and therefore feel compelled to offer drinking water to their guests, the drinking water has become a saleable commodity in recent times, and that too, at a price almost equivalent to that of milk.

Drinking water being the fluid that keeps the body and soul in the living beings together, the governments should ensure that it is available either free or at a nominal price affordable by everybody. The task of uninterrupted provision of disinfected pathogen-free drinking water to the masses of population of the country has to be taken up in the form of a revolution or movement, just like the governments take up pulse-polio programme for the new-born. Disinfection of water to the level of potable grade by pre-treatment and filtration, followed successively by other treatments such as coagulation, flocculation and sedimentation, and *total disinfection* with the aid of domestic or government-sponsored community level UV-ray electronic machines would be a pragmatic solution to ensure that all the people in the country, including those in villages, get clean drinking water free from pathogens. That would serve as a 'group health insurance' for the masses of population.

The UV-ray based electronic disinfectant machines should be manufactured and sold in the country in such large numbers as comparable to those of television sets. Advertisements in mass media encouraging people to buy these machines need to be launched in aggressive manner. The Union Government, on its part, has to extend subsidy in its annual budgets to the manufacturers of UV-ray water disinfectants, for a limited period, say ten years, so that these machines become popular and affordable for purchase by the middle- and low-income segments of population. Besides, the State Governments should consider extending sale-tax exemption to the buyers of these water-disinfection machines. We have to ensure that we provide pathogen-free drinking water through innumerable outlets everywhere in cities, towns and villages, such as kiosks, drinking water-points and multiple taps in bus stations, railway stations and other common places either free of cost or at a nominal cost. That would relieve the common man and those living below the poverty line, of *buying* drinking water at a premium price, as also the government can save itself from the embarrassment of allowing the price of drinking water to go at par with that of milk.

Approach to water policy

From an enviable status of possessing excess or adequate water resource in major parts of the country's terrain, India has slid down to the level of a water-deficient country due to a combination of reasons. We have badly gone wrong in managing our water resources in recent times when we grew into a nation of one billion-population, and when we were simultaneously faced with errant monsoons and also under stress to produce more foodgrain and industrial products to face global competition. We have failed to classify the quality of water required for our industries, for our personal requirements in various activities, for building construction, for recreation and swimming pools and for maintaining the urban green belts. We kept on using the raw water up-gradable to potable water as well as the treated potable water, for direct oral consumption alongside for washing railway wagons and automobiles. Even today, same quality of water is used in most part of the country for cooking, for washing cloths and for flushing toilets. Such misuse of water did not matter when the population of our country was just half of the present size. Drastic steps are the need of the hour, as human activities need to be classified keeping in view the desirable quality of water in different human activities, and allocating appropriate water for those activities.

As population increases, as people's expectation for comfortable lifestyle increases, as water resource becomes scarcer, as less quantity of essential commodity like water has to be shared with *more* people, and as all these things happen *simultaneously*, we necessarily have to adopt sound techniques in managing our resources by changing our priorities and adapting ourselves to new habits. Water resource is not an exception to this rule. Therefore, budgeting and management of water resource in the country has gained top priority in the beginning of the new millennium. We have to pick up the signals of alarm transmitted by Climate Change, Global Warming, El Nino, errant monsoons, truant rainfall, retreating glaciers and droughts (all these aspects are discussed in detail in different chapters of this book) in time and act fast. Lest, we may ourselves gradually end up living amidst geographic conditions that are alien to those of the present times and be heading towards the Lost Civilizations.

If we classify the human and industrial activities and assign different quality of water to different activities, and substitute the recyclable water to such activities where water quality is not an issue, it would mean that we have taken the first step towards effective management of water resource. Treatment of wastewater and generation of treated recyclable water has the twin advantage of (i) augmentation of water resource for use in limited human activities and certain industries where water quality is not a serious matter; and (ii) restoration of past glory of our rivers and lakes by channelling treated water into them. Paper and pulp industry, and servicing centres for railway wagons and automobiles are, for example, two places where treated wastewater can be substituted for groundwater or surface runoff water.

A chain of desalination plants along the East Coast of India, employing distillation or membrane techniques can produce adequate volumes of potable grade water. Seawater desalting using desalination plants is capable of quenching the thirst of people residing not only in coastal belts, but those in the hinterland and far off-coastal areas too. In the event we launch aggressive campaign to augment the water resource by technological means by installing several distillation or reverse osmosis plants and wastewater treatment plants, we need to generate a lot of electricity too. Justification is made in this book as to why we have to resort to bolstering our power generation using nuclear reactors. In addition to augmenting the power and water resources, the wastewater treatment plants, the desalination plants and the nuclear power plants would demonstrate our technological capability, besides providing employment to a few millions of our youth.

The people of India have always risen to contribute or offer finance to right causes—be it in wartime or at the instances of natural calamities. The Indian bourses could mop up a mind-boggling Rs.35,000 *million* in one single public issue of shares of just one firm in early 2004, and it only reflects the financial soundness of the country. Indian citizens are not averse to finding a lasting solution to the vexed and repetitive issue of the scarcity of potable water in our cities, towns and villages, and they would welcome technological missions such as desalination plants along the coastlines and wastewater treatment plants, which would enhance the country's image, besides providing water for their daily needs. If only the governments and the self-proclaimed environmentalist organizations are serious about finding lasting solutions to the scarcity of potable water as well as pollution of rivers, lakes and reservoirs, funds have to be raised for installation of desalination plants and wastewater treatment plants evenly distributed all over the country and channel the treated water into the water bodies.

Pilot projects of these technologies were indeed started in the past in India, but due to several extraneous reasons, neither many of these projects could be sustained nor new projects could be initiated. It needs to be emphatically stated at this juncture that the technologies cannot be faulted if a project fails due to the failure of a system or incapability of a group of individuals to deliver anticipated positive results. Thorough introspection by all concerned is necessary to analyse why these technologies have not yielded success or could not be sustained in India, when they had made tremendous impact in several other countries. Developmental programmes, such as nuclear power plants, potable water resource augmentation, wastewater treatment to save our rivers, lakes and reservoirs, should move on at any cost. Execution of such projects requires leaders, bureaucrats, technical personnel and grass root level workers whose '*character ethics*' can not be doubted or challenged.

Information on climate, rainfall and causes of droughts will be profitable when we study why we should start making sustained and planned long-term investments in water technologies, both in terms of money and human resource. Therefore, a few chapters describing these aspects are included in this book. Many of the 18 chapters in this book can stand independently, and therefore they can be read and understood individually. Only some chapters are interlinked. It is nevertheless suggested that the book be read chapter-wise to maintain continuity. Those who have background of various technologies discussed in this book can directly read without any difficulty the chapters presenting overviews of those technologies. Complementary information can be obtained from the references and websites listed out in the end.

A recollection of the first stanza of our National Song composed in Sanskrit by Bankim Chandra Chatterji is very much in context at this juncture. Our National Song starts thus: "*Vande Mataram! Sujalam; Suphalam; Malayaja Shitalam; Shasyashyamalam; Mataram*!" The translation in English of our National Song, as given by Sri Aurobindo, is as follows: "*I bow to You, Mother! Richly Watered; Richly Fruited; Cool with the winds of the south; dark with the crops of the harvest; The Mother*!" The address as The Mother refers to The Motherland. Significantly, the very first adjective while describing our motherland says it is "*richly watered*". Our efforts should be directed to regain the riches of water resources we had in the past and replenish them in our Motherland. Every time we chant the National Song, we have to passionately resolve that our policies are focused and efforts dedicated to make our country eternally "*richly watered*".

Chapter 2

As a pre-requisite to know about water resource, this chapter on climate deals with:

- Important aspects on the fundamentals of Climate and Biome.
- Climate in different parts of India.
- Monsoon and Rainfall in India.

CLIMATE

Climates are the characteristic conditions created on the Earth's surface by atmosphere under the influence of the Sunrays and the moisture derived from oceans. Atmosphere is the gaseous envelope surrounding the Earth, and it is a mixture of essentially nitrogen and oxygen, with small quantities of water vapour, carbon dioxide and inert gases such as argon. The climatic patterns, the wind systems, the clouds, the rainfall and the snowfall over large geographic regions are, in general, determined to a large extent by atmosphere. Climate is the long-term weather prevailing in a particular geographic location, and it includes the general pattern of seasons, weather conditions, rainfall, droughts and weather extremes such as thunderstorm, hurricane, typhoon and cyclone.

Two important factors determine an area's weather patterns and rainfall, and therefore, its climate; they are—Air temperature and Precipitation. In turn, these two factors are *partly* linked to the surface temperatures of the landmass and the oceans, and the moisture-content transmitted into the atmosphere. The climate mainly determines the type of plants that grow in a geographic region and the animals that inhabit it. All the three components—climate, plants (flora) and animals (fauna)—are interwoven to create the fabric of, what is known as, a *biome*.

Biome is a large geographic area bearing distinctive plant and animal groups, which are adapted to that particular climate and environment. The climate and geography of a region determines what type of biome can exist in that region. Major biomes include grasslands, tundra, desert, forest and a large variety of aquatic environments. Each biome consists of several eco-systems whose floral and faunal communities have adapted to minor variations in climate and environment inside the biome. The biomes that exist all over the world are controlled by the climate. Thus, the biomes, the ecology, the climates and the atmosphere of the planet Earth are intricately and delicately balanced aspects created by the Nature; if one of these aspects is tilted, the others too get disturbed with unpredictable consequences.

It is well known that the Earth rotates about its axis, which is tilted at an angle of 23.5°. This tilt coupled with the Sun's radiation results in *seasons* that are experienced on the Earth. Sun transmits its rays and they impinge upon the Earth's surface at different angles. The Sunrays reach Earth's equator in the areas lying between 23°N and 23°S latitude, at a direct angle. The intensity of the Sun's radiation that reaches the atmosphere within this area (*i.e.*, between 23°N and 23°S latitudes) is more compared to the radiation intensity received by the areas located away from those latitudes. The temperatures in this area are in higher range, compared to those in the regions away from it. In other areas, the Sunrays arrive at an angle to the Earth's surface and are, therefore, less intense. Such areas tend to experience cooler temperatures. The closer is a place to the poles of the Earth, the smaller are the angles at which the Sunrays reach that place; hence, the less intense is the Sun's radiation. The coolest temperatures are therefore obviously experienced at and near the Polar Regions of the Earth.

The climate system of the Earth is based on the location of these hot and cool air-mass regions, and the atmospheric circulation created by the wind systems, known as the *trade winds* and the *westerlies*. The trade winds north of the equator blow from northeast, and in the areas south of the equator, they blow from southeast. The trade winds of the two hemispheres meet near the equator and cause the air to rise. When the rising air becomes cool, raindrops develop from the moisture-content of the air, and the rain is precipitated on the Earth's equatorial regions. The resulting cloudy and rainy weather near the equator creates *tropical condition*.

The westerlies blow from southwest on northern hemisphere, and from northwest in the southern hemisphere. The westerlies steer the storms from west to east across middle latitudes. Both westerlies and trade winds blow from the 30° latitude belt. The surface winds are light over large areas situated at 30° latitude. Therefore, the air slowly descends to replace the air that blows away. The moisture that the air contains evaporates in the intense heat prevailing on the Earth. The tropical deserts exist under these regions.

The climatologist Vladimir Koppen had divided the Earth's surface into various climatic regions that generally coincided with the patterns of vegetation and soils all over the world. This classification is known as Koppen Climate Classification System and is employed for classifying the world's climates (Strahler, 1984). This system recognizes five major climate types based on the annual and monthly averages of temperature and rainfall. Each of these five types is indicated by capital letters such as A, B, C, D and E.

Type A is the *Moist Tropical Climate* known to occur with high temperatures round the year, as well as high amounts of rainfall in most parts of the year. Type B is the *Dry Climate* characterized by scanty rainfall and a wide range of diurnal temperature variation. Type B has two sub-groups namely—S-type, characterized by *Semi-arid* or *Steppe*; and W-type characterized by *Arid* or *Desert*. Type C consists of *Humid Middle Latitude Climates* with dry summers and cool wet winters. Type D is described by *Continental Climates* experienced in the interiors of landmass; weak to moderate rainfall and widely variable seasonal temperatures are characteristic of Type D climate. Type E is the *Cold Climate* experienced in the areas of permanent snow and in *tundra* (the level of treeless plains, as that of the arctic regions).

There are sub-groups in the above-mentioned types. These sub-groups are indicated by a second letter of lower case annexed to the main types (indicated in capital letters, A to E). Thus, we have Af or Bf or Cf Climates, which are *Moist* with adequate rainfall in all months and no dry season. The Am Climate is the *Rainforest Climate,* in spite of short dry season in the monsoon-type cycle. The Bs and Bw climate types denote *dry seasons* in summer and winter respectively. They are otherwise known as 'high-sun' and 'low-sun' seasons, in the respective hemispheres of the Earth.

Apart from the Koppen Climate Classification System, the climates are recognized under three basic climate groups, along with biome characteristics (Strahler, 1984). They are—Group I: Low-latitude Climates; Group II: Mid-latitude Climates; and Group III: High-latitude Climates.

Group I: The Tropical Moist Climates (Af) as in the Rainforests, the Wet-Dry Tropical Climates (Aw) as in the Savanna, and the Dry Tropical Climate (Bw) as in the Desert constitute the Low-latitude Climates of Group I. In the *Tropical Moist Climates* (Af) of *Rainforest,* the humidity varies between 75 to 90 percent, the temperatures remain about 27°C, and the rainfall is heavy in all the months of the year. The high temperature coupled with high humidity causes the formation of clouds in the afternoons on most of the days round the year. Heavy rainfall takes place both in the summer (due to the tropical air-masses) and winter seasons. The annual rainfall is more than 250 centimetres.

Savanna is characterized by *Wet-Dry Tropical Climates* (Aw), wherein the seasonal changes occur between wet tropical air masses and dry tropical air masses. These contrasting conditions result in a very wet season alternating with a very dry season. Cool dry season will be followed successively by extremely hot summer, and wet rainy season.

In the *desert* biome, *Dry Tropical Climates* (Bw) typical of desert climates are experienced. The moisture in the wind is subjected to evaporation by intense heat. The air masses seldom penetrate the area and hence, the rainfall is scarce. This results in extremely dry and pronouncedly high temperatures. Most of the low latitude deserts are concentrated near the tropics of Cancer and Capricorn.

Group II: The climates in the Mid-latitude zones are influenced by the tropical air masses that move towards the poles, and the polar air masses, which move towards the equator. Either of these air masses dominates the area. In this group, there is *Dry Mid-latitude Climate* (Bs type) prevalent in the *Steppe* characterized by grasslands. This climate is typically of semi-arid type. Depending upon the rainfall, the area may vary as *arid desert* when the area receives less rain, and as tall-grass prairie when the rainfall is adequate.

The *Chaparral* biome experiences *Mediterranean Climate* (Cs) characterized by wet winter and dry summers. These are caused by the sinking air of the subtropical 'highs' and may last for about five months.

In the *Grasslands* biome, *Dry Mid-latitude Climates* (Bs) prevail. The ocean air masses are blocked by the mountain ranges, allowing the polar air masses to dominate in the winter months. In summers, continental air masses are dominant, yielding some rainfall. The temperatures range widely, with warm to hot summers and cold winters.

Moist Continental Climate (Cf) operates in the *Deciduous Forest* biome. This climate is in the polar front zone where the polar and the tropical air masses tend to meet. The seasonal changes between the summer and the winter are pronounced in this climate. Daily temperatures change frequently. Abundant rainfall occurs throughout the year. The tropical air masses bring more rainfall in the summer season.

Group III: High-latitude Climates dominate in the Polar and arctic air masses. In the northern hemisphere, Canada and Siberia are the two air mass sources in this group. There is no such corresponding air mass in the southern hemisphere. Under this group, a typical biome by name *Taiga* is recognized. The climate that exists in this biome is known as *Boreal Forest Climate* characterized by long cold continental winters and short cool summers. The temperature ranges from as much as 41°C to minus 25°C. Small amounts of annual rainfall take place in the summer months, at an average of 30 centimetre. The boreal forest climate varies as humid, sub-humid and semi-arid types.

The *Tundra Climate* experienced (along the arctic coast) in the *Tundra* biome is a long extremely severe winter season due to the influence of the polar and arctic air masses, and a short mild season, which is not a true summer. However, the ocean winds moderate the severity of the low temperatures.

The *Alpine* biome experiences *Highland Climate*, which is cool to cold as in the mountains and high plateaus (example, Alps and Himalayas). The climate changes rapidly on the mountains, becoming colder in high altitudes. The temperature range is very large from about 30°C in summer to minus 10°C in winter. Interestingly, the climate of a highland area is closely related to the climate of the surrounding biome. This can be explained as, the highlands have the same seasons as well as dry and wet periods as in the biomes they are situated in. The mountains work as natural water storage areas. Snow is preserved till the spring season, but in summer, water is released in the surface drainage (perennial rivers) by slow melting of the snowpacks.

CLIMATE IN INDIA

An understanding of the climates in various parts of India is essential from the point of view of knowing the distribution of the water resources in different terrain segments of the country. Further, it would offer an insight into the water resources of the various parts of the country for the effective management of these resources. Having got familiarized with the general climatic types of the world, let us now take an overview of the climate, climate zones and biome types that exist in India.

Three geographic aspects influence the climate in India—(i) the Himalayan mountain range in north, (ii) the desert in northwest India and (iii) the Indian Ocean in the south. The Himalayan range forms a barrier and prevents the dry icy winds of Tibet, north Asia and Siberia, from blowing into the plains of India. Further, this geographic set up creates a sub-tropical climate with warm terrestrial winds and pronounced diurnal temperature variation. Indian Ocean in the south generates a humid and warm monsoon climate with plenty of atmospheric moisture, characteristic of tropical zones. The desert is largely responsible for the movement of the moist wind system (discussed later) in different segments of the terrain. Cyclonic moist wind systems develop every year over the bay (Bay of Bengal) on the eastern side of the peninsula, due to which seasonal rainfall occurs along the eastern coastal belt, central India and the interior landmass adjacent to the east coast. As a matter of fact, this rainfall is the lifeline for a majority of people living in the said thickly populated areas of the country, both for instant use and as replenished groundwater. The said three geographic entities cumulatively create an overall typical sub-tropical to tropical climate that characterizes the major part of the country's terrain situated to the south of the Himalayas.

Four distinct *seasons* prevail during a year in most part of the country. They are: (i) Dry and cold weather season from December to early March. (ii) Hot weather season between early March and mid June. (iii) Rainy season from mid-June to the end of September. (iv) Dry weather interspersed with occasional wet spells during October to mid December. Being a peninsular landmass gives India an advantage of significant variations in weather patterns in these four seasons and the type of flora in different parts of the terrain, such as the coastal belts, midlands and Sub-Himalayan belt.

Further to this, the presence of hill chains both in the western and eastern sides, a little away from the coasts, known as the Western Ghats (*Sahyadri* in Sanskrit and other Indian languages) and the Eastern Ghats respectively, creates distinct local variations in the weather patterns and flora. These local variations in the climate have set up distinctive types of biomes too within the Indian peninsula. The climates and biomes of India, like the country's ethnicity, culture and languages, show significant variation across the country, from north to south, from east to west and from northeast to northwest. The climates vary from *perhumid* in the northeast segment to *arid* or *desertic* in northwest (Rajasthan). *Semi-arid* climate prevails in the interior parts of India. Whereas the west coastal belt, situated to the west of the Western Ghats experiences a *humid* climate, the rain-shadow zone east of the Western Ghats, Deccan plateau and central India are characterized by *semi-arid tropical* climate.

India can be classified into seven climatic regions, based on the pattern and quantum of rainfall, weather, biome characteristics and geographic locations of different terrains. They are: (i) Tropical Rainforest Climate Region; (ii) Tropical Savannah Region; (iii) Tropical Steppe Region; (iv) Sub-tropical Steppe Region; (v) Tropical Desert Region; (vi) Humid Sub-tropical Region; and (vii) Mountain (or Highland) Climate Region. These climatic regions are similar to Koppen Climate Classification System, but modified by Trewartha on the basis of temperature, flora and geographic set up (in other words, biome). Accordingly, four major climatic regions namely A, B, C and H are recognized (Mamoria, 1979). Descriptions of these climatic regions and biomes (as in the Koppen Climate Classification System) are already summed up in the earlier part of this chapter. The adjectives such as 'tropical' and 'sub-tropical' are essentially qualifying terms added to give a vivid picture of the local climatic conditions.

(i) Tropical Rainforest Climate Region: The West Coast, the Western Ghats, and the northeastern States of Meghalaya, Nagaland, Manipur, Mizoram, Tripura and the low altitude areas of Arunachal Pradesh in northeast India come under the Tropical Rainforest Climate Region. These regions are characterized by consistently high temperature of above 18°C. The months of May and June are hot, but July and August are the most humid and hottest months. Copious rainfall, to the extent of over 250 centimetre, occurs during July and August. The terrain supports dense evergreen forests with tall trees and very thick undergrowth, including dense grass and a wide variety of ferns. These parts are somewhat similar to the Af-type *rainforest* biome described earlier; but there are minor variations in the Indian rainforests, mainly in terms of, and due to, the geographic set up of India. Nevertheless, the tropical rainforest climate in these regions is almost analogous to the climate that exists in the rainforest biome.

(ii) Tropical Savannah Region: This is the dominant region in peninsular India and includes a major part of Deccan plateau, eastern Gujarat, Madhya Pradesh, Chattisgarh, south Bihar, Jharkhand, Orissa, northern and central parts of Andhra Pradesh, eastern Maharashtra, coastal Tamil Nadu and rain shadow zones situated to the east of Western Ghats in parts of Karnataka and Kerala. The climate consists of long dry summer spells, up to 46°C in the day time of peak summer season, short spell of dry winter with day temperatures of not less than 22°C, and annual monsoon rainfall of about 100 centimetre. This climate is very similar to the Wet-Dry Tropical Climate (Aw) type experienced in the Savannah biome earlier described.

(iii) Tropical Steppe Region: This region is characterized by the semi-arid type climate and encompasses the rain shadow zones situated to the far-east of Western Ghats in the interior parts of Karnataka and Tamil Nadu, off-coast Tamil Nadu and Andhra Pradesh and central Maharashtra and minor parts of southern Madhya Pradesh. Dry weather with daytime temperatures of about 30°C prevails for a major part of the year. Hot and dry summer in April and May with daytime temperatures of over 46°C, short spell of monsoon period with about 75 cm rainfall and dry moderate winters with minimum night temperatures of about 14°C for short durations characterize this region.

(iv) Sub-tropical Steppe Region: Vast tracts of terrain in Punjab and Haryana to Kuchch, Saurashtra, western Rajasthan and northern Gujarat constitute this region. It is characterized by arid weather and it experiences extremely hot summer with temperature hovering around 48°C in the months of April to June, weak monsoon-induced rainfall of 50 to 75 cm, and dry winter with night temperatures of as low as 5°C exists in this region.

Both tropical and sub-tropical steppe region of India are very much akin to the dry mid-latitude climate (of Bs-type) with minor variations, in the steppe or grassland biome mentioned earlier.

(v) Tropical Desert Region: The desert of western Rajasthan and the sandy plains of Kachch together constitute a Tropical Desert Region. This region experiences extreme conditions both in summer and winter; the daytime temperature in the summer months of May and June reaching almost 50°C and the night temperature in winter is about 4°C. Rain is extremely scarce (about 10 cm in a year) in this region. This climate is exactly similar to the dry tropical climate (Bw) type prevailing in the desert biome as described previously.

(vi) Humid Subtropical Region: This region is in the sub-Himalayan plains and is spread over the foothills of Himalayas, eastern Rajasthan, plains of Uttar Pradesh and Bihar, northern Bengal and Assam, and southern Arunachal Pradesh. Monsoon rain occurs simultaneously with summer in July and August when the day temperature of about 40°C. Winters are dry. Dense subtropical *rainforests* exist in this region. The eastern segment of this region, i.e., sub-Himalayan Bengal, Assam and Arunachal Pradesh experience more rain than the western segment due to the activity of the Bay of Bengal arm of the southwest monsoon (described later).

(vii) Mountain Climate Region: The entire Himalayan mountain chain constitutes a Mountain Climate Region characterized by moderate summer day temperatures of about 20°C and sub-zero night temperatures in winter. In summer, the daytime temperatures may go up to 30°C or even more. Snowfall occurs in the winter. Rainfall is scanty on the northern slopes and Trans-Himalayan plateau where cold dry desert-like weather prevails in summer. The southern and western slopes experience markedly excessive rainfall of over 250 cm during the monsoon period. In the eastern part of the Himalayan chain in Sikkim and Arunachal Pradesh, the rainfall is far in excess than the average rainfall in the western Himalayan segment. Overall, the Mountain Climate of India may have a notable similarity to the earlier mentioned Highland Climate that prevails in the Alpine biome.

RAINFALL IN INDIA

The climate of a geographic location has a direct link to the rainfall it receives. In India, areas of very heavy rainfall exist on the windward parts of the Western Ghats, the Meghalaya plateau and the eastern parts of the sub-Himalayan belt. On account of elevation, topography and rainfall factor, many of the major rivers or their tributaries originate from these three regions. The average rainfall over India's territory is about 120 cm, which is highest in any part of the world. The rainfall over the areas other than the above-mentioned regions is highly variable. The occurrence of rain in the entire Indian terrain is totally controlled by the annual weather-related activity known as the '*monsoon*'.

Monsoon

The monsoon (meaning 'season' in Arabic) is a wind system, controlled by the convection of moist winds in the atmosphere. India experiences the annual phenomenon of weather known as the *southwest monsoon* that occurs over a major or almost the whole of peninsula, and a subdued *northeast monsoon*, which is active only in the southeastern coast in Tamil Nadu. The activity and movement pattern of the wind system linked to the southwest monsoon are very complex and somewhat unpredictable, yet its arrival time is mostly punctual, and the resulting yield of rainfall quantity, both spatially and temporally, is highly skewed.

The intense summer heat during March to the end of May over the north and northwest segment of the Indian terrain comprising the arid zone and the sandy desert causes a low-pressure zone along a northwest-southeast axis from Rajasthan to central India, roughly along the Indo-Gangetic plains. This low-pressure zone, which is also known as the 'monsoon trough', drags the southwesterly moist wind system that originates in the Indian Ocean, causing the *southwest monsoon* over the Indian peninsula. This monsoon first hits the landmass in the West Coast of Kerala by the end of May, and soon bifurcates into two branches, namely the Arabian Sea Branch and the Bay of Bengal Branch (Mamoria, 1979). Together, these two branches of the monsoon pick up over ten thousand billion cubic metres of moisture from Arabian Sea and Bay of Bengal, and precipitate about a quarter of it over the Indian subcontinent, mostly in India's peninsular region and small parts in its neighbourhood like Bangladesh, Myanmar and Pakistan. The southwest monsoon is associated with cyclonic storms and rainfall over the peninsular India. About eighty percent of total rainfall in India is caused by the southwest monsoon.

The Arabian Sea Branch of southwest monsoon proceeds northwards and by the middle of June, it spreads over the coast of Konkan and Maharashtra, and then to Saurashtra and Kachch. The Bay of Bengal branch, in the meantime, migrates northwards and hits Bengal and the northeastern States of India. Thereafter, it merges with the Arabian Sea Branch due to the barrier caused by the Himalayas and becomes active over the Gangetic plain and major parts of northern and central India by early July. The southwest monsoon is active all over India during July to September and brings rainfall to almost the entire country. The southwest monsoon is a natural phenomenon that unifies the whole country, breaking the barrier of languages.

The southwest monsoon starts retreating by the end of September and brings about a marked weather change from rainy spells to dry cool weather over the interior parts of the peninsular landmass. By the end of September, the southwest monsoon winds are drawn up to Punjab, by mid-October up to central India and by early November up to southern India. Therefore, from the September end to November, the southwest monsoon winds seem to withdraw in stages; hence, this season is known as 'retreating southwest monsoon season'. In the beginning of October, a low-pressure zone is centred over the northern parts of Bay of Bengal and by early November, this low-pressure zone moves southwards. Under its influence, precipitation of residual rain intermittently takes place during October to December in the coastal areas of Andhra Pradesh and Tamil Nadu.

In December, a low-pressure zone develops over the Bay of Bengal, while high-pressure conditions prevail due to rapid cooling of the land in the northern plains of Punjab and Himalayan foot-hills belt. Therefore, offshore winds start blowing from high-pressure zone of northern plains towards the low-pressure zone in the northern tip of the Bay of Bengal. Thereafter, these offshore winds pick up moisture from the Bay of Bengal, move from northeast to southwest and precipitate rain on the southeastern Indian coast. The rainfall due to these 'northeast trade winds' or *northeast monsoon* is restricted to coastal zone of Tamil Nadu and Andhra Pradesh.

The average rainfall over the Indian terrain due to the monsoon is about 120 cm, with a variation of about twenty percent deficit or excess every year. As such, the rainfall in different parts of the country is widely variable from about 300 cm in Meghalaya plateau and on the western side of Western Ghats, about 160 cm in parts of Assam valley, about 75 cm in central India and about 20 to 30 cm in most parts of Rajasthan and Saurashtra.

Tropical Depressions in monsoon

The activity of the monsoon season and the resultant rainfall in different parts as discussed above gives just a generalized annual pattern. The duration and quantity of rainfall that the southwest monsoon system brings to different parts of the country hinge on the complex natural process of the development of local low-pressure pockets called *tropical depressions*. These depressions originate in the Bay of Bengal and move landward in a circular pattern in the west-northwesterly direction. The rainfall occurs mainly to the south of these depression tracks. Furthermore, the moisture-content in the monsoon wind system, the intensity of the wind circulation, the clockwise or anti-clockwise pattern in the circular movement of the depression and the temperature of the sea relative to the land temperature are the other important factors that control the rainfall quantum as well as the area where the precipitation occurs. During some monsoon seasons, as a result of a single or a combination of factors mentioned above, the depressions may be less intense or do not develop at all. In that condition, the rainfall becomes scanty in the area that comes under the influence of the Bay of Bengal arm of monsoon. It is likely that the activity of the monsoon as well as the rainfall brought by it are complicated locally by excessive human interference, such as deforestation and release of the methane, carbon dioxide and other industrial gases in the atmosphere.

Even if the depressions are of normal intensity, the movement direction of the depression track controls the rainfall distribution in the north and central Indian terrain. If the depression follows a northwesterly direction, there will be excessive rain in the Gangetic plain and north-central India. If the movement direction of the depression is along west or southwest, the south and south-central India receive normal to excessive monsoon rains, but the Gangetic plains and northern India will be affected by paucity of rain. The depression tracks always weaken during its passage and eventually terminate at Kachch or Rajasthan, depending upon their intensity; the region where the depression terminates also receives residual rainfall. Thus, the monsoon depressions are important factors that manoeuvre the rainfall in major parts of Indian peninsula.

Cyclone

The coastal belts and some parts of southern peninsular India situated proximate to the coasts come under the influence of *cyclones* and associated storms and tidal waves during the southwest monsoon. The cyclones are tropical storms with winds blowing at over 120 kilometres per hour. Similar storms are known in other countries as hurricanes or typhoons. Though the cyclones originate both from Arabian Sea and Bay of Bengal, those originating from the Bay of Bengal are often devastating. In the past, when facilities for predicting the cyclones were absent, there was no way either to anticipate the probable area that would be hit by the cyclone. As a result, tidal waves associated with cyclonic storms frequently caused havoc in Andhra Pradesh and Orissa coasts and repeatedly brought about extensive damage to the agricultural crops and took very heavy toll of human life and livestock. There were several such cyclones in record until the recent past, for example, the brutal killer cyclone that hit the Andhra Pradesh coast in November 1977. Excess rains to the extent of 25 cm in a day in some pockets of area cause heavy floods, inundate the low-lying areas and destroy the standing crops. Paradoxically, even when some parts at and near the coast come under the influence of cyclones and receive rainfall in normal or in excess quantity, the adjoining interior areas of Orissa, Tamil Nadu, Andhra Pradesh, Karnataka and central India simultaneously come under the grip of dry weather conditions due to paucity of rainfall in the rain shadow zones.

In recent times, the satellite imagery are monitored onscreen continuously to study the cloud patterns, track the cyclones and to predict the coastal zones that are likely to be hit by the cyclones. Such cyclone prediction, information on the area likely to be hit by the cyclone and precautionary evacuation of the population from the danger zone are possible nowadays on a 'real time' basis, thanks to the continuous availability of satellite imagery for meteorological purposes. This has considerably reduced the loss of life due to cyclones in recent times, though the destruction of standing crops often goes unchecked. Occurrence of the super-cyclone (hurricane of Category-5) of Orissa, associated with gale winds of speed ranging from 260 to 300 kilometres per hour during 29th to 31st October 1999 that claimed inestimable human lives is an unavoidable and freak manifestation of a furious weather pattern rather than a deficiency in interpretation of the satellite imagery for cyclone prediction.

Western Disturbances

Cold wave conditions associated with dense clouds, rainfall in the plains of northern India and snowfall on the western Himalayas occur during winter months (December to March) due to typical rough and unpredictable weather phenomenon called as the *western disturbances*. The origin of the western disturbances is in the area situated northwest of Indian peninsula such as Iran and Afghanistan, but their influence is felt in Indian terrain from the geographic region situated to the west. Cold wave conditions frequently extend up to central India and Assam, under the influence of the western disturbances.

Variations in Climate

The climate as described above is a generalized or an average pattern in the entire geographic region of India. Variations in the climatic pattern each year in different regions of the country are more a rule rather than exception. These variations are both of spatial and temporal nature, and they include: (i) time duration of different seasons getting either shortened or prolonged, (ii) quantity of rainfall, (iii) non-seasonal rainfalls, (iv) marked increase in temperature and (v) wide fluctuations in the temperature. In addition, the weather, cyclonic storms, and the timing and quantum of rainfall and snowfall are becoming more and more unpredictable gradually over a period of time. These variations and unpredictability in climate are recorded not only in India, but also in the entire globe, and this is attributed to a phenomenon called variously as the *'Climate Change'*, *'Global Warming'* and *'Enhanced Greenhouse Effect'*. This phenomenon is discussed in a later chapter. The fallout of the 'climate change' is manifested in the form of occurrences of more and more *'extreme weather events'* in the entire world (Brown, 1996). Floods, cyclones, heat waves, droughts and sudden cold waves would be globally more frequent due to the process of 'climate change'. In extreme cases, the polar ice would start melting, which would lead to sharp rise in the sea level; this would mean that current small islands might face the danger of partial or total submergence in the seawater. Some of the low-lying coastal areas are also under threat of submergence. Deserts would continue to grow, new arid zones might be transformed into deserts, and rainfall would turn totally unpredictable or unreliable.

On account of excessive invasive acts of the human beings on the Nature by way of his interference with it, such as mowing down of the forest canopy, and the continuous emission of excessive carbon dioxide, toxic and 'greenhouse' gases (discussed later) into the atmosphere, the 'climate changes' are likely to be more frequent and more catastrophic in future. In this scenario, we, the humans scattered all over the globe, have two alternatives: One, to overindulge on the Planet Earth, alter the atmosphere above us by adding toxic gases into it and, as a consequence, live on the edge under hostile 'climate changes' and 'extreme weather events'. Two, to alter our own life patterns, improve the overall environment of the biome in which we live, and live in peace and harmony with the Nature.

Chapter 3

In this chapter, brief discussions on

- Different types of water resources,
- India's rainwater resource and
- Glacial snowmelt water resource are presented.

WATER RESOURCES

The water resources of a place are divisible into two major categories namely, the *Surface Water Resource* and the *Groundwater Resource*. In turn, the surface water resource can be sub-divided into the *Rainwater Resource* and the *Snowmelt Resource*. Though the groundwater is a significant water resource, large amounts of human efforts and peripheral facilities (such as mechanical and/or electric powers) are required to tap this resource. The readily available source of water comes as the surface water resource, and this is mostly confined in the river system of a place. Logically, thus, rainfall becomes the one and only contributor to both the surface water resource and the groundwater resource of a terrain.

India is one of the richest countries in the world in terms of the water resources that are cumulatively available in the whole country, and are also seasonally replenished year after year. Further, a dense network of rivers, streams and tributaries in the Indian terrain facilitates the tapping of the surface resource an easy exercise. From the point of view of the annual occurrence of monsoon and the architecture of the terrain, India is one of the few countries in the world to have bountiful water resource that flows in the river system. The geographic and climatic parameters, however, vary very widely from region to region of the country. Hence, the rainfall, which is linked to these parameters, is also variable in different parts of the country. The wealth of water resource is unevenly distributed in various segments of the country. Assessment and quantification of a nation's water resources are extremely complex exercises, further complicated by the erratic rainfall, periodic losses due to evaporation and transpiration, uneven percolation into the groundwater system and the lack of equity in distribution patterns to the groups of population as also human activities. The Indian Council of Agricultural Research, ICAR, (2002) indicates that the water resources of a region are influenced by three factors namely—1. Climatic Factors, 2. Physiographic Factors, and 3. Geological Factors, but primarily, the rainfall and climate are pointed out as the main parameters.

Despite the complexities associated with the exercise of determining the volume of the water resources, various figures are currently available as estimated water resources of India. The 'total flow in all the river systems of India is assessed at 1,645 thousand million cubic metres' (Rao, 1975). This water is said to be distributed in three groups of rivers—the first group comprising fourteen major perennial rivers with catchment areas each of 20,000 square-kilometres and above; the second group of forty-four rivers each with a catchment areas in the range of 2,000 to 20,000 square-kilometres; and the third group of ephemeral and desert rivers, each of them having a catchment area of less than 2,000 square-kilometres.

Quoting Dr. A.N. Khosla, the ICAR (2002) has indicated the mean annual natural runoff (or surface water resources) of the country as 167.23 million hectare-metre (mham) distributed in six water-resource regions of the country. These six regions are the river basins of Indus, Ganga, Brahmaputra, Luni-Ghaggar, west-flowing rivers and east-flowing rivers.

The Research, Reference and Training Division (2002) has recorded the average runoff in the river system of the country as 1,869 cubic-metres. Of this, the 'utilizable portion by conventional storage and diversion is estimated as about 690 cubic-metres...and the replenishable groundwater potential is estimated at 432 cubic-metres'. But, the disturbing piece of information furnished by the Research, Reference and Training Division (2002) is that the 'per-capita availability of water at national level has reduced from about 5,177 cubic-metres in the year 1951 to the estimated level of 1,869 cubic-metres in 2001 with great variation in water availability in different river basins'. This trend does not augur well for an agrarian country like India where 70 percent of the people are one way or other dependent on the agriculture, and 25 percent of the country's gross domestic product (GDP) is contributed by the agriculture sector. This also would not augur well for a nation, which has a vision to grow as one of the major economies of the world and compete with the other countries with big economy (otherwise known as 'the developed countries') in the year 2020.

When the population of the country during the period 1951-2001 has roughly doubled and breached the one billion mark, the per capita availability of water at the national level has nose-dived to get reduced by more than two-and-a-half times. This is in stark contrast to the outlay or expenditure incurred towards the major and medium irrigation projects, which in the Annual Plan for the year 2000-01 stood at Rs.10,976 crores and for the Eighth Five Year

Plan (1992-97) period at Rs.21,669 crores, whereas for the First Five Year Plan (1951-56) it was just Rs.376 crores (Research, Reference and Training Division, 2002). This points to the fact that the availability of water for human needs is becoming less and less by every year. When on one side the very economy of the country hangs precariously balanced and linked to the water resources and in turn, to the rainfall, on the other side, the day-to-day activities of the country's population are likely to go haywire, if further reduction takes place in per capita availability of water.

Resource from the Rainwater

The atmospheric wind system that constitutes the southwest monsoon attracts huge quantities of moisture in one single season just preceding the monsoon from the Indian Ocean, Arabian Sea and Bay of Bengal, all of which together encircle the peninsular Indian terrain. It is estimated that the moisture carried by the Arabian Sea Branch of monsoon between June and September is about 7,700 billion cubic-metres and that picked up from the Bay of Bengal Branch is about 3,400 billion cubic-metres (Mamoria, 1998). Subsequently, this wind system precipitates just over a quarter of the moisture, i.e., about three billion cubic metres, over the Indian subcontinent, including Bangladesh, Pakistan and Myanmar. About eighty percent of the total rainfall in India is caused by the southwest monsoon. Therefore, the southwest monsoon and the rainwater it precipitates over the Indian terrain are pivotal to the quantity of surface water resource and also for the annual replenishment of groundwater.

A part of the rainwater that reaches the Earth's surface infiltrates into the shallow sub-surface soil. This infiltrate may continue to flow laterally at shallow depths along the surface slope within the pervious soil layers and eventually join the streams. This flow of the water is known as *sub-surface runoff*. Yet another part of rainwater percolates deep into the sub-surface and becomes part of the groundwater system. Some fraction of the groundwater may reach the stream and become the base flow of the stream. This part of the rainwater is known as *groundwater runoff*. Apart from the rainwater infiltration, other surface waters collected in reservoirs, ponds, stagnation in the agricultural fields, canals, lakes and rivers too contribute to the groundwater resource. When the infiltration of rainwater ceases after the rainwater saturates soil and groundwater, the overland flow of the rainwater starts and it is known as the *surface runoff* (described later). The surface runoff, sub-surface runoff and groundwater runoff together constitute what is known as the '*total runoff*'.

A delicate balance exists between the inflows (from rain) to the water resources of a region and the outflow components (usage in the human activity). Under healthy and effective conditions of water resource management, the outflow component should not be allowed to exceed the inflows. But this never happens, especially during weak or failed monsoon seasons, resulting in the outflow of water resources for human activities far exceeding the inflows in the total runoff. This upsets the balance and the region acquires deficiency in terms of its surface water resource. The prudent water resource management under the condition of less inflow (due to deficient rainfall) should necessarily consider cutting down the intensive water-linked human activities like usage of water for agriculture or water-intensive industries. Alternatively, efforts should be made to identify recyclable water resources to ease the stress on the surface water resource that is hit by the deficient inflows.

Likewise, there is always a strong inter-relationship between the surface water resource and the groundwater resource of a region, and therefore, they too exist in perfect balance. The surface water resource is the one and only contributor to the groundwater resource by recharging through streams, reservoirs, ponds, lakes and so on. It is obvious then that when there is deficient condition in surface water resource due to scarce rainfall, the recharge to the groundwater is also affected to a major extent. Hence, the management of groundwater in a rainfall-deficient period should ensure that the groundwater resource is not plundered in intensive water-linked human activities.

The annual average rainfall in the Indian terrain is a healthy 120 centimetres, much more than the world average rainfall of 100 centimetres. The paradox, however, is the geographic nature of India, which shows a large variety of climates and variations in rainfall, characteristic of distinctive biomes, as earlier discussed. This variation is exemplified by the fact that while Meghalaya in north-eastern India receives over 1000 centimetres of rainfall in a season, several parts of Rajasthan in north-west gets a paltry 20 centimetres rain in the four-month monsoon period. Likewise, when the areas west of the Western Ghat (Sahyadri) range of hills, along the coastal parts of Kerala, Karnataka and Maharashtra receive as much as 600 centimetres of rainfall, the rain-shadow zones of the Western Ghats range, situated at about 75 kilometres away to the east, receive just about 70 centimetres or so of rain. On east coast, the rainfall is high in the coastal zones, but in the inland to the west, the rainfall decreases gradually farther and farther from the coast. Further, nearly 75 percent of the total rainfall is received in the four-month period of monsoon from June to September. Rainfall in other seasons is generally scarce or totally absent.

Thus, in a paradoxical situation, Indian terrain receives almost the whole quota of rain for the full year in a short period of about four months. That indeed is the reason for water resource management becoming a complex affair in India, for, the rainfall that is received during this short period has to be distributed for a variety of activities (discussed in a later chapter) throughout the year. We have been successful in this exercise most of these years, but also have not been that successful in some years.

If we try to depend only on the rainwater resource for all our activities, we may be playing a risky game with the nature on which we have no control. It is a debatable topic whether our success in providing food to a billion-strong population is attributable to the successful researches in producing high-yielding hybrid seeds or to successful management of water resources developed largely by rainwater source. At the same time, one fact that would not evoke debate is that our surface water and groundwater sources have certainly depleted over a period of time on account of the extreme stress we had mounted over these two resources. Steep fall in groundwater table levels especially in the urban and semi-urban places, necessity to abandon cropping seasons due to non-availability of water, vast tracts of lands turning into semi-arid to arid regions bereft of surface and sub-surface water, and seawater incursion in the coastal aquifers (discussed later) are just strong indications to it. We have abused and plundered the pristine rainwater resources for purposes and activities in which water quality is not an issue. We cannot afford to continue this way for long in future. We have to find out ways to identify suitable substitutes for the rainwater that is now misused.

Nature alerts us in many different ways to mend our ways of living. We should be prudent enough to understand and infer the silent messages conveyed by the Nature, and alter our attitude towards it, lest we would have to face the consequences of extreme conditions impacted by us on the Nature. That is what the lost civilizations of the world teach the humanity of the current times.

Glacial snowmelt water resource

Whereas the precipitation of rain due to monsoon activity is a vital source of water for almost the whole of Indian peninsular terrain, the snowmelt from the glaciers of the Himalayan belt is a significant contributor to the water source to the valley parts and the foothill region of Himalayas, which supports a very large population of India. The Himalayan glaciers have enormous amount of water contained in the form of snow- or ice-cover or what are known as '*snowpacks*'. These snowpacks melt in summer season, contributing considerable volume of water in the perennial rivers flowing through the Himalayan mountain chain and its foothill belt.

The significance of this bounty of the Nature is that the snowmelt feeds the drainage basin in a gradual manner commensurate to the gradual increase of temperature over a period of time in summer season and eventually flows into the rivers that flow along the sub-Himalayan regions. Paradoxically, though the Lesser Himalayan belt as well as the sub-Himalayan belt receive considerable rainfall during the monsoon season as also blessed with the Nature's bounty in the form of glacial snowmelt as water source, several Himalayan habitats and population face extreme hardships in managing water resource for irrigation of agricultural lands, due to the rugged nature of the terrain. The scarcity of good quality protected water in the Himalayan and sub-Himalayan towns such as Dehra Dun, Almora, Itanagar, Gangtok, Darjeeling and Roing is also noteworthy in this regard and it gives an impression to the observers that the sustainable management and development of water resources appears to be our least priority.

The amount of water resource from glacial snowmelt or snowpack of Himalayas is not assessed in an organized way so far. An approximate input of water resource, which gets released in Sutlej and Beas Rivers in peak seasons, is estimated for the dam management authorities as a model for glacier melt runoff quantity (for eg. Kulkarni *et al*, 2002). The water boards of riparian States such as Delhi, Punjab, Haryana and Himachal Pradesh seek similar information to plan the water resource in lean periods. Continuous monitoring of the thickness of glaciers in various seasons is done by the agencies of Central Government, but such exercise has limitations due to inaccessibility of the terrain and hostile weather condition in most part of the year. It is in this context that the photogrammetry and remote-sensing techniques have come handy to project the inflows of water in the Himalayan Rivers, based on algorithms and models.

It is important to note at this juncture that the findings of many agencies on the study of glaciers have revealed significant reductions in the thickness as well as the retreat of the snout portion of several Himalayan glaciers. This has led to considerable reduction in the release of snowmelt and resultant inflow of snowmelt water into the rivers. The study of the remote-sensing scientists by temporal (time-related) monitoring, over a period of several years, of the equilibrium line of Gangotri Glacier in Uttarkashi district, Uttaranchal can be cited as an example in this connection. This study has revealed that the Gangotri Glacier is now about 26 km long occurring at an altitude of approximately 4000 m, and fed by seven lateral glaciers. However, in its peak and prime of growth in the past, it was 60-65 km long with overall thickness of glacial ice in the accumulation zone approximately 100 m more than the present thickness (Tangri *et al*, 2001).

The phenomenon of global warming is attributed as a cause to this reduction in the size of glaciers and ice-sheets all over the globe, but this cause is yet to be proved in clear terms beyond doubt. The process of global warming spreads over a geological period comprising few thousands of years, and therefore, its impact on the glaciers and the snowmelt will be a debatable issue in abstract science. However, as far as the issue of water resource management is concerned, there is no debate that the snowmelt from the Himalayan glaciers has considerably reduced in recent times. This aspect also needs to be seriously taken into account by those who feel that linking of the rivers will grossly mitigate the water scarcity in the central and southern peninsular India, because linking of rivers is going to involve a few thousand crore rupees and such a project should consider availability of water in rivers on a long term basis.

Chapter 4

This chapter is aimed at taking reviews on:

- The phenomenon of Drought and its impacts.
- The causes of drought, including Global Warming and El Nino.
- The drought-prone areas in India.

DROUGHT

Some of the most important points that emerge from the discussions so far can be summarized as follows: Rain is the major source of water that charges the surface water and groundwater resources of India. The snowmelt water from Himalayan glaciers, the only other source, feeds small quantities of water to the Himalayan Rivers such as Ganga and Yamuna. The exact quantity of snowmelt water that accretes to the water resources of India is not exactly known, but it is small compared to the total resources available. Further, even if the snowmelt water adds to the water resources, the terrain south of the Yamuna River Basin does not receive the snowmelt. Therefore, rainwater is the whole and sole source of water for a major part of India. The rainwater received by the peninsular Indian terrain comes from the southwest monsoon rains. Rains of residual nature due to northeast monsoon add small quantities of water resource to the southeastern coastal India in Tamil Nadu and Andhra Pradesh.

The southwest monsoon being a natural phenomenon, to a large extent, the Nature's vagaries control the quantity of rainfall, the surface runoff and the quantity of rainwater that percolates to the sub-surface. Moreover, the phenomena of southwest monsoon and the tropical depressions are themselves so capricious that when there are excessive rains and resulting floods in one segment of the Indian peninsula (especially the coastal area that comes under the influence of cyclonic storms), the interior rain-shadow parts of central and southern peninsular India (parts of eastern Uttar Pradesh, Madhya Pradesh, Chhatisgarh, Jharkhand and off-coastal areas in Maharashtra, Andhra Pradesh, Orissa, Karnataka and Tamil Nadu States) get only local and scanty rainfall and, therefore, these areas frequently come under the grip of dry weather conditions. The quantity of the rainfall itself being a variable parameter, the water recharge into the groundwater system in a location hinges on the intensity of the monsoon activity in that area. Paradoxically, very often many interior parts of peninsular India, such as Deccan plateau, southern and central India, come under the grip of dry spells even during the monsoon season itself due to scanty rainfall. Such areas, which are gripped by weak monsoon activity and its attendant effects such as scanty rainfall, low surface runoff and low infiltration into the groundwater system, and impaired agricultural activity are said to be drought-affected areas. In scientific jargon, *drought is a phenomenon in an area where the atmospheric moisture-content drops sharply in such a way that the precipitation of (sufficient) rainfall is affected, resulting in a dry spell over a considerable period of time.*

Studies conducted in the last five decades or so have indicated that the drought is not the result of a single cause, but a cumulative effect of many causes. The drought manifests as sharp depletion of surface water and groundwater resources for a prolonged period, say two to three years. Not only the availability of water for irrigation of the agricultural lands is acutely hit, but also the day-to-day human life is subjected to inconvenience due to the shortage of water for personal use.

According to the Indian Meteorological Department, an area is declared 'drought-affected', if the annual rainfall is less than 75 percent of its normal. If the rainfall is less than 50 percent of the normal, that area is declared to have come under 'severe drought'. If the droughts get repeated for two or three consecutive years, not only agriculture and the nation's economy are hit, but also the very lifestyle of the people itself totally changes for worse. The immediate impacts of the drought, besides the failure of agriculture, are migration of population, loss of livestock and scarcity of food and fodder. Owing to the creation of adequate buffer stocks of foodgrain by the State and Union Governments, in the recent times, the food situation in the drought-affected areas in India is brought under control by timely and adequate supply of grains. But, the cattle bear the brunt of water and fodder shortage and perish in large numbers every year in the drought-affected areas. Despite advancement in technologies such as weather forecasting with the aid of satellite imagery interpretation and the study of past weather cycles and patterns to study the causes that were responsible for previous droughts, it has not been possible to precisely forecast the imminent drought in a particular place, because of the vagaries associated with the Nature and its manifestations such as the monsoon and the quantity of rainfall it is capable of bringing to us.

History has chronicled several droughts in India, and no part of India has escaped dry spells due to the failure of monsoon, and the resultant drought.

Abandonment of villages and migration of population were the common impacts due to droughts in the past. Studies in sociology in recent times have revealed that the crime rates such as robbery and looting steeply increase in the drought periods, as a sequel to the migration of people affected by drought looking for a decent life, but taking to unsocial activities in the absence of job opportunities. The water supply, an essential civic amenity provided by the government-controlled authorities in the urban centres, is drastically cut during droughts and it affects the day-to-day life of the people; the worst affected are those living as community in multi-floor apartments. The common man turns rebellious on account of the non-availability of water for their day-to-day minimum activities. Mass movements are organized by the vested interests in politics under the issue of water supply (or the lack of it). We see that if the general elections are held in a year affected by dry spell due to the lack of adequate monsoon rainfall, the ruling political party loses the election.

In recent times, sharing of river water for agriculture in the post-monsoon period by riparian States has become an issue between several State governments. What is paradoxical is that the States, which have long coastline, have not taken any steps all these years to mitigate the water scarcity by means of technologies and the people continue to suffer due to inaction during the unimaginative rule of the consecutive governments. Ironically, all those who cry hoarse about the inadequacy of water during the dry and peak summer seasons, including the worst affected human segments, become silent the moment the first spell of rain occurs in the beginning of the monsoon. The cycle repeats in the next and then the subsequent years of summer, without an iota of improvement in the availability of water over the previous summer seasons. The situation cannot forever continue like this in future.

If the pressure exerted by the growing population on land for habitat and agriculture continues to escalate at this rate, and consequently the available water resources are utilized as now, even under the conditions of normal monsoon and rainfall, by the year 2020, there will be very acute shortage of water resources in the country even for the human beings' day-to-day use, apart from the need of water for agriculture. The crisis would become more acute and the problem compounded if monsoon periods do not bring sufficient rainfall over large areas of the country consecutively for two years or so, to replenish both the surface water and the groundwater resources. Needless to emphasize that concrete steps and visible measures towards drought and dry-weather management need to be taken and implemented at any cost as a matter of Water Policy, by way of developing our water resources by alternative means. In our efforts to build a strong and developed

India by providing infrastructure for sustainable growth in agriculture and industry with a vision for healthy situations in those two sectors in the year 2020, we should start planning for the development of ample water resources in the whole country.

CAUSES OF DROUGHT

Though the root cause of drought is repeated deficient or failure of rainfall during rainy seasons, the very reasons for the deficient or failure of rainfall is linked to several atmospheric parameters. The cause of drought can be attributed to two fundamental meteorological parameters—temperature and precipitation—which together control the climate of a geographic location on the Earth's surface. Logically, therefore, the same parameters that control the climate cause the droughts too.

The Sunrays that impinge upon the surface of the Earth raise the temperature. The high temperatures created on the land and the oceans are primarily responsible for loading water vapours or moisture in the atmosphere. The amount of the moisture in the atmospheric air is expressed as *humidity*. The parameter known as *relative humidity* (expressed in terms of percentage) refers to the amount of water vapour or moisture the atmospheric air holds, compared to how much it could hold (in other words, the level of saturation) at a certain temperature. The meteorologists, however, indicate the moisture in the air in terms of *dew point*, which is the temperature at which the relative humidity reaches one hundred percent. Water is transferred from the land and ocean surfaces to the atmosphere by two natural processes known as (i) evaporation and (ii) transpiration.

The *evaporation* of water can be defined as the process in which the water in liquid form is transformed into a gaseous or vapour state, without necessarily reaching the boiling point. The evaporation in geographic sense refers to the transfer of water from the Earth's surface to its atmosphere as moisture. It is estimated that the oceans contribute approximately 80 percent of humidity resulting from evaporation, and the remaining 20 percent comes from the inland water bodies and vegetation. The wind systems play their own role in transporting and shifting the evaporated water in the atmosphere around the globe, influencing the humidity in the atmospheric air around the Earth's surface. Most of the evaporated water exists as a gaseous cover outside the clouds. Expectedly, evaporation is more intense in the geographic locations that experience warmer temperatures, as that of the Peninsular Indian landmass.

Transpiration, in geographic sense, refers to the process of the transfer of water from the Earth's surface to its atmospheric levels, through the medium of plants and vegetation that drape the Earth's surface. This transfer of the Earth's surface moisture to its atmosphere takes place through the plants' stomata, which are small openings found on the underside of the leaves and are connected to the vascular tissues of the plant. Transpiration is an extremely slow process that is largely controlled by the humidity of the atmosphere and the moisture-content of the soil. Some of the plants that survive dry conditions have the remarkable ability to open and close their stomata. This adaptation is necessary to limit the loss of water from the plant tissues. Without this adaptation these plants would not be able to survive under dry weather condition and drought.

It is often, or always, difficult to distinguish between evaporation and transpiration. Evaporation from the landmass is more intense due to warmer temperatures, such as warm equatorial regions. High temperatures promote the process of transpiration also. The volume of water lost due to transpiration can be very high from the agricultural fields of equatorial regions. A composite term namely *evapo-transpiration* is employed to indicate the water loss or transfer from the land surface of the Earth to the atmosphere, owing to the effects of evaporation and transpiration combined. The process of evapo-transpiration uses up parts of both surface runoff water and the soil water occurring up to certain depths to transfer to the atmosphere as moisture. Thus, deficient or absence of rainfall coupled with the simultaneous water loss from surface of the landmass in the equatorial, tropical and sub-tropical regions may create extremely dry weather conditions. Such prolonged dry conditions in the semi-arid and arid regions considerably bring down the atmospheric moisture content, and therefore, droughts and drought-like conditions.

Meteorological scientists study the short-term weather patterns almost on a twenty-four hour—seven-day-week basis. Modern remote-sensing tools such as satellite imagery are widely generated with appropriate spectral, spatial and temporal resolutions and employed specifically for forecasting the weather and rainfall patterns. Current researches in remote-sensing-based meteorology are directed to measuring the surface temperatures of land and sea, applying specifically worked out algorithms for on-screen computer-aided studies. Besides, anomalies in the patterns of temperature and wind circulation are monitored almost on a continuous basis, and possible forecasts are also made with the aid of computer animation by application of simulation models and changing the algorithms and/or input data. Short-term weather patterns are predicted with satisfactory accuracy by these exercises. The results of even such rigorous continuous monitoring of the relative temperatures of land surface vis-à-vis ocean/sea surface, and the atmospheric moisture-contents to forecast the weather and rainfall have been frequently inaccurate, subjective and hotly debated. They have yielded hit-or-miss results in predicting possible long-term anomalies in the pattern and quantity of rainfall during a rainy season, and therefore the droughts.

If that is so for short-term weather pattern, one can imagine the constraints in predicting the climate and possible occurrence of droughts. Historical records are of help to know the climate of a region. Largely, droughts occur in the rain-shadow zones and in the areas where the monsoon clouds do not possess adequate moisture contents to trigger precipitation. To a great extent, anomalous atmospheric circulation and the absence of tropical depression during monsoon periods deprive an area of rainfall during monsoon season, thus inflicting drought. Predicting and identifying the cause of droughts are linked to so many variables of climate—temperature, rainfall, relative humidity, evapo-transpiration, air circulation, changes and anomalies in wind patterns to name a few—and therefore, indeed, a complex matter.

The immediate cause of drought is the predominant sinking motion of air, which results in compressional warming or *high pressure*. These processes inhibit cloud formation and cause lower relative humidity and considerably less precipitation of rain. Regions that come under the influence of high pressure for a major part or almost permanently round the year are the deserts. Besides, other climatic regions such as arid and semi-arid regions also experience high pressure to varying degrees intermittently. Droughts for prolonged periods occur when anomalies in atmospheric circulation patterns persist for long durations, say several months.

Researches in meteorological studies in the last fifteen years or so have pointed out the occurrence of strange atmospheric phenomena over continents and oceans. The rainfall pattern, the temperature of the Earth's surface, droughts, floods and the quantity of rainfall that occurs over a continent are linked to these atmospheric phenomena, which produce two characteristic conditions. They are: (i) Global Warming and (ii) El Nino. The meteorological researches have attributed both droughts and flash floods due to sudden heavy downpour, to these two phenomena. Let us know some of the important facts about these two phenomena.

GLOBAL WARMING

We have understood that the drought is the major fallout of a complex process related to the climatic changes, which in turn are linked to the interplay of the intricate processes taking place in oceans and atmosphere. In the past, before the Industrial Revolution to be precise, the climate changes occurred as a natural phenomenon. The intense human activity in the post-Industrial Revolution era began altering the climate by bringing about drastic changes in the ecosystem of the biomes world over. This is largely a consequence to the modification of the agricultural and industrial practices to suit the increasing demands of the human society.

On one side, there was excessive deforestation carried out by the growing population to meet the demands of wood for fuel, produce heat for protection against cold weather, house construction and a host of other activities that required wood as a main input. The loss of the canopy of vegetation has considerably brought down the evapo-transpiration and as a result, the moisture content in the atmosphere. On the other side, the gases such as carbon dioxide, methane and nitrous oxide, known as the '*Greenhouse Gases*', are released continuously in atmosphere during intense human activities. There is a belief currently that the agricultural fields with standing paddy crops release great volumes of methane, which gets concentrated in the atmosphere. In the recent past and current times, the excessive use of automobiles by the growing population around the globe has led, and has been leading, to burning of fossil fuels like petrol and diesel, and as a sequel to this, volumes of carbon dioxide are added to the atmosphere. In addition, a mixture of industrial gases is released from the surface of the Earth and these gases too have brought about large-scale alteration in the composition of the atmosphere.

The atmospheric research scientists are unanimous in their opinion that the emissions from the automobiles are the greatest pollutants of the atmosphere, and therefore, the countries that have high density of automobiles are solely responsible for releasing large quantities of carbon dioxide and adding them to the atmosphere (see Brown, 1996). Powerful greenhouse gases that are not naturally occurring in the atmosphere, such as hydrofluorocarbons, perfluorocarbons and sulphur-hexafluoride, which are generated in a variety of industrial processes, are also released in the atmosphere where they get trapped, concentrated and remain for several decades.

The energy from Sun heats the Earth's surface. In turn, Earth radiates the energy back into the space. The atmospheric natural greenhouse gases, such as water vapour and carbon dioxide, trap the heat that the Earth radiates back into the space. This natural process is familiarly known as the '*Greenhouse Effect*'. Without Greenhouse Effect, the temperatures on the Earth would have been much lower than they are now, and the life on Earth would not have been possible. Therefore, this is known as 'Natural Greenhouse Effect'.

However, problems arise when the concentration of greenhouse gases increases in atmosphere. By increasing the concentration levels of the greenhouse gases and other industrial toxic gases in the atmosphere as described above, the human activities have multiplied the Earth's natural greenhouse effect manifolds. Introduction of great amounts of additional carbon dioxide, methane and a host other greenhouse gases has caused extra warmth at the surface of the Earth. To distinguish this stepped-up greenhouse effect from the 'Natural Greenhouse Effect', the term *'Enhanced Greenhouse Effect'* was introduced (Brown, 1996). Consequent to the *enhanced* greenhouse effect, the Earth's surface temperature has risen considerably in the past century. This increase in surface temperature of the Earth is what has come to be known as *'Global Warming'*. Several forums such as the World Climate Conference (1990), the Earth Summit of Rio (1992) and the Climate Conferences of Berlin (1995) and Geneva (1996) have taken place to deliberate on this issue, and a blend of scientists and representatives of the governments from all over the world participated in the deliberations. In subsequent meetings where the results of scientific research were highlighted, it was felt that 'a doubling of carbon dioxide concentrations in the atmosphere would cause global warming of between 1.5°C to 4.5°C' (Brown, 1996). Further, an estimated temperature increase at the rate of 0.3°C per decade as the possible speed of global warming was highlighted, though it was felt that this estimate was on a higher side. A lot of debate had been going on for long as to who is to blame, whether the 'developed' countries who emit profuse carbon dioxide by excessive use of automobiles, or populous countries such as India and China, which emit greenhouse gases, especially methane, from extensive agricultural fields. For overall information on this aspect, as well as the global political implications of the global warming, the interested readers may refer to Agarwal and Narain (1991). All the countries across the globe are unanimous in accepting the role of excessive carbon dioxide and its deleterious effect on the atmosphere and climate. In a meeting in 1987, an international treaty to cut back on the expected rate of release of greenhouse gases to reduce the dangers of excess warming was suggested.

High surface temperatures and accelerated warming were recorded during the past two decades, both in the northern and southern hemispheres of the globe and over the oceans. A catastrophic drought in the American Midwest in 1988, a freak hurricane in the English Channel in October 1987 and the Hurricane Gilbert, which caused widespread damage in the Caribbean Islands are cited as possible causes of global warming. Further, in the South Pole, a 160 km long and 40 km wide Antarctic iceberg broke. Nearer home, the decrease of snow cover and dimension of many glaciers due to the speedy melting of snow in the case of Himalayan glaciers are cited as the evidence for global warming.

One important aspect is that the greenhouse gases emitted by the human activities remain in the atmosphere for long periods ranging from a few decades to centuries. Therefore, the global warming process as a result of enhanced greenhouse effect would also continue as long as the human activities such as most needed agriculture, burning of fossil fuels by automobiles and thermal power plants, and industrialization go unchecked.

The extent of global warming due to the human activities since pre-Industrial Revolution times is difficult to gauge. The Intergovernmental Panel on Climate Change (IPCC) has stated in its Third Assessment Report (2001) that 'there is new and stronger evidence that most of the warming observed over the last 50 years is attributable to human activities'. The IPCC has cautioned that the complex systems, such as the 'global climate system, would respond in non-linear ways and produce surprises'. These surprises in the warmer Earth could lead to more frequent, more intense and more fierce storms, cyclones and hurricanes, with intervening spells of severe droughts. Likewise, the warmer oceans raise the sea levels, affect the global climate systems, damage corals, disturb the phytoplanktons, chlorophylls and fish distribution, and interfere with the marine ecosystem.

The humanity all over the globe, therefore, has to get itself prepared to face hurricanes, floods, droughts and other freak climatic events in the years to come, and Indians are no exception. It would be prudent on the part of those formulating national policies on health and environment, forest and ecosystem, and above all the country's water resource management, to realize that the global warming *is a fact*, and it is our own making. They have, therefore, a role in facing and tackling the long-term effects of global warming, particularly the dry spells and possible droughts, which are going to be recurrent phenomena in future, due to the large amounts of greenhouse gas cocktails we have already pumped into the atmosphere above us. They would also do well to find out new and innovative ways to minimize the gas emissions in the atmosphere, in the light of the greenhouse gas contributing factors discussed here.

EL NINO

The global warming appears to have an effect, though it is not proven, on the phenomenon called as *El Nino* (in Spanish it means The Little Boy). El Nino is a disruption of the oceans' atmospheric system in the tropical Pacific, but making tremendous impacts on the weather around major parts of the globe. The atmospheric phenomenon linked to El Nino is termed the '*El Nino Southern Oscillation*' (ENSO), which involves exchange of air between the eastern and western hemispheres centered in tropical and sub-tropical latitudes. Among the consequences of El Nino and ENSO are increased rainfalls across the southern part of the United States and in Peru that has caused destructive flooding, and drought in the West Pacific areas (Indonesia and Australia). Sometimes, it is associated with devastating brush fires in Australia. The National Oceanic and Atmospheric Administration (NOAA), United States of America has observed that the effects of *El Nino,* though has its origin in Pacific Ocean, can have effect on the weather around the globe and modify the weather from time to time.

El Nino was originally recognized by the fishermen off the coast of South America as the appearance of unusually warm water in the Pacific Ocean, anomalously in the beginning of the year. In a recent study, Rodionov *et al* (2003) have found out that the North American winter of early 2003 was impacted by a typical El Nino event as evidenced from the distribution of sea-surface anomalies in the North Pacific, with a strip of anomalous warm water along the west coast of North America. Concurrently, an extensive pool of anomalous cold waters occupied the central part of the North Pacific. This sort of unusual or anomalous cold ocean temperatures in the equatorial Pacific region is termed as *La Nina,* meaning 'the little girl' in Spanish.

GLOBAL WEATHER PATTERN AND DRY WEATHER OR DROUGHTS IN INDIA

Empirical relationships have been demonstrated to exist between precipitation and the ENSO events in the tropical zones of the Earth. Nevertheless, the impact of El Nino on the Indian monsoons, rainfall and dry weather patterns is not established. Further, the possible influence due to the variations in the global weather patterns, climate change and global warming on the monsoons, humidity in the atmosphere over India's landmass, and the drought-like conditions in India is not at all understood. It may be possible that the anomalous rains in some parts and drought-like dry weather conditions experienced in some of the semi-arid zones of Indian sub-continent in recent times could be the probable fallouts of El Nino, though this cannot be concluded with absolute certainty. The monsoons of Indian sub-continent are wind systems and are controlled by convection of moisture generated from Indian Ocean. Therefore, any variation that takes place in the Pacific Ocean would have some impact on Indian Ocean too. It would, therefore, be logical to ascribe the development of the moisture contents, atmospheric circulation and tropical depressions that are key factors in controlling the rainfall as well as the high pressure conditions in some of the semi-arid zones in Indian peninsula during the recent monsoon periods, to the weather pattern that occurs in the adjoining areas, such as Pacific land mass that are affected by the effects of El Nino.

Global Warming and El Nino phenomena are key factors in manipulating the moisture contents in the atmosphere over the Indian Ocean. These phenomena are likely to alter the climate, the monsoons and the rainfall in the Indian sub-continent in the decades to come. If the dry weather pattern in the peninsular Indian terrain in the recent past and the present times is any indication, in all likelihood, the frequency of such anomalous dry weathers and drought-like conditions would be more, and their intensity more severe in future. Therefore, we have to be geared up to face such frequent failures of rainfall, dry weather patterns or drought-like situations.

DROUGHT-PRONE AREAS IN INDIA

The Bay of Bengal arm of the wind system constituting southwest monsoon gets impoverished in moisture by the time the monsoon clouds reach the northwestern part of India (in western Rajasthan) due to precipitation in the western sub-Himalayan areas. It is because the rainfall takes place along the windward sides of the Eastern Ghats belt, the Satpura hills, the Vindhyan hills and the Aravalli hills. The monsoon clouds that pass over the Aravalli range, therefore, do not possess adequate moisture to trigger precipitation. As a result, arid climate characteristic of typical desert biome, called Thar desert, is created in the northwestern terrain of India, mainly in Rajasthan. This desert terrain is known for chronic droughts.

The intriguing aspect defying the Nature and the logic, however, is the drought-prone terrain belt of Kalahandi and Koraput (Orissa) in eastern India, geographically situated *not* very far away from the East Coast. Though it is a fact that this belt is situated partly in the rain-shadow zone of the Eastern Ghats and thus explains the paucity of rains, it is difficult to explain the cause for perpetual drought-like conditions in this terrain segment. Excessive human interference in, and exploitation of, the Nature may be cause for this. Adequate corrective measures, such as development of new thick forests, soil conservation by binding the soil with the aid of new grasslands, and reclamation of wastelands using drip and sprinkler irrigation are bound to improve the prevailing drought condition in that area. These measures may lead to alteration of the type of the existing biome in this terrain segment to more congenial for humans, flora and fauna.

The area aligned roughly in a north-northwest—south-southeast direction in the rain shadow zone situated to the east of the Western Ghats annually receives about 75 centimetre average rainfall with local variations. But this rainfall is highly erratic. Frequently, the rainfall plays truant in this area, which extends from parts of Tamil Nadu in the south, through Andhra Pradesh and Karnataka, up to Maharashtra. Drought-like conditions prevail in this zone for most part of the year. Of particular notoriety for drought-prone conditions are the areas such as North Arcot (Tamil Nadu), Anantapur and Chittoor (in Andhra Pradesh), Bellary and Bijapur (in Karnataka), and Osmanabad and Aurangabad (in Maharashtra). Vast tracts of land in the Anantapur area are perpetually drought-affected.

It should be noted that most of the areas in this rain-shadow zone is densely populated with hard-working, knowledgeable and enterprising people. There are several towns having more than 1,00,000 population, besides innumerable villages situated in this zone. Extremely dry and arid conditions, with daytime temperatures of over 45°C for about 75 days in year, sparse vegetation, pediment type of wastelands and deep levels of groundwater table make the life somewhat difficult in these places. Wastelands dominate over the cultivable fields, and in the cultivable lands too, just one crop is grown in a year. Proper water resource management in this terrain zone is capable of transforming the arid to semi-arid conditions into highly productive areas. Irrigation by canals has been resorted to in the recent past. It would appear that the areas in this belt would be ideal to resort to drip irrigation and sprinkler irrigation to reclaim the drought-affected areas.

A few more pockets of chronic drought-prone areas are—1. Ramanathapuram and Thirunelveli Districts, Tamil Nadu. 2. The linear terrain between Coimbatore and Palakkadu (Palghat). 3. The Saurashtra and Kachch regions of Gujarat. Arid to semi-arid climate with less than 75 centimetre of average annual rainfall characterize these areas. Despite the fact some of these areas get significant rainfall to the extent of 50 to 75 centimetre, it is the truant or erratic nature of the rainfall, which makes these areas drought-prone.

K. L. Rao (1975) has assessed that in India 'the total area affected by inadequate rainfall is a little over one million square-kilometres'. Further, the area experiencing less than 40 centimetres of rain is about 12 percent. Thirty-five percent or a little more than one-third of the country's area receives less than 75 percent of rainfall. It is further indicated by K.L.Rao (1975) that some 56 million hectares of cultivated area of the country experience inadequate rainfall. These figures would have obviously gone up in recent times, on account of repeated severe summers and inadequate rainfall in the first three years of this century. There are reports about water scarcity of varying acuteness even in areas that receive more than 125 centimetre of average rainfall, such as parts of Kashmir Valley, the cities like Guwahati, Ernakulam, Visakhapatnam and Mumbai, and the areas located in Ganga, Yamuna, Mahanadi, Krishna, Godavari and Cauvery River Basins. One wonders whether more and more areas in India are turning into drought-prone. This is a matter of serious concern.

In India, in current context, the areas that—(i) receive less than 40 centimetre of average annual rainfall, (ii) experience prolonged dry weather and (iii) repeatedly erratic and scanty rainfall may be indicated as drought-prone areas. Of these, the latter two categories of area may well be situated in the semi-arid climatic zones that experience rainfall in the range of 75 to 100 centimetres. In recent times, it has been becoming excessively difficult to draw a line of separation between the drought-prone areas and the areas affected by water scarcity due to inadequate rainfall during monsoon seasons. Because, under dry and arid weather conditions, when the scarcity of water affects the human needs and agriculture alike, it gives an impression that the area is affected by drought. It is, however, not so. As a matter of fact, long dry spells and erratic or truant rainfall, coupled with population pressure and requirement of large amounts of water for irrigation needs to step up the agricultural production have, of late, brought about conditions akin to arid to semi-arid climatic regions even in the areas that receive more than 120 centimetre of average annual rainfall. Reports of the arable lands situated slightly farther away from the flood plains and the deltaic regions of Ganges, Godavari, Krishna and Cauvery Rivers coming under the grip of water scarcity for agricultural activities are becoming more and more frequent in recent times. Locations like Kodaikanal, Udhagamandalam (Ootacamund) and Nilgiri Hills in Tamil Nadu, and Dehra Dun and Mussouri in Uttaranchal are examples for excessive human interference leading to unsustainable urban development and decadent imbalance in the availability of water resources in areas otherwise known for copious seasonal rainfall. Besides building a stress on the availability of potable water both within and the areas adjoining such urban centres that have undergone unsustainable development, a fragile eco-system unsuitable for the humans, the flora and the fauna alike is created in the vicinity of the locations of the cities which have thus grown wildly. That precisely is the reason why sustainable management practices for the available water resources needs to be introduced as a matter of Water Policy.

Chapter 5

In this chapter, we discuss:

- Water Resource Management in drought control.
- Identification of effective Water Resource Management methods.
- Organizations in India that handle Water Resource Management to fight droughts.

WATER RESOURCE MANAGEMENT IN DROUGHT CONTROL

Huge funds are currently invested and transacted in India in modern areas such as telecommunication, information technology, cinematography, entertainment, cricket, advertising, tourism, sports, beauty pageants, fashion design and betting. Among these, barring the information technology, the other activities are virtually *'new systems of wealth creation'* as described by Toffler (1990). These new systems have provided unconventional types of employment to a few millions of people of the country, directly or through links. The wealth generated, transacted and spent in these areas is of mind-boggling proportions. Huge investments in these activities *cannot* be resented, because these are areas of modernity and we have to be progressive in our outlook as well as we have to keep pace with the grueling speed at which the other modern countries vie with each other in these fields. What is deplorable, however, is that towards our efforts to achieve greater heights in these sectors, we appear to have lost sight of the most important aspect of life, for that matter the very lifeline—*The Water.*

The one vital sector—perhaps only next to the country's Defense—that requires the attention of the planners for the twenty-first century, relates to the augmentation and sustainable management of the country's water resources. Though this may sound as an over-statement, only in the events of capricious monsoon, less-than-average rainfall over vast terrains, and the water-taps endemically going dry in the cities and towns in summers, as had happened in the years 2002 and 2003, we realize the inadequacy in our efforts to tide over the crises of water, as also near-drought conditions prevailing in major parts of the country. Even the media, be it electronic or print, focus the inadequacy of water resources only during acute water scarcity, without projecting the existing technologies in water resource augmentation. The paradox is that when the monsoon breaks every year and the rainfall arrives in trickles, all the ordeals of the summer are forgotten. Nobody, including the media, appears to feel concerned about the excessive water resources that were spent in the preceding summer and the ways and means to replenish it for use later.

The most powerful medium of current times, *viz.* the electronic medium, has to share a lot of blame for its failure to educate people on the water-related affairs and technologies. Though multiple news channels churn out same news of marginal importance to the society through different correspondents, they hardly come to an understanding to focus society-oriented subjects such as educating the masses on the water augmentation technologies. This has been going on for the last two decades, as a result of which the common man thinks that the monsoon rainfall is the only source for us to rely upon, unaware of the fact that there are several mechanical means currently available to boost the water resources. Consequently, the water resources in our country have come down heavily without replenishment.

Our vision for a strong India in the year 2020 has to begin with building up a mammoth stock of agricultural products far in excess of our own needs. Repeated drought-like conditions at this juncture would certainly be impediment in proceeding ahead with our agenda for the year 2020. We have to fight the droughts and water crisis at any cost and go ahead with mass agricultural production. That would facilitate export of our agricultural produce at competitive price to the industrialized countries and the new economies such as Japan, Korea, Taiwan and Malaysia. It should not be forgotten that India could tide over the economic tailspin that hit the South and Southeast Asian nations and Russia in 1997-98 on the strength of our agriculture. Though India was marginally affected due to that global economic caving-in, our economy had shown resilience by bouncing back and making significant growth in the years that followed the dramatic 1997-98 economic coup. Our agricultural stock had largely saved us from that gloomy situation, because the then India need *not* had to buy foodgrain at a premium with a comparatively weaker currency at a time of economic meltdown, as did Thailand and Korea.

We cannot afford to remain complacent on our successful encounter with the 1997-98 economic slide-down that the other countries in our neighbourhood had suffered. It should not be forgotten that India had experienced excellent monsoon in the years preceding that 1997-98 economic fiasco of South and Southeast Asian nations. Barring a few sporadic patches of agricultural lands in Madhya Pradesh, Rajasthan, Gujarat and Andhra

Pradesh, most of the arable lands in the other parts of the country were adequately rainfed during those monsoons, which resulted in agricultural production in plenty.

Our efforts to raise bumper crops year after year hinge on our ability to fight drought-like conditions. This would be possible only by making provision of ample water for agriculture. If we make our water resources available for the mounting needs of our population, building construction, growth of industries and other activities that support daily human lives, we not only instantly consume the surface runoff water, but deplete the groundwater resources too. We should remember that these two resources are the main source of water for agricultural activity. Therefore, if we are serious about drought control and not letting more agricultural fields go parched and dry, we have to find alternative source of water to support the other activities of human life, allotting the two above-mentioned resources in bulk for agriculture-related activities. That would be our first step towards water resource management for drought control.

This statement does not mean that currently water resource management practices are not adopted in the country. It would be apt to quote the Research, Reference and Training Division (2003) at this juncture: '*Planning and development of water resources need to be governed by national perspectives. Since 1987, a number of issues and challenges have emerged in the development and management of the water resources sector. As such, the National Water Policy needed to be reviewed. The (Revised) National Water Policy was adopted by the National Water Resources Council in its Fifth Meeting held on 1 April 2002 at New Delhi. To achieve the desired objectives of the policy, each State has to formulate its own State Water Policy, backed with an operational action plan...*'.

Soil conservation: Successful agricultural practice is hinged to effective tackling of the extremely arid as well as drought-like conditions and combating the complex natural phenomena, which are responsible for making the terrains drought-prone. Two of the critical factors for these complex natural phenomena are: (i) soil erosion and degradation, and (ii) non-availability of water resources.

Just like there is a depletion of water resources all over the world, the available soil resources too have been undergoing depletion on a global scale. It is estimated that in the beginning of the current century, more than one third of the productive lands on global level are affected at varying degrees (Ahern and McMichael, 2002). The soil in these lands are eroded and washed away, or affected by degradation due to compaction, salination, water-logging and addition of chemicals. The use of chemicals has reduced the organic contents of the soils and resulted in contamination with heavy metals, such as mercury and lead, and other pollutants.

A study on the causes for the soil degradation coupled with the measures to tackle it and the soil conservation methods are vast themes of research, and a discussion on these aspects may not be within the scope of the present topic of discussion. Soil conservation and treatment are continuous processes to keep constant vigil on the soil quantity and health. Nevertheless, it should be noted with contentment that several organisations in India are striving in the areas of soil research and the soil nutrient replacement in the agricultural fields, with a good measure of success. Of particular importance are the efforts of a number of Soil Conservation Research, Demonstration and Training Centres founded by the Government of India. These centres, spread all over India in the beginning, were amalgamated into a single organization called Central Soil and Water Conservation Research and Training Institute (CSWCRTI), Dehra Dun and brought under the Indian Council of Agricultural Research (ICAR) in 1967. This Institute has been conducting researches all over India through its Regional Stations on the soil (and water) conservation problems all these years (Dhruva Narayana, 2002). Almost all the State Governments of Indian Union too have formed local centres to tackle the soil degradation and conservation as well as to replenish the soil nutrients. In recent years, remote-sensing technique is widely used in soil surveys and to study the stress on the soil due to moisture- and nutrient-contents. Besides, several organisations have come up with an objective to look after the health of the soil in India's dryland and semi-arid tropical zones. Loss of soil cover due to erosion as well as the soil degradation due to excessive use of fertilizers, pesticides and other chemicals is a problem faced by the entire world and India is no exception. It is encouraging, however, that thanks to the concerted efforts on the soil treatment carried out by the above-mentioned organisations in India, the agricultural activity is still moving on uninterrupted.

The same thing cannot be said with certainty about the conservation of water resource. Agricultural activities in vast tracts of arid, semi-arid and drought-prone zones are affected in the beginning of this century due to wide gaps in the management of our water wealth. Water resources management over a large terrain such as India's vastness and complexity is, indeed, an intricate matter and it requires a multi-pronged approach involving the people at large with active support of the government machinery. The aspects that are important for the water resource management in the arid regions and in drought control may be broadly outlined as follows:

1. *Surface runoff management.*
2. *Watershed-linked rainwater management.*
3. *Management by locating water-bearing geomorphologic features.*

4. *Groundwater recharging.*
5. *Rainwater harvesting.*
6. *Alternative Methods.*

In an attempt to give an overview of these management methods, each of these aspects is discussed in as much details as possible in the forthcoming chapters. That would give us an opportunity for self-analysis on the pitfalls of, and inadequacies in, our current approaches to water resource management and to adopt appropriate corrective measures in our future approach. The Alternative Methods are dealt with in great detail, as this segment of management needs a greater impetus.

In a fair analysis, it needs to be indicated pre-emptively that one method in isolation or more of these methods in different combinations are currently employed as water resource management practices in various places of India. Yet, four out of six of the above methods, namely surface runoff management, watershed-linked rainwater management, groundwater recharging and rainwater harvesting are directly linked to the quantity of precipitation in an area and therefore directly linked to the rainfall. In the event of a deficit rainfall due to a weak monsoon, the water resource management in a particular region or State or a district, is thrown out of gear. The entire life as well as the agricultural and industrial activities are hit. As regards the other two methods, one of them, namely attempts to locate water-bearing geomorphologic features, is in practice in several parts of India either by the State governments or the Central agencies or by both. The alternative methods, however, is not practiced at all anywhere. In this respect, our prudence lies in adopting a common, integrated and holistic approach to the management of the water resources. What is required is a uniform policy on the water resources management to be enunciated by the Governments, both in the Centre and the States.

Currently, the Central Water Commission (CWC), attached to the Ministry of Water Resources in the Union Government, is the organization in India that monitors the water resources. There are several other high-powered Central agencies too that are trying to find out ways for water management. The CWC looks after initiation and co-ordination of schemes for control, conservation and utilization of water resources throughout the country in consultation with State Governments, for the purposes of flood control, irrigation, navigation and hydroelectric power projects. Since its inception in 1945, CWC has several major achievements both in India and abroad especially in the hydroelectric power projects.

Despite several drought-like conditions, complex weather patterns, shortage of water resources, arduous terrain conditions and disputes between the riparian States over sharing of river waters, today India is a power to be reckoned with in the field of agriculture. Besides the CWC, a host of other organizations have continually been striving to keep India on top in the field of agriculture (and power) and to successfully tide over crunchy situations with regard to the availability of water resources. Both in the annual and the Five Year Plan periods, hefty funds are allocated to create *Major and Medium Irrigation Projects*. A *Command Area Development Programme* has been in vogue since 1974 to improve the irrigation and optimize agriculture production. The *Accelerated Irrigation Benefits Programme* was launched to give loan assistance to the States to help them complete some of the unfinished projects. There are programmes exclusively for *Minor Irrigation* and *Flood Management*. A *Hydrology Project* is now in progress in nine States with credit assistance from the World Bank. The Central Ground Water Board looks after the exclusive task of *Groundwater Development*. The high-power apex body known as National Water Resources Council functions from 1983 and evolves national level policies for development and use of water resources in conformity with the highest national interest. This Council has the Prime Minister as the Chairman, Union Minister for Water Resources as Vice-Chairman, and the State Chief Ministers among others as Members.

There are also Public Sector Undertakings such as Water and Power Consultancy Services (India) Limited (WAPCOS) and National Project Construction Corporation (NPCC) Limited, actively engaged in the water resource-related works. In 1982, an organization known as the National Water Development Agency (NWDA) has been instituted to look into the developments of the Peninsular and the Himalayan Rivers. An exhaustive review of the functions and achievements of the said organizations and programmes is not within the scope of this book. For this review the reader may refer to the Research, Reference and Training Division (2003) of the Government of India.

There have, indeed, been significant achievements year after year in the field of water resources management, in spite of various complex situations, as a result of the committed service to the Nation by these organizations and the programmes implemented. This is evident from—(i) the huge buffer stock of foodgrain we possess as a result of irrigation practices in agriculture, and (ii) the availability of large quantities of hydel power. But in the process, we have exhausted almost all our existing water resources as well as reserves in the form of groundwater. Now it is time to think of ways to augment the water resources and to build up comfortable reserves of water for the future.

Erratic rainfalls, coupled with the ever-growing population of India and the never-ending needs for water for a host of human activities, dilute the programmes implemented through the great efforts put in individually and collectively by all the above-mentioned organizations. The efforts of all these organizations, without exception to that of National Water Resources Council, National Water Development Agency and Central Water Commission, in the development of agriculture and infrastructure that is needed to support the lives of a billion-strong population of the country, need to be constantly accelerated every now and then. As observed by the Research, Reference and Training Division (2003), the *'National Water Policy, indeed, needs to be reviewed'*. Effective water resource management is the only remedy for the control of drought-like conditions and all the six methods mentioned above need to be holistically employed as a matter of Water Policy continually on a long-term basis over the entire terrain of India, to achieve the desired results. An overview of the six methods outlined above at this stage would provide information on the utility and advantage of each of the methods.

Chapter 6

We consider following points in this chapter:

- Management of surface runoff.
- Parameters that control/affect the efficiency of surface runoff.
- Simple time-tested techniques in management of surface runoff.

SURFACE RUNOFF MANAGEMENT

The moment rain is precipitated on ground the rainwater flows on the natural ground slopes and then reaches the first order streams. The vegetal cover instantly appropriates a part of the precipitated rainwater to itself. Yet another part of the rainwater fills up the depressions on the Earth's surface. A major portion of these two parts is eventually lost in evapo-transpiration. A part of the precipitated rainwater percolates down to the Earth's sub-surface and reaches the groundwater system. Further, a considerable part of the rainwater subsequently flows into the second, thence into the third and successively higher order streams before it reaches the channels of major rivers.

When the rainfall intensity at the surface of the Earth exceeds the infiltration in the soil and percolation into the groundwater system, the flow of the rainwater over land begins and the water is led into the active channels of the rivers to become the *surface runoff*. The surface runoff is that part of rainwater and/or floodwater, which gushes over the earth's natural slopes and lower order streams, and quickly transported into the streams' channels and thence to the drainage basin of the highest order of the river in that region. The surface runoff eventually gets emptied into the seas as surplus. Since the rivulets as well as the lower order streams and tributaries follow the maximum slopes in that location, a part of the surface runoff gushes along these drainage channels too. Effective drought control in rainfed areas is linked to prudent ways to harness the surface runoff before it is released into the seas through major river channels, and putting it to use in irrigation and also as a means to recharge the groundwater system.

Despite frequent and long periods of drought in central and south India, several million cubic metres of fresh rainwater goes into the Bay of Bengal as surface runoff during almost every monsoon season through perennial rivers such as Godavari and Krishna Rivers in Andhra Pradesh and Mahanadi River in Orissa. Ironically, major parts of several districts situated proximate to Mahanadi River, such as Kalahandi and Koraput districts in Orissa, large parts of Nalgonda district in Andhra Pradesh through which Krishna River flows, and many parts of Karimnagar and Khammam districts in Andhra Pradesh which are drained by Godavari River and its tributaries, face drought-like condition year after year. This only suggests that the measures taken to harness the surface runoff are far from adequate.

The above fact can be corroborated by the following statistics furnished by the Research, Reference and Training Division (2003), and to quote it, '*The average run-off in the river system of India has been assessed as 1,869 cubic-kilometres. Of this, the utilisable portion by conventional storage and diversion is estimated as about 690 cubic-kilometres...the replenishable groundwater is estimated at 432 cubic-kilometres*'. The remaining quantity working out to about 40 percent of the runoff is lost in various natural processes such as evapo-transpiration and loss due to draining into the sea through the rivers. Effective water resource management has to address methods to harness this 40 percent of fresh water runoff of the river system, which goes as waste, and diverting it for human and agricultural needs. One of the effective ways achieving it is through watershed development, which will be discussed later.

The proportion of the surface runoff to the part of the rainwater that percolates into the groundwater system depends largely on the quantum of rainfall in an area, the nature of the soil or rock cover, and the geological structures (which function as conduits to the groundwater system) such as foliations, fractures and joints that are present in the rocks. During the monsoon season, the wind system gathers a few thousand billion cubic-metres of moisture from the oceans and then precipitates this moisture as rain in the areas of low pressure called tropical depressions on the landmass. The quantity of rainfall that takes place during a given period of time is an important factor that controls the surface runoff.

Normally, the monsoon wind system covers over several tens of thousands of square-kilometre land area, and yields rain at a slow rate in intermittent spells, on an average of ten to fifteen centimetres in each spell, covering about twenty-four to thirty-six hours. This situation is ideal as far as the natural process offering remedial measure against drought, as it saturates the soil that has gone dry in the preceding summer, with adequate water. Further, under this condition, efficient percolation of water into the groundwater system takes place, as also the flow of floodwater and excessive surface runoff into the sea is averted. However, complex weather patterns are locally created on the landmass during the monsoon seasons as a result of a host of parameters such as enhanced levels of greenhouse gases in the atmosphere and small pockets of low-pressure zone over the landmass. Under such conditions, intense tropical depression builds up over a small area of land, and heavy to very heavy rainfall occurs there within a short duration of time, which results in floods. Though such features are regular in peninsular Indian terrain just prior to the onset of monsoon (therefore known as pre-monsoon downpours), similar phenomena take place right through monsoon season too. Paradoxically, under these conditions, while floods affect the area that receives heavy rainfall during a short period, it is often observed that the area in the periphery of the flooded area comes under the grip of drought, until and unless rainfall takes place subsequently. The reason is that due to the lack of ample time for the floodwater to saturate the parched lands and the depleted groundwater system, it just gushes as surface runoff and is emptied within a short time into the sea through major river channels. It is in paradoxical instances like this that the importance of tanks and bunded structures is realized, for they would function as receptacle and storehouse for the floodwater.

Impervious soil cover, such as beds of clay and fine silt as also shale and fine-grained mudstone, impedes the percolation of rainwater and therefore, the surface runoff is greater in the terrain covered with these soils. Jointed and fractured gneissic or granitic terrain allows percolation to some extent, but the sheet-like exposures of these rocks facilitate rapid runoff. The basaltic rocks of Deccan Trap terrain, occurring in most part of Deccan plateau in western and central India, are well-jointed rocks. Further, by virtue of being a plateau in terms of topography, the Deccan Trap terrain displays more or less flat or gentle slopes restricting rapid surface runoff to a great extent, allowing moderate percolation. The top layers of the Deccan Trap basalt are frequently altered over extensive areas by chemical weathering process into laterite, which subsequently on physical weathering yields lateritic soil. Both laterite and the lateritic soil being of loamy or spongy nature and the pores being interconnected, the terrain covered by laterite and lateritic soil allows moderate amounts of percolation and restricts rapid runoff. Percolation of water into the groundwater system and arrest of excessive and rapid surface runoff in the areas comprising clay, shale, mudstone, silt and other impervious layers of soil is promoted by growing moderate cover of vegetation such as orchards and horti-agricultural plant support.

Harnessing of the surface runoff water by construction of check dams or bunds or barrier walls at locations that are about one kilometre away from the break in slope, using impervious earthen materials, has been an age-old water resource management practice observed in India. A probing examination of the topographic maps offers insight to the ways in the past when the surface runoff is collected and stored in a series of tanks in locations slightly (about 500m) away from the foot of hill ranges of the Eastern Ghats, and/or wherever the steeper slope of hill chain/cluster ends. The logic is that the rainwater precipitated on the hills is carried along the slopes, and the moment it reaches the plain area, it is impounded. The bunds of the serial cascading tanks constructed along river courses are arranged one bund behind the other downstream. In striking contrast, the bunds of the surface runoff tanks are arranged characteristically sideways to impound the rainwater to harvest it for use later, particularly in dry summer months (Figure.6.1). These tanks are provided mostly with arc-shaped bunds or barrier walls that show gentle concave face toward the water storage side.

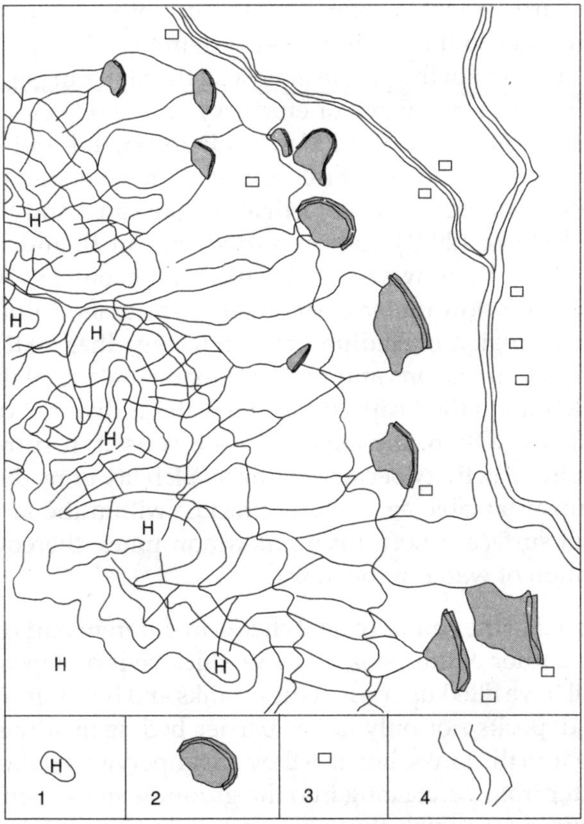

Figure 6.1. Geomorphologic map of a location in Andhra Pradesh, interpreted from topographic map, showing bunded tanks located sideways a little away from fool-hill to impound surface runoff rainwater. Some of the tanks show a cascading pattern to collect the water overflowing from the tank located upstream.

Index: 1. Contour lines indicating positive topography or hills. 2. Bunded tanks. 3. Villages in pediplain. 4. Active channel and flood plain of a stream.

In the western and northwestern part of Indian States such as Rajathan, such bund or barrier is known as *johad*. At some places, similar bunds are constructed on the down-stream side too, as seen in Figure 6.1. Thus, a train or train-cum-cascading system of tanks punctuates the topography of the area where surface runoff practice is actively resorted. During active monsoon, the runoff directly accretes to the train of tanks situated at the break in slope area, on the upslope. When the first tank gets filled up, the runoff flows over the bund and follows the slope or the channel of the stream, to reach the second tank on the down-slope. The process continues and all cascading tanks thus get filled up over a period of time in the rainy season.

The water thus harnessed is irrigated to the arable lands situated close by these tanks. Furthermore, a significant amount of water under storage in the tanks percolates to the groundwater system too. This practice appears to have been common in the past in almost all the States in peninsular India, comparatively at a more degree of efficiency, in the off-coastal areas of all southern States, Maharashtra and Madhya Pradesh. It is sad that in recent times, most of those bund structures, if not all, have gone defunct due to disuse because of modern ways of irrigation like canal and lift irrigations. Rajankar and Dolke (2001) have reported that in Bhandara district of Mahatrashta alone, there were 43,381 tanks built about 250-300 years ago, but their existence is now under great threat. The situation is not different in the Rayalaseema region of Andhra Pradesh (Gangadhar, 2001). Further, in the event of weak monsoons and resultant inadequate rainfall, these tanks do not get filled up to their capacity, as the water is diverted directly to the agricultural fields. Often, the tank bed itself is used in current times for intense agricultural activity because of the availability of ready water in the tanks. This unsustainable agricultural practice within the tank had led to poor inflow of surface runoff during monsoon times, thereby prohibiting the accumulation of water in the tanks.

It has been noted in the course of field checks in different parts of the Deccan region by the author of this book that very thick beds of deposits of silt and fine sandy soil have filled up most of these tanks and bund structures. These fine-grained deposits not only act as barrier bed against the water from accumulating into the tanks, but also they are impervious sediments that do not allow water from percolating into the groundwater system. Therefore, the very purpose for which these bunds or *johad* are constructed during different times in the past is defeated in current times due to the disuse of these water-bearing structures, mismanagement and indifferent attitude of the authorities and modernization of irrigation methods. Adoption of modern methods in any field needs to be encouraged, but not at the cost of the existing time-tested methods, which are eternally of tremendous value.

Barring a few committed non-governmental organisations (NGOs) in some parts of the States, which conduct sporadic desilting programme of a few tanks in some parts of a district just before the onset of the monsoon once in a while, there is no organised effort by government or local administrative bodies to conduct such desilting programmes. Some of the State governments have formulated some crude temporary programmes in the recent past with an objective of desilting of these tanks, but the efforts are pathetically inadequate as indicated by the overall results achieved under those programmes. Desilting and deepening of the tanks to attract adequate surface runoff needs to be taken up as an annual event of mass movement nature with the whole-hearted participation of the rural youth, a force to be reckoned with.

Antala (2001) has reported that *"inspired by a voluntary organization....farmers in the Saurashtra-Kachch region have recharged more than 300,000 wells and tubewells"*. In this effort, runoff rainwater was diverted by canals and/or tubes into the open wells and to the locations close to the tube-wells, in order to recharge them during monsoon season. The result of this effort was so significant that the groundwater of the whole area got recharged substantially. With more focus and orientation towards the objective, such movements are capable of producing results, since the local people are indeed serious about augmenting water resources in the vicinity where they live and therefore there is strong motivation in them towards this objective. Importantly, gently undulating terrains with small hillocks, hummocks or steep slope alternating with low-lying areas, as existing in the Deccan region and central Indian terrain are excellent locations where surface runoff can be harnessed with ease and least efforts, in tanks and irrigation reservoirs. The cumulative area possessing the geomorphologic feature as mentioned would be a few million hectares, and therefore, the benefit that would accrue if this practice were revived in these regions would be unthinkably tremendous. What is currently needed is local leadership of undoubted integrity showing never-say-die spirit against heavy odds and on the face of opposition from vested interests to guide the people toward achieving the goal of collecting the seasonal rainwater that flows as runoff.

The crucial factor for harnessing the surface runoff in the tanks, however, is the very availability of surface runoff, which is expected to be sufficiently high enough to initially saturate the parched land and subsequently to accrete to the tanks to fill them up. *"You cannot solve a problem on the same level on which it was created. You have to rise above it to the next level"*, observed Einstein. These prophetic utterances have profound meaning in current situation on the water resource management. Several million hectares of agricultural lands across the States of Indian Union, as well as the urban and suburban regions of the country occupied by human habitation are affected by repeated deficient rainfall for long periods. Alternatively, these areas are drought-hit, or are over-exploited in terms of available surface water and groundwater resources, or are deprived of percolation of water to the groundwater system. These terrain segments require far more surface runoff than the quantity of water available under normal climatic conditions and rainfall that occurs in India annually. Even if it presumed that the monsoons in the future years would be timely and the seasonal rainfalls adequate, several millions of hectares of the north-western, western, central and south Indian terrains, which supports human habitat and agriculture would require far excessive, but slow surface runoff, to make its surface saturated with water and replenish the depleted subsurface soil and groundwater resource too.

Protection of surface runoff water is a challenge. It may be possible to protect a reservoir from major human activity such as supply of potable water, but, in the case of a river, protection may be possible, if at all, only over a limited segment of the river. Advanced technologies such as wastewater treatment (discussed in a subsequent chapter) come handy to protect the rivers and reservoirs by letting in cleaner influent waters. Surface runoff management is the easiest method of, and the starting point for, the water resource management, as also the direct and most effective way for rainwater harnessing. If we are not able to build new tanks and reservoirs, we have to take steps to at least maintain, preserve and use the existing ones by repairing the bunds, barriers, check-dams and *johads*. We also need to build new tanks and reservoirs at the foothill regions and wherever there is a natural slope in the terrain, so that the runoff rainwater is most effectively harnessed, managed and administered to the Indian masses. Further, we have to revive the method of building tanks and reservoirs evenly in all regions, especially in the rain-shadow parts of central India and Deccan region.

Figure 7.1. Enhanced false colour composite subset of IRS 1D-LISS 3 imagery (RGB-3,2,1) of a watershed.

Chapter 7

We now take reviews of:

- Watershed concept and water resource management at watershed level.
- Remote-sensing methods in watershed delineation.
- Limitations in the management of watersheds.

WATERSHED-LINKED RAINWATER MANAGEMENT

Watershed is a terrain comprising a natural drainage system of streams, which join and drain at a particular point in the downstream side. Watershed is a mappable unit of a terrain (mappable on any scale) in which the runoff waters essentially flow through the streams' channels and drains through a particular location within the drainage basin (Figures 7.1 and 7.2). Murty (1994) calls watershed as *'an area draining the rainwater into a stream'*, and has presented a few fundamental aspects on the watershed management in India. Occasionally, the term watershed offers the same or similar connotation as that of 'catchment area' of a reservoir or a 'drainage basin', but it has a notable difference from the latter in that several watersheds constitute one drainage basin. Watersheds are small units of terrain, which can be administered with ease in terms of gauging, monitoring and distribution of the water resources available or accrue in the streams that drain those terrain units, by means of effective management mainly for irrigation of arable lands and a source for potable water.

Logically, therefore, planning for effective water resource management of a large terrain constituted by a drainage basin should start with planning at the watershed levels. Furthermore, watersheds being essentially rainfed, the management of water resources in the watersheds would mean that the rainwater in the smaller terrain units is effectively handled for a variety of human activities. In the case of surface runoff, the rainwater is liable to skip a considerable number of streams or tributaries in a watershed by following the Earth's natural slopes to create floods in the same or one or more nearby watersheds. Contrarily, in the case of the watershed-linked rainwater management, the rainwater is made to saturate the parched stretches of land situated within the watershed area. Further, the water that drains in

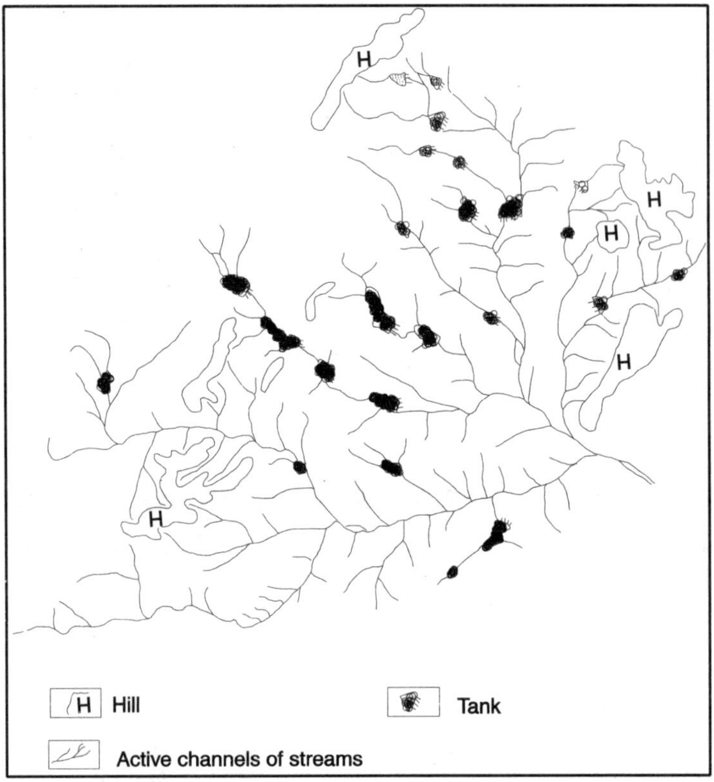

Figure 7.2. Map of the watershed interpreted from the imagery shown as Figure 7.1.

the lower order streams is frequently impounded in tanks and bunded reservoirs for irrigation purpose, as well to facilitate replenishment of the groundwater resource.

WHO (1993) recommends that the watersheds should be protected from human activities. This is because the quality of water over large areas depends on the cleanliness maintained in the watersheds, and the cleanliness, in turn, is hinged to the human activities. This could include isolation of the watershed and control of polluting activities in the area, such as dumping of hazardous wastes, mining and quarrying, use of fertilizers and pesticides, and the limitation and regulation of recreational activities.

These facts underline the necessity to delineate the watersheds of appropriate sizes in terms of area in square-kilometres, in all parts of the country. That would help in mapping them in great detail using various technologies, linking them together, and eventually bring out the compiled maps of the drainage basins. The availability of these maps of watersheds and drainage basins would facilitate arriving at appropriate decisions by the concerned authorities in the government with regard to locating the sites for reservoirs and tanks by construction of dam or bund or *johad* across the streams that carry the rainwater in a watershed, and if necessary, diverting that water to the water-deficient sectors of the same or other nearby watersheds.

Remote-sensing technique in mapping watersheds

India is a mighty force in the world to be reckoned with in the fields of satellite launch, satellite imagery generation and the interpretation of the satellite imagery data for multiple purposes, otherwise known as Remote-sensing. The Indian Remote-sensing Satellite (IRS) imagery are world class in terms of spatial and spectral resolutions. The IRS imagery are currently in use in a host of advanced and off-beat scientific applications such as Precision Farming, agriculture, horticulture, forestry, environmental studies, urban planning of several cities and towns, laying of telecommunication network, seismic and landslide studies, Ocean Colour Monitoring for the purposes of Fishery, strategic defense planning, and the all-important water resource identification and the mapping of watersheds. If carried out with prudence by experienced scientists, the remote-sensing technology using satellite imagery is the empirical means to delineate and map the watersheds in most speedy, accurate, logical, scientific and modern means, which is verifiable by anybody, anytime and from any part of the globe. Systematic delineation of the watersheds of any appropriate area in terms of its size and location, and suggestion for proper measures for the management of the watersheds thus delineated, should form one of the thrust areas in scientific research in the Five Year Plans, as well as Scientific Action Plan with a vision for developed India in the year 2020. These studies can also be used to identify the prospective locations for bunds, dams and *johads* within the watersheds demarcated.

The organizations that carry out remote-sensing by interpretation of the IRS imagery have already generated a huge wealth of data pertaining to the watersheds in several segments of Indian terrain. These data are accessible only in bits and pieces for isolated locations of the country. Further, generation of these data almost lack a uniform and systematic approach. The data on watersheds have to be made available to any work group of planners and policy makers on contiguous basis in terms of drainage basin or river basin, in order that the reasons for the root causes for dry conditions and poor surface runoff can be pointedly understood. This is necessary for implementing appropriate and meaningful management practices for the water resource development from the watershed levels. In the absence of data on watersheds on contiguous basis, the planning and implementation would go totally haywire and the purpose for which the watershed management exercise was undertaken would not yield the desired results. Further, a plethora scientific data generated by these organizations for many watersheds tend to be repetitive, lacking uniformity in scientific contents. The aspects that are conspicuously lacking in most of these studies are the empirical field validation of the imagery-interpreted data, meaningful synthesis of the watersheds in terms of effective management, and authoritative information on meteorology, geomorphology and geologic structures like faults that are likely to have a bearing on the seismicity. All these information of fundamental nature on each watershed are expected from the remote sensing scientists to facilitate the Governments of the Indian Union and the States to arrive at a comprehensive Water Policy.

River Water Disputes: Construction of multi-purpose dams across high order rivers to impound the water and build reservoirs might appear to be an easy way to collect surface runoff, and use it for irrigation in lean summer seasons. The major perennial rivers of the country are inter-State rivers. The growing demands of water sometimes lead to disputes between the riparian States. Such disputes are resolved through negotiations amongst the river basin States with the assistance of the Central Government. Adjudication through the appointment of water dispute tribunals is also resorted to as and when warranted (Research, Reference and Training Division, 2003).

The disputes over (i) the sharing of Cauvery River water, (ii) Almatti Dam on Krishna River, (iii) the proposed Paragodu reservoir across the Chitravati River and (iv) Sardar Sarovar Project on Narmada River are the focal points to highlight that construction of dams for large reservoirs may not be the ideal way in the present Indian context to harness the surface runoff and watershed-linked rainwater harvest. The Central Government has appointed five Water Dispute Tribunals, namely Godavari, Krishna, Narmada, Cauvery, and Ravi-Beas Water Dispute Tribunals, to resolve the issue of water sharing by the riparian States (Research, Reference and Training Division, 2003).

Further, the construction of dams has three more undesirable effects. They are:

(i) Loss of vast of agricultural lands due to submergence;

(ii) Displacement of a large number of population;

(iii) Seismicity that are natural and Reservoir-induced.

Submergence of land and displacement of people

Several hectares of fertile agricultural lands would be permanently lost due to their submergence by the water impounded by dams. Whereas the wastelands, which are not currently under cultivation and situated somewhat proximate to the reservoirs might be reclaimed for agriculture by irrigating from the reservoirs, there is no doubt that the small and medium farmers would be severely hit by submergence, as their land holding would be permanently lost under the reservoir water cover. Though they would certainly get alternative land as means for their livelihood, it might take several years for a settlement of this issue. With such uncertainty looming large, the people who face displacement from their settlements and the farmers who are likely to lose their land-holding, organize themselves to protest against the proposal for the construction of the dams. Efforts to resolve this issue generally take decades.

Seismicity

The Himalayan Rivers, as discussed earlier, carry large quantity of water, both surface runoff during monsoon seasons and snowmelt water from the large quantity of Himalayan snowpacks. Therefore, the Himalayan Rivers are ideally suited as sites for multi-purpose dams related to mega-projects of hydroelectric power generation and irrigation. The big challenge, however, is related to the safety of the dam against seismicity and earthquakes. It is well known that the whole chain of Himalayan Mountain is criss-crossed by several faults, the geological structures that make an area prone to seismicity and earthquakes. Most of the rivers and natural surface drainage pattern in the Himalayan mountain belt are structure-controlled as they usually follow major faults. Therefore, major engineering structures such as dams may become vulnerable to damage in the event of seismicity and the resultant earth tremors. The existing dams in the Himalayan belt have withstood the fury of several earthquakes in that belt because of the enormous amounts of exercises that had been carried out, specifically on the Himalayan dam projects, to draw up earthquake-resistant designs for them. On account of the intrinsic element of risk due to the seismicity faced by the Himalayan dams, the cost of these dam projects becomes very high, often prohibitive, as compared to those of the dams outside the Himalayan belt to conform to the stiff standards of seismic-resistant design. For this reason, micro-hydroelectric projects that can be operated using small reservoirs and much smaller dams as compared to the dams of peninsular India are currently in vogue in several locations in the Himalayan belt, apart from a few major projects.

In view of a series of earthquakes that had rocked the peninsular Indian region in recent years, such as Latur, Jabalpur and Bhuj seismic events, as also micro-seismic activities (earth tremors of small magnitude and short duration) that are recorded in several places, extreme precautions are currently taken in the dam design and construction, and as a result, the cost of the dam projects to harness the watershed-linked rainwater and surface runoff water in the peninsular India has also gone exponentially high in recent times.

Reservoir-Induced Seismicity (RIS)

The impoundment of surface runoff water in reservoirs by constructing large dams has yet another drawback. There are evidences to link the occurrence of seismic events to the development of large water reservoirs with the aid of dams. Such seismic events are termed as Reservoir-Induced Seismicity (RIS) or Reservoir-Related Seismicity. There are subtle difference between these two, but a discussion on it would be currently irrelevant.

The RIS has a potential to cause damage to the dams as also lead to loss of life and property, due to the cumulative effect of seismicity and damage to the dam on account of the seismicity. In the seismically active areas, a reservoir may facilitate or advance the occurrence of earth tremors, possibly leading to more frequent but smaller magnitude events. The relationship between the water impoundment and the seismic events, and the mechanisms by which a reservoir could cause earthquakes are not wholly understood and there is a lot of debate on this subject. The Reservoir-Induced Seismicity is caused by a host of parameters such as,

(i) the geological structures in and around the reservoir area,

(ii) ambient stress field conditions,

(iii) mechanical resistance offered by the rocks to the water column in the reservoir,

(iv) hydraulic properties of the crustal rocks below the water column,

(v) the pressure exerted by the water column (called hydrostatic pressure) on the dam and on the floor of the reservoir, and

(vi) the water level fluctuation in the reservoir.

Nevertheless, it has been suggested that large reservoirs may induce increased seismic activity through:

(a) Change in the normal effective stresses in the underlying rock due to increased pore pressures.

(b) Transmission of hydrostatic pressure through discontinuities in the underlying rock.

In the case of Acu reservoir in northeast Brazil, it was found that a combination of factors such as geological structures, seismic anisotropy of the rocks beneath the water and a two-dimensional pressure diffusion within the fault planes beneath the reservoir are responsible for the Reservoir-Induced Seismicity. Observations and monitoring have revealed that the Reservoir-Induced Seismicity can be attributed primarily to the water level fluctuations in the reservoirs, as in the case of a few case histories in Australia. In the Indian scenario, seismicity is known around the Koyna Reservoir, Mahatrashtra, where scientific investigations have been going on for over two decades on the Reservoir-Induced Seismicity.

The means for the application of safety measures against the Reservoir-Induced Seismicity is no different from that of the seismicity caused naturally by the Earth's intrinsic dynamic system. Provision of heavy reinforcements in the design of the dam at the time of construction, to make the dam and its peripherals earthquake-resistant, is the only way to safeguard the dam and reservoir from the Reservoir-Induced Seismicity. Therefore, for impounding the surface runoff water in the watersheds, developing large reservoirs by construction of the dams, which are earthquake-resistant as well as safe against Reservoir-Induced Seismicity, would remain substantially expensive in the years ahead. Besides, the wisdom to go ahead with those plans would draw flak and remain debatable and unresolved for decades without any solution, thereby involving more investments for carrying out research studies and repairs as a sequel to those researches.

Limitations in watershed approach

In a nutshell, in terms of the cost involved, the mega projects to impound the surface runoff water exclusively for the purposes of human consumption and irrigation of the agricultural lands would become less viable exercise in future, unless the projects are envisaged as of multi-purpose nature with a hydroelectric power project complement in it. Even in the event of our decision to go ahead with the expensive exercise of dam construction, we may have to face stiff resistance from those nurturing vested interests, ill-informed but powerful groups of the so called 'human activists' and self-proclaimed environmentalists. The matters concerning relocating masses of humanity that would be displaced by the proposed reservoir, the demand for adequate monetary compensation involved in their relocation in alternative sites, and the pleas for environmental and/or wildlife protection in the area of submergence are likely to drag on for decades. That would undoubtedly eat into the vitals of the economy concerning the project and stall the very purpose for which the project was planned. Such instances are not new in the Indian scenario, as the past experiences should have already taught us a lesson or two, and if we do not learn from past experiences, we have only ourselves to blame.

The time-tested and age-old method of water resource management by construction of several small bunds or *johads* across lower order streams for harnessing the surface runoff water still appears to be the most effective way of collecting runoff water in a series of small cascading tanks before it joins the higher order streams. This way, the surface runoff water gushing instantaneously into higher and highest orders of streams/rivers and eventually getting wasted by flowing into the sea during the peak monsoon seasons and local flood periods can be effectively impounded in the watersheds for use in lean summer season. Most importantly, this measure should be complemented by frequent desilting exercise to be carried out in the cascading tanks, so that they can accommodate more runoff water flowing into them during the monsoon and flood times. This way, we would start our realistic and pragmatic water resource management practices in watersheds, by preserving, reviving and nurturing the olive trees of age-old customs taught and handed over down the line to us by our forefathers who, in fact, are more realistic and practical for their times than we currently are. What else would be a better and more fitting way of paying a tribute to them than following the ways taught by them?

Chapter 8

In managing water resource, we attempt locating new sources of water. In this chapter, we:

- Review geomorphologic processes and features where water is the agent.
- Get familiarized to geomorphologic features favourable as source of water.
- Try to know means and ways to identify those geomorphologic features.

LOCATING NEW SOURCES OF WATER

'Geomorphic processes leave their distinctive imprint upon landforms, and each geomorphic process develops its own characteristic assemblage of land forms'. So goes one of the fundamental concepts of Geomorphology, the science of the planet Earth's morphology, or simply the *'science of the landforms'*. Just like each and every plant or animal has its own diagnostic characteristics, the landforms too have their individually distinguishing features that are developed by the geomorphic processes. The geomorphic processes are all those physical and chemical changes, which effect modifications on the Earth's surface. These processes are carried out by any natural medium that is capable of accumulating and transporting the Earth's material. Thus, wind, running water (including the surface runoff and the stream water), groundwater, glaciers and movements within the water body are strong agents for the geomorphic processes. These geomorphic processes work together in continuum, both spatially and temporally, to create a wide variety of landforms on the Earth's surface.

The study of geomorphology is an important constituent in the scientific pursuits to locate the favourable landforms that bear significant quantities of water within them. For this reason, the geomorphic study ought to be one of the major inputs in our approach towards the water resource management. It is a strong logic that the water resources associated with the Earth's surface (as surface runoff in the watersheds, drainage basins and reservoirs) as also in the sub-surface (groundwater) are contained within specific landforms. Hence, in all purposeful attempts towards the water resource management, the study of geomorphology or landforms becomes an integral part. The lack of an in-depth analysis of geomorphology in water resource management might lead to a situation wherein the desired results would be difficult to come by.

The study of geomorphology in conjunction with the landslide-related and seismological investigations is important in the feasibility studies for the proposed river-linking projects too. The possible fallout in the case of ignoring the geomorphic study might vary from geological hazards to the projects to escalation of the cost of the projects to unimaginable proportions to overcome the possible hazards. Further, for proposing sites for the check dams, large and small reservoirs, canal alignments and locations for groundwater exploration, the studies of geomorphology and landforms become a crucial factor, lest the results obtained become disastrous, with the whole exercise becoming uneconomical and/or unsustainable.

Geomorphology, by itself, is a vast topic of study and research. Every geomorphic agent namely, wind, water and glacier carves out an array of geomorphic features. Further, the geomorphic features created by these agents are divisible into two distinct processes—namely, the *erosion* and the *deposition*. The water resource management concerns the geodynamics related to the rivers and streams, or what is known as the *fluvial* processes. Therefore, we shall now confine to the study of some of the landforms associated with water activity and developed due to erosion and deposition under the fluvial or river system, wherein water is the geomorphic agent.

The geomorphic processes can be broadly classified as Endogenetic and Exogenetic (Thornbury, 1969). The *endogenetic geomorphic processes* originate from the dynamic forces that operate from *within* the earth's crust, such as diastrophism and volcanism. The endogenetic processes elevate or build up portions of the earth's surface, thereby preventing the earth's surface from eventually getting reduced to the sea level. The diastrophic processes are of two types—*Orogenic*, when mountain-building activities (such as the formation of Himalayas) with deformation of the rocks occur, and *Epeirogenic*, associated with regional uplift of landmasses without significant deformation of the rocks. *Volcanism* is the movement of the moulten rock or magma onto the earth's crust. These moulten rocks may be extruded through centralized vents to result into volcanoes, or through linear fissures as mass eruptions. The effect of igneous intrusions are dome-shaped hills, or intrusion of magma into older rocks, which when exposed by erosion, give rise to landforms different from those developed upon the rocks enclosing the volcanic rocks. The endogenetic processes continually attempt to build up or restore areas, which have been eroded away by the exogenetic processes.

The *exogenetic geomorphic processes* result from the forces that act *externally* from the earth under the influence of the above-mentioned geomorphic agents and bring about weathering, mass wasting and erosion of the crustal materials, to create new landforms. The exogenetic geomorphic processes create the landforms that are related to the water resources, and thus, they are of our immediate concern. Let us now take a brief review of these geomorphic processes and landforms created by them.

Erosion: Erosion is a comprehensive term applied to various ways by which the mobile agencies like water and air obtain and remove rock debris. There are four aspects to most of the processes of erosion by the rivers:

1. The *hydraulic* action by the river water by which loosening and removal of materials takes place.
2. The *corrasion* action comprising wearing away of bedrock, sides and floor of the rivers through the impact of boulders, pebbles, sand and silt on transit.
3. The *attrition* process in which mutual wear and tear of rock particles take place while they are on transport, due to collision, rubbing and breakage.
4. The *corrosion* action is the solvent and chemical action of the stream water on the soluble rock material—mostly limestone, carbonate and calc-silicate rocks—with which it comes into contact.

Erosional geomorphic forms mostly possess greater relief and they show more striking landforms. The geomorphic agents such as water always try to reduce these landforms into sediments or debris and deposit them whenever the velocity and energy of the running water drops, or if the supply of the debris and sediments exceeds in a way that the stream water can no longer move or transport them.

The most common erosional landforms generated by the river are valleys. Contrast to the glacial valleys, which show typically U-shaped cross sections, the valleys cut by the rivers display V-shaped sections. The development of a river valley depends on the original surface slope, the quantity of rainfall received in the area and the underlying geological structure, which determines the varied resistance to erosion offered by the rocks. Where a newly emergent land provides an initial seaward slope, the rivers that flow down the slope and the valleys that these streams excavate are said to be *consequent*. The valley faces by themselves form the secondary slopes down which tributaries can develop; these tributaries and their valleys are called *subsequent*. In succession, other generations of tributaries form. The main river and all its tributaries constitute a river system, and the whole area from which the river system derives water and rock-waste is the *basin* of that river. The term *drainage pattern* refers to the particular plan or design, which the individual stream courses collectively form.

It is pertinent to mention here that a river system or a drainage basin of a consequent stream comprises many watersheds, which are made up of several subsequent streams. Further, the river basin has a totally different connotation than the groundwater basin and they are totally independent of one another. A groundwater basin may be fed or recharged by more than one consequent stream and, therefore, it may spread over more than one river basin. Contrarily, some of the watersheds belonging one consequent river basin, which feeds or recharges a groundwater basin, might charge a separate groundwater basin situated adjoining to it without physical link or continuity.

Weathering of rocks continuously supplies rock-waste, which falls or is washed away by rain into the nearest stream. The stream carries away the debris contributed to it, and at the same time acquires more by eroding its own channel. *Valleys* develop by the removal of material, all of which is carried away by the streams that drain them. Rivers are, by far, the chief agents in excavation of valleys, not only because of their own erosive capability, but also because of their tremendous power of transportation. Under the same erosional activity by river or water, which involves the development of valleys, several residual landforms such as peaks, ridges, isolated pinnacles of resistant rocks and inselbergs result. The consequent and subsequent streams dissect the original uniform topography gradually transforming it over a geologic period into a varied and uneven landscape. Sooner or later, as rock-waste continues to erode away by streams from every part of the area, the valleys are widened and the intervening divides are reduced, until the region may be worn down to a low-lying surface of residual relief, which are called *peneplain*. This sequence of transformation takes place during a long geologic period and is known as *cycle of erosion*, which is divided into three stages referred to as *youth, mature* and *old age*.

Erosional landforms: The development of *pot-holes* in the rocks of the valley floor is an example of erosion due to the stream by corrasion action. The pot-holes develop from the gaps, hollows and depressions of rocky channels of the streams where boulders and pebbles act as erosive tools and rapidly swirl around by *eddies* or spiral motions triggered by moving water. In the humid and semi-arid regions such as peninsular India, as the precipitation of rain takes place, a mechanical effect of the water on the surface of the Earth takes place, and the water washes down the loose particles to the lower levels of the topographic relief. Eventually, the rain-wash is swept into *rills* (small brooks separated by roughly equal distances along slopes) and streams. In the event of a heavy downpour, a large volume of floodwater as a sheet of flowing water is produced at a local level, which sweeps the underlying soil leaving a long gash in the hillside. The gash gradually deepens into a gully by subsequent rains, and soon the gully may tap the water table, to carry both the surface runoff and groundwater to transform itself into a rivulet. This is one of the ways in which the tributaries originate. Besides, in the semi-arid regions, where the rainfalls, especially pre-monsoon and summer rains, frequently turns violent, the gently sloping ground is dissected by vigorous rain gashes to produce intricate forms of innumerable *gullies* and small *ravines*, separated by *spurs*. The ridges that intervene the gullies are further dissected into ribs and trenches. Such landforms are known as *badlands*. Frequently, along slopes of hills, slow downward movement of soil, a phenomenon called *soil creep* takes place. A direct field evidence of this is the tree-trunks curving at the ground towards down-slope. In the gorges of youthful rivers of Himalayas and Western Ghats, *debris-slide* or *rock-avalanche* often occur in the similar mechanism as that of soil creep, but on many times at vigorous magnitudes. Furthermore, similar conditions favour *landslides* on much bigger scales, wherever slumping or sliding occurs on the sides of slopes, precipices and cliff, or of road and railway cuttings, particularly where dense rock formation overlies weak formations liable for soaking by water.

As the river continues to flow from one side to the other side of its basin, it undercuts the bluffs (steep bank overlooking the river) wherever a bend impinges upon them, and so the widening of the valley floor proceeds. The freely forming bends along the river course are called *meanders* (Figure 8.1).

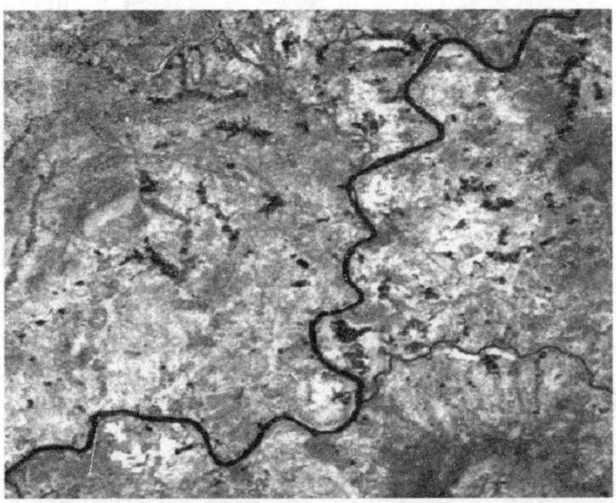

Figure 8.1. Subset of an IRS 1D LISS 3 imagery in band-3, showing meanders of a high order river in peninsular Indian terrain.

The *valley flat*, a bedrock erosional surface, is the fundamental landform produced by the lateral erosion by the stream. The development of a valley flat involves elimination of the spurs that occur between the stream meanders by lateral erosion, as the stream flows against the upstream and downstream sides of spurs in going around the meanders. Continuation of this process reduces the width of the meander spur, until only a narrow *meander neck* separates the stream on the two sides of the meander. Continued impingement of a stream against the two sides of a meander neck would result in the stream cutting through it and forming a *meander cutoff*. The development of meander cutoff in three stages is illustrated in Figure 8.2. A remnant of the meander spur will then be left out as a *meander core*, and the abandoned route around it becomes a *meander scar*. The stage that precedes the formation of a detached meander core is that in which a stream flows across a meander neck beneath a natural bridge or arch.

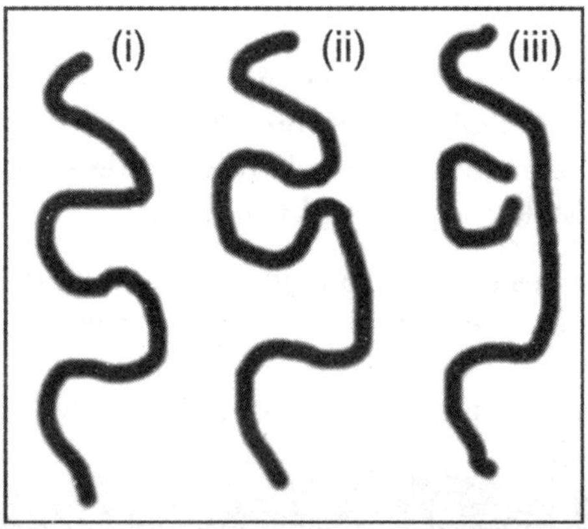

Figure 8.2. Meander cutoff development, shown in three stages. A meander (i) effects lateral erosion of the banks of its channel and produces a meander neck (ii). Continued erosion results in isolation of meander neck, resulting in meander cutoff (iii).

In the course of erosion, if an underlying bed happens to be less resistant or non-resistant, it is gradually worn away by undercutting, and the overlying resistant rocks collapse along joints. The resistant rocks eventually begin to stand out as prominent ledge, which subsequently becomes steep to become *escarpments*. The escarpment and its dip slope on the other side of the escarpment together form *cuesta*. If the beds dip at high angles on either side, the resulting feature is a *ridge* or *hogback*. If the beds are horizontal, the erosion leads to the development of *mesa* or tableland. By continued erosion and weathering of mesa produces an isolated small flat-topped hillock called *butte*.

Figure 8.3. IRS 1D LISS 3 imagery (false colour compositeRGB-3,2,1) subset showing channel fills (seen as dendritic pattern) in an area dissected by low order streams.

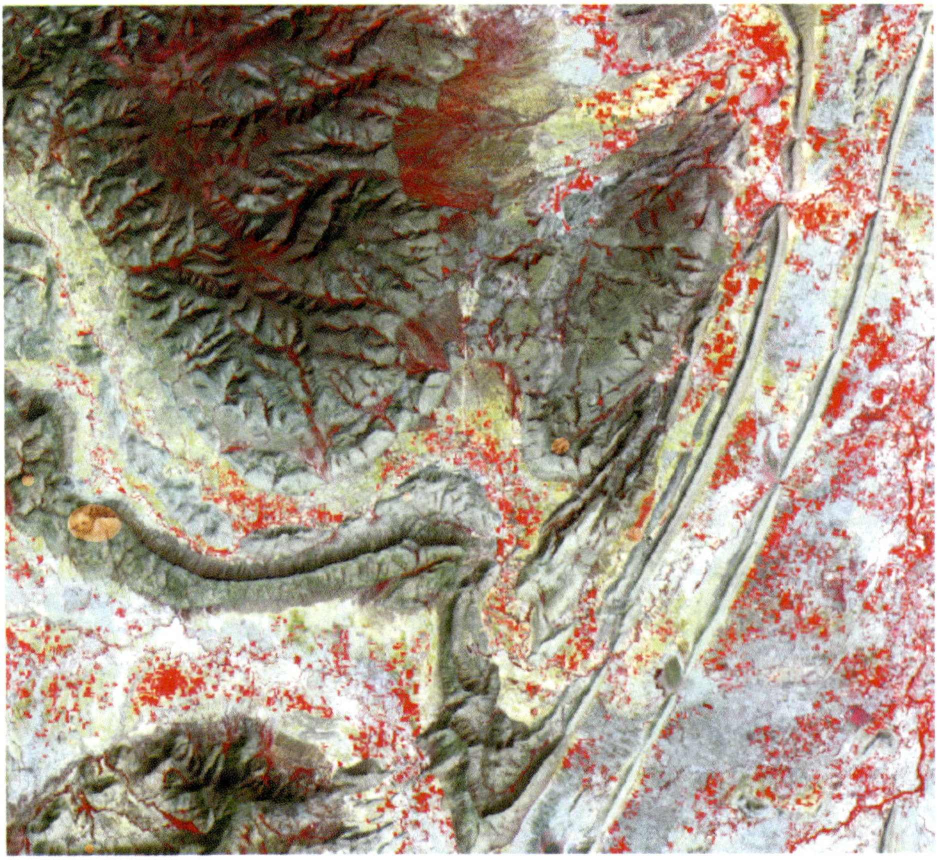

Figure 8.5. IRS 1D LISS 3 linear-stretched imagery (false colour compositeRGB-3,2,1) of structural valleys filled with the sediments derived from the enclosing structure-controlled hills. There is intense agricultural activity in these valley-fills on account of the availability of water in plenty.

Deposition: Deposition is the process that succeeds erosion and it is the process of dumping of sediments or debris that is transported by the river/ stream water within the river channel. Deposition is the complementary process of erosion as these two processes together are responsible to the general leveling of the earth's surface.

Depositional landforms: If a stream is heavily laden with coarse debris from nearby mountains, its minimum gradient and rate of flow are necessarily greater than for one carrying finer material. As floods subside, the energy of the flowing water drops, and the initiation of *flood-plain,* the deposition of alluvium over the valley flat, begins. Just as the valley is the fundamental feature resulting from stream erosion, the flood-plain is essentially the product of deposition. The flood-plains consist of two types of deposition materials. The *channel deposits* or *channel fills* are those which fill the river channels by deposition of alluvium and debris (Figures 8.3 and 8.4), otherwise known as river-bed load sediments. The *overbank* deposits are those laid down on the surface of the flood-plain when the stream overflows its banks.

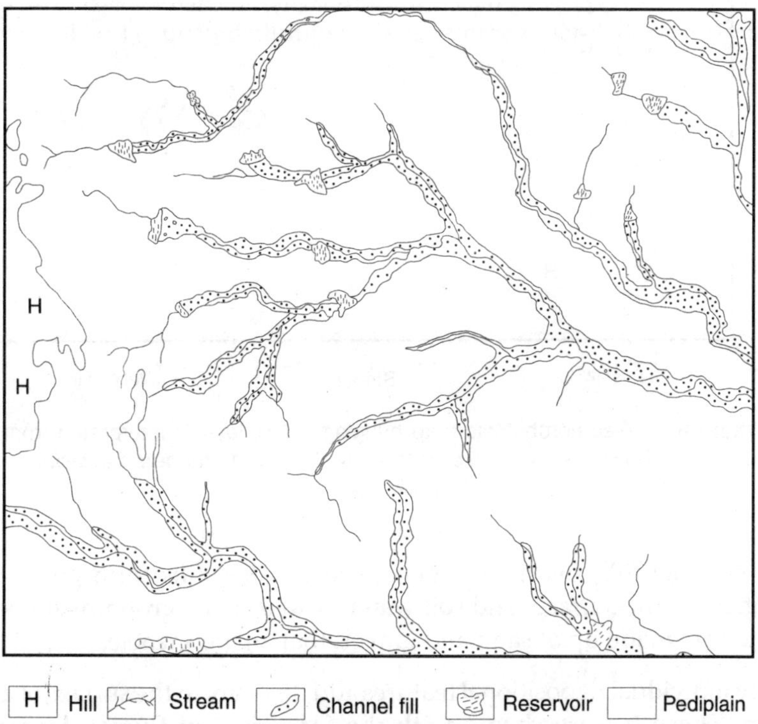

Figure 8.4. Geomorphologic map interpreted from the imagery shown as Figure 8.3, delineating channel fills. (*Assistance of Mr. U. N. Satapathy in preparation of this map is acknowledged*).

Whenever the stream overflows its banks, the water currents are checked at the margin of the channel, and the coarsest of the debris is dropped there. Thus, a low embankment known as *levee* is built on the either side of the flood-plain. Gradually, the deposition of debris carried by the river from the adjoining hills progresses and large quantities of those sediments in the form of sand and gravels along with boulders fill up the valley to build up what are known as *valley fill* deposits (Figures 8.5 and 8.6).

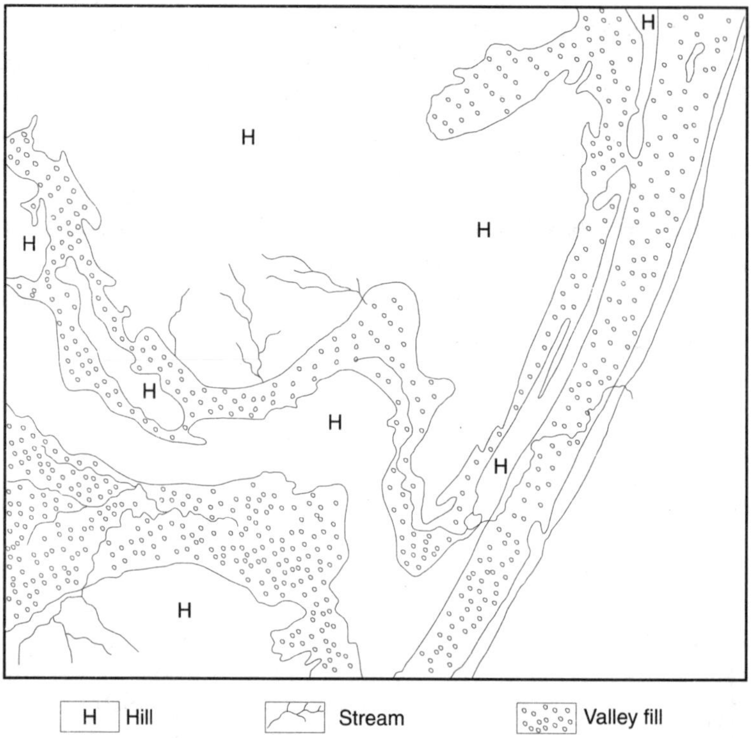

Figure 8.6. Geomorphologic map bringing out valley-fills interpreted from the imagery shown as Figure 8.5. (*Mr. U. N. Satapathy rendered assistance in preparation of this map*).

The channel-fills, overbank deposits and valley fills form good quality aquifers for the surface- and soil-water as well as the groundwater system, as they are made up of sand and gravels that are permeable.

Many individual depositional features add diversity to the surface of a flood-plain. These features begin with the formation of fluvial depositional landforms known as *bars*. *Point bars* (or meander bars) grow by individual increments of deposits outward into the meander curve. *Channel bars* are ellipsoidal or lens-shaped deposits that are located in the stream course, and these are responsible for the sinuous course as well as the branching of the river channels. Figure 8.7 shows a few point bars and channel bars in the course of the Godavari River.

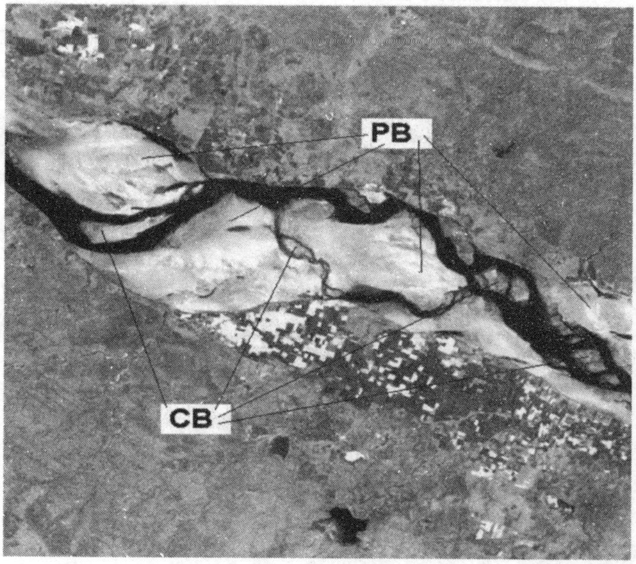

Figure 8.7. Point bar (PB) and channel bar (CB) along a part of Godavari River.

When a river reaches the sea, much of its load may be quickly deposited, partly because the current is checked and partly because the salt water coagulates the fine particles of the sediments. The sediments thus settle more rapidly than in fresh water. Waves and currents may be sufficiently strong to sweep away the material and prevent the mouth from being silted up. As the deposition progresses, the channel of the river becomes choked and obstructed, as a result of which the river can no longer carry the water, and, therefore, the river channel breaks. This process is repeated several times, until branching channels called *distributaries* are formed. The resulting fluvial landform is a combination of fine alluvial terrace deposit punctuated by distributaries, and it is known as *delta*.

Along with flood-plains of rivers, the deltas have provided sites for mankind's civilization, both in the ancient and modern times, in India and elsewhere, for the reason that they form excellent aquifer, the life-line, as well as fertile soil source for agriculture.

The features similar to delta, but occurring in the interior land away from the sea, are known as *alluvial fans*. Where a sediment-laden stream emerges from hills or mountain onto a lowland there is a marked change in gradient with resulting deposition of alluvium, apexing at the point of emergence and spreading out in a fan-like form onto the lowland. If the slope of the surface is steep, the feature is called as *alluvial cone.* A series of fans may coalesce to form an extensive *piedmont alluvial plain* or *bajada* (Figure 8.8).

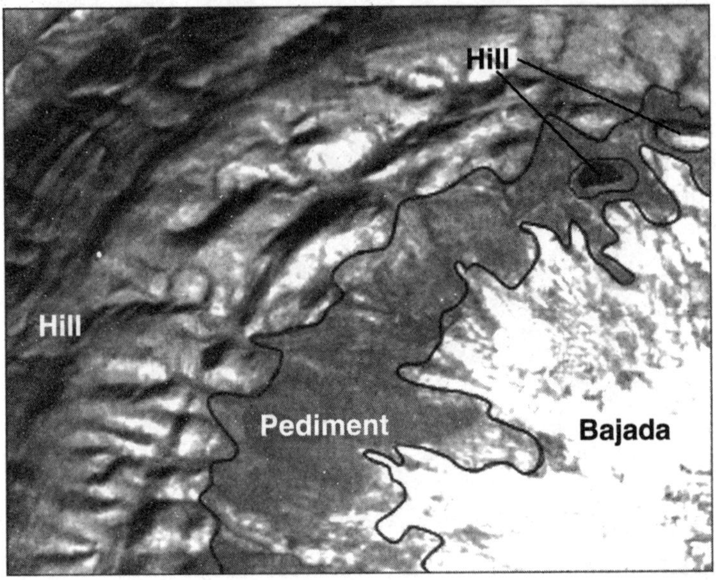

Figure 8.8. IRS 1D-LISS 3 imagery bringing out geomorphic features like pediment, bajada and hills.

In the arid and semi-arid areas, the hilly landforms are mostly bordered by slopes, which extend downward to neighbouring river basin floors. This area of slope consists of two parts—a lower part comprising *bajada*, and an upper part, which is an eroded bedrock surface veneered with alluvium. The upper part is called as *pediment* (Figure 8.8), whose slopes vary between about one to seven degrees, whereas the mountain slopes, which they abut are much steeper, ranging from 15 degrees to sub-vertical. In the arid and desert regions, the term pediment is frequently used to designate the rock-floored plains adjacent to, or interpenetrating, the mountain masses, which are developed by planation of the mountains at altitudes dependent upon those of the interior basins into which the mountains drain. A series of coalescing pediments constitute *pediplain.* Pediment zones mostly do not make up aquifers, but at places, unconfined aquifers may be present in pediment zones. Pediplain allows infiltration of surface runoff due to its plain surface without any slope; therefore, pediplain has good potential for developing huge groundwater reservoirs and basins.

The Earth has always been a dynamic planet. The orogenic and epeirogenic processes in the diastrophic movements, as well as the structural and tectonic disturbances continually accompany the evolution of the Earth during the geological times, including during the Quaternary period (since approximately 1.8 million years from now). These processes often cause gradual upliftment of the landmass and result in the disturbance of the surfacial drainage courses. Such disturbances are recorded on the surface of the Earth as changes or migrations in the course of the river channels and the abandonment of the river channels. These are brought about due to terrain upliftment, leading to sudden changes or modifications in the slope angles of the terrain or choking of the river channels as a result of sudden deposition of huge quantities of sediments and debris. These modifications of the Earth's surface lead to the river channels becoming disused, though the original course of the rivers/streams/tributaries get preserved as relics. The relic courses of rivers are what are known as the *paleo-channels*, and they may not currently show any fluvial activity in the form of active channels, flowing water, erosion of sediments, but the activity of the relic fluvial activity may be discernible as dried-up river channels and deposition of fluvial sediments in the remote-sensing tools such as satellite imagery (Figure 8.9) and aerial photographs, as also gravel beds in field. The existence of the mythological

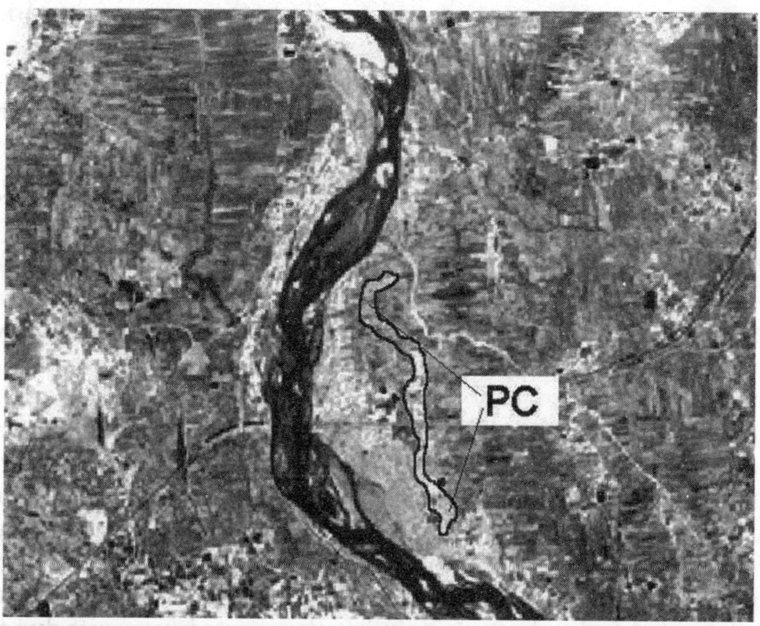

Figure 8.9. Paleo-channel (PC) of a major river, demarcated in an IRS 1D LISS3 band-3 imagery.

Saraswati River in the northwest India has been inferred from a few segments of its paleo-channels. Under somewhat similar conditions of diastrophic and tectonic movements as well as choking of channels due to excessive deposition, major rivers and its tributaries might abandon their present channels and migrate to follow a slightly different course. These channels are known as *abandoned channels.*

Study of geomorphology or landforms in water resource management

The channel-fills, valley-fills, paleo-channels and abandoned channels are efficient aquifers as they most often or always contain beneath them large quantities of water that would have infiltrated through them over a period of centuries of years, as well as connate waters (the water in the groundwater system that is entrapped simultaneous with the deposition of the sediments), both of which would have together built huge groundwater reservoirs and basins that might contain substantial reserves of groundwater. Further, over a period of geologic time, the geomorphic features in the meander system (such as meander cutoff, and even the meanders) are filled with sedimentary deposits due to fluvial processes or diastrophism or a combination of both. These deposits also hold tremendous potential for groundwater. Therefore, water resource management as well as efforts towards augmentation of water resources would necessarily have a geo-scientific input in the form of locating these geomorphic features on ground. The exercise of locating these landforms or geomorphic features on ground is, indeed, an arduous task. For this purpose, one needs to develop extraordinarily sound scientific logics and a flair for withstanding extremely rigorous conditions of field life in the course of their efforts in identifying these geomorphic features.

It is heartening at this stage to note that a powerful alternative mode of technology in the form of *Remote-Sensing* is currently available for quick results in a minimum time duration and least human labour for this exercise. Remote-sensing with the aid of satellite images is an inexpensive, speedy and non-invasive technology that offers insights into the efforts to locate the landforms such as channel-fills and paleo-channels. The meander system units of the present and the geological past too display characteristic patterns in the satellite imagery. Remote-sensing, therefore, can be applied prior to the invasive and expensive procedures such as drilling for locating the aquifers and detailed exploration for the water resources. The satellite imagery in digital format are currently used in the interpretation by computer-aided technologies in locating the groundwater reservoirs and the geomorphic features.

One should, nevertheless, remember that remote-sensing is only a technological tool to minimise the arduous search for the landforms that are favourable for water resources, but would not constitute a holistic approach towards this endeavour. Without validation using the empirical field data on geology, soil-type, bedrock and geophysical parameters, the remote-sensing technique would not stand scrutiny on its own, and therefore, the remote-sensing scientists need to concurrently develop proper skills to recognise the field parameters that diagnose the channel-fills and paleo-channels. In other words, delineation of geomorphic features and landforms that overlie the aquifers using remote-sensing as beginning, followed by validating the thus delineated landforms using empirical field data would constitute the healthy practice in the study for water resource by remote-sensing.

The efforts of water resource management, therefore, would be *inter alia* aimed at identifying the valley fills, channel fills, meander cutoffs, pediments, abandoned-channels and paleo-channels of rivers in all parts of the Indian terrain. This is not a difficult task given that India has the entire wherewithal in the form of aerial photographs, satellite imagery and the geo-scientists who can apply the proven technology in a systematic way so as to cover the whole country by this exercise. Just as the team of Indian space scientists constitutes a mighty force to be reckoned with in the field of satellite launch and space technology, the Indian scientists who can apply the space data also constitute highly effective and powerful groups of remote-sensing scientists in various organizations. They are involved in the interpretation of satellite imagery both in the hard copy format and the computer-compatible digital format, for a large variety of purposes, the most important being the water resource management. This is evident from a large volume of data on water resource that is generated, presented and hotly debated in several scientific meets and published in innumerable journals. A large number of scientists employed by the Government of India, various State Governments as also private entrepreneurs have generated a host of data in the form of thematic maps, such as isopach maps, geomorphologic maps, landform maps, drainage maps, geo-hydrological and geological maps. These data are, however, spread over widely as bits and pieces. Therefore, a meaningful synthesis by focusing them to a single theme and orchestrating them for water resource management in respect of the whole country as well as to arrive at a decision on crucial matter such as Water Policy remains a veritable challenge. A national-level consortium of organizations governed by the Government of India had taken a significant step forward in this direction to create what is known as National Spatial Data Infrastructure (NSDI) for the whole country in the form of multi-thematic layers with the aid of GIS, the Geographic Information System. (A brief account of GIS is given later). The data generated by remote-sensing with field validation on the occurrence of the above-mentioned geomorphic landforms suitable for the surface and groundwater resources in the whole Indian terrain should form one of the constituent layers in the NSDI, in order to facilitate formulation of a concrete, unambiguous Water Policy for the entire country. Generation and compilation of data on the water resources development on the national level should be given the topmost priority in the creation of the NSDI with a farsighted plan and development for the year 2020. India, as we are fully aware, is a mighty force in both remote-sensing and information technology and our youth contribute significantly to the global operators of these technologies. Even if a fraction of the talents and knowledge bank available with our youth is channelized and tapped for projects like water resource-linked information related to the NSDI, we would not only be able to make tremendous impact in locating new sources of water, both of surface and sub-surface levels, but also creating new job opportunities to the millions of our generation next.

Chapter 9

This chapter is devoted to studies on:

- Fundamentals of Groundwater System and groundwater resource management.
- Aquifers, their characteristics and seawater ingression into aquifers.
- Groundwater scenario in India.
- Methods of groundwater recharging.

MANAGEMENT OF THE GROUNDWATER RESOURCE

In recent times, there is a tremendous awareness generated among the country's masses on the importance of groundwater recharging, thanks to the vigorous campaigns conducted by some of the State Governments of the Indian Union. This awareness, however, could not be so far translated to any substantial water wealth replenishments to the groundwater system. Nevertheless, a large number of urban dwellers have started practising the groundwater recharging, and therefore there have been local rise in the groundwater levels. This rise, however, is not sufficient in areas of groundwater overexploitation. Prior to understanding the groundwater management and augmentation of groundwater resources, it is necessary to review the concepts and the other scientific aspects of the groundwater, since, until and unless these concepts are fully understood, appropriate approach to the optimum utility and the recharging of groundwater cannot be employed effectively. Therefore, important scientific aspects pertaining to the groundwater are summarised in the beginning of this chapter. Further, modern approaches to groundwater resource augmentation are also briefly discussed.

FUNDAMENTALS OF GROUNDWATER

Groundwater refers to the water beneath the surface of the Earth, occupying voids, fractures, joints, pores, interstices between the grains of sediments, and cavities within the rock sequences that constitute geological formations. Groundwater is a part of the Hydrologic Cycle, diagrammatically illustrated in Figure 9.1. The geological rock formations of the Earth's crust act as conduits for transmission and as reservoirs for storage of water.

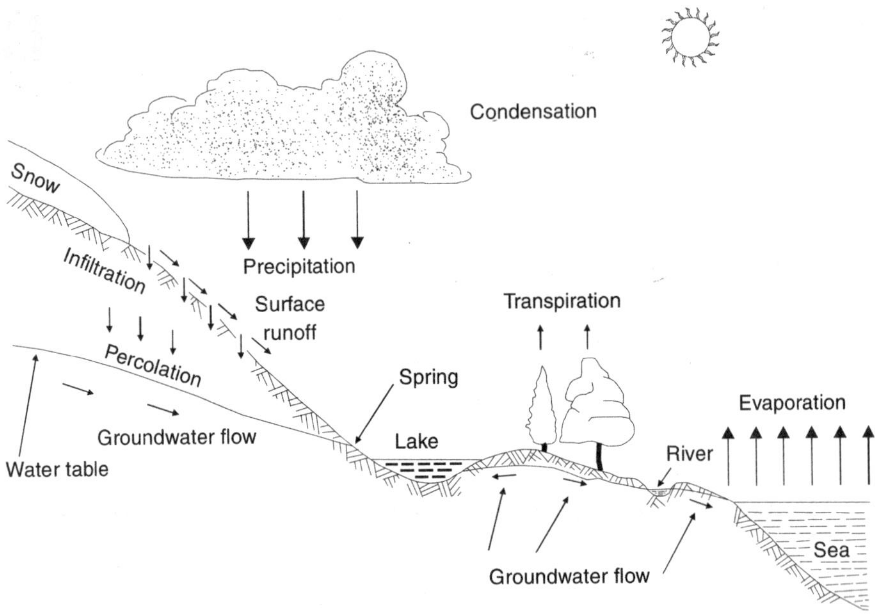

Figure 9.1. Hydrologic Cycle.

Groundwater originates initially as surface water, which owes its source mainly from precipitation. Some contribution in a few areas at certain altitudes and latitudes accretes to the groundwater resource from snowmelt too. Water at times gets entrapped in the interstices of the sediments at the time of deposition; such water is known as connate water. These two sources form the original or primary recharge to the groundwater system.

Water enters these formations from the Earth's surface runoff or from surface water bodies and preserved in underground reservoirs. The water thus accumulated in the subsurface reservoirs over several years or decades or, at times, even centuries provides large resource of water, and it is known as groundwater resource. A part of the subsurface water, however, travels slowly using the natural slope, until it reaches the surface by action of natural flow or human activity. Groundwater that emerges into the surface stream channels aids in sustaining the flow of the stream under the condition of depleted surface runoff. Apart from the rainwater, significant quantities of water contributed by artificial recharge from the sources of excess irrigation of agricultural lands, seepage from canals, lakes, ponds and several other forms of human activities enter the groundwater. Water can enter the groundwater system even from the sources of sea, creeks, backwaters and lagoon, if the hydraulic gradients slope downward in an inland direction. The quality of the groundwater, therefore, depends on the cumulative effect of all these natural and artificial sources of water that gets recharged into the groundwater system.

Release of groundwater as *springs* occurs when water re-emerges to the ground surface where the subsurface level of groundwater intersects the ground levels, such as the locations of deep erosion. This water flows as streams, or gets accumulated in large depressions on the surface as natural lakes. Frequently, groundwater reaches the ground surface through porous media of rock or soil at a slow rate to be known as *seepage*, and forms small surface water bodies like ponds. In contrast, *spring* is a concentrated discharge of groundwater occurring as a current of flowing water. The springs of groundwater origin are known as *gravity springs*, which result from water flowing under hydrostatic pressure.

There are four varieties of gravity springs:

1. *Depression springs* are formed where the land surface intersects water table.
2. *Contact springs* are created by permeable water-bearing formation overlying less permeable formation, which intersects the ground surface.
3. *Artesian springs* result from the releases of water under pressure from confined aquifers either at an outcrop of the aquifer or through an opening in the confining bed.
4. *Tubular* or *Fracture* springs release groundwater from rounded channels such as solution channels or fractures in impermeable rock connecting with a groundwater supply.

Thermal or *hot springs* are of non-gravitational type of groundwater associated with volcanic rocks or deep-sited fissures or fractures or faults, and therefore warm. Part of these waters may be juvenile, related to magma, but get mixed up with groundwater on its rise from deep inner parts of the Earth to the ground surface.

The groundwater near the surface returns directly to the atmosphere by evaporation from within the soil and transpiration from the vegetation. The effect of evaporation is high in arid and semi-arid regions and transpiration is phenomenal in subtropical rainforests. Jointly, these types of climatic conditions prevail over a major part of India. Therefore, the loss of groundwater from shallow levels due to the combined effect known as *evapo-transpiration* is considerably high in the temperate weathers of India. Nevertheless, the water thus reaching atmosphere soon becomes a part of the Hydrological Cycle and returns to the surface of the Earth.

AQUIFERS

The movement of groundwater in the subsurface is governed by the *permeability* of aquifer and the *hydraulic gradient*. In order that the geological formations can bear groundwater, these rock formations essentially need to be permeable, a trait which is expressed in terms of *coefficient of permeability* (Todd, 1959). The coefficient of permeability is defined as the flow of water at 60° F in terms of gallons per day through a medium having a cross-sectional area of one square-foot, under a hydraulic gradient of one foot to one foot.

The permeable geological formations, which contain groundwater, are known as '*aquifers*'. Also, a rock formation, which has the capacity to host or yield significant quantities of groundwater, is defined as aquifer. An impermeable geological formation, which can bear water, but is not capable of permitting water to trickle through it, such as clay bed, is known as '*aquiclude*'. Impermeable rocks or formations neither containing water nor allowing water to pass through them are '*aquifuge*'. Massive, unfractured hard rocks such as gneiss, granite and Deccan Trap basalt can form aquifuge. The groundwater occurs in the subsurface either in the *zone of saturation* or the *zone of aeration*. In the zone of saturation, all the interstices in the rock are filled with water.

The portion of the rock or soil that is not occupied by mineral matter or cement is filled by water in the groundwater system. These vacant spaces are known as voids, interstices, pores or pore-space. Primary interstices and voids are created by geological processes such as sediment deposition, which govern the formation of the rocks. There are also secondary interstices and voids developed after the formation of the rock due to deformation of the rocks, and those are known as joints and fractures. Further voids such as solution cavities in the terrain comprising calc-silicate and carbonate rocks, and limestone also accommodate the groundwater. The interstices may be termed, on the basis of their sizes, as *capillary, super-capillary* and *sub-capillary*.

Capillary interstices are sufficiently small so that surface tension forces would hold water within them. *Super-capillary* interstices are those larger than capillary interstices. *Sub-capillary* interstices are so small that water is held primarily by adhesive force. Depending on the connection of the interstices with one another, they may be either *communicating* if they are inter-connected, or *isolated* if they are not connected. The porosity of a rock or soil is a measure of the interstices, and it is expressed as the percentage of void space to the total volume of the mass. Porosities of the rocks range from zero to above fifty percent, depending on the degree of compaction of the rocks, irrespective of whether the rocks are sedimentary, igneous or metamorphic.

The zone in which the groundwater occurs may be the *zone of saturation* or the *zone of aeration*. In the zone of saturation all the interstices, joints, fractures and voids of the geological formation are filled with water. The zone of aeration consists of the geological formations whose interstices are occupied partially by water and partially by air. The upper surface of the zone of saturation is known as the *water table*. The zone of saturation extends downward up to a depth where an impermeable stratum such as a bed of clay or the massive bedrock without voids occurs. The term groundwater actually refers to the water that occurs in the zone of saturation. The water occurring in the zone of aeration is known as *suspended* or *vadose* water. These aspects are diagrammatically illustrated in the Figure 9.2.

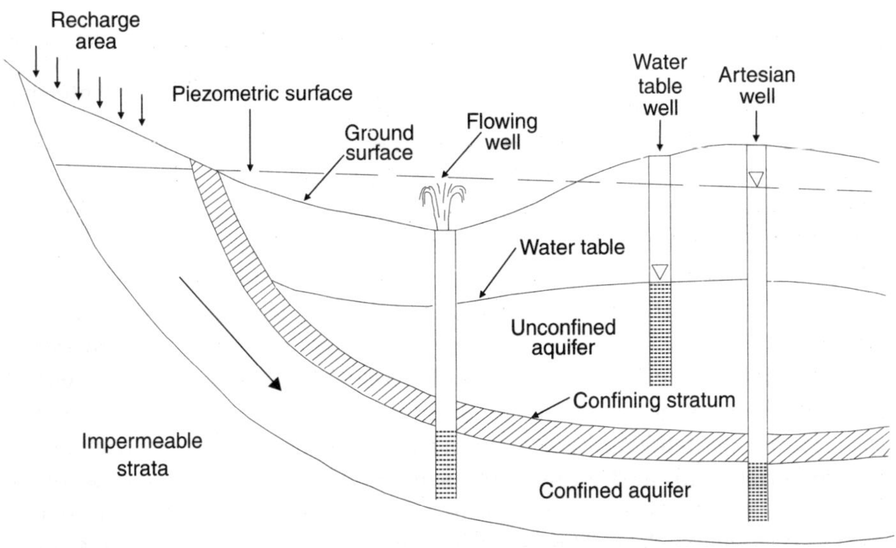

Figure 9.2. Various components of Groundwater System. (Adopted from Todd, 1959).

Zone of Aeration: The zone of aeration is sub-divided into (a) Soil Water Zone, (b) Intermediate Zone and (c) Capillary Zone (Todd, 1959). In the soil water zone, water exists at less than saturation, except temporarily when excess water reaches the ground, as for example, from rainfall or irrigation. The soil water zone extends from the ground surface to the zone of major roots of the plants. The soil in the soil water zone, in places of high humidity, tends to be *hygroscopic*, meaning that it has a tendency to absorb moisture from the atmosphere. The thickness of the soil water zone varies with the soil type and the types of vegetation cover.

The study of the soil water zone has a tremendous bearing on the soil resource management and agriculture, as these activities get main support from the soil water zone. The intermediate zone extends from the lower edge of the soil water zone to the upper limit of the capillary zone. The thickness of this zone is highly variable from zero, when the bounding zones merge with a high water table approaching the ground surface, to several meters under deep water table conditions. The water in the intermediate zone is considered to be non-moving or pellicular that is held in place due to the hygroscopic force from the zone above, and the capillary force from the underlying capillary zone. The water in excess from the soil water zone moves down under the influence of gravity to the intermediate zone, and therefore, a part of the water in this zone is also called as *gravitational water*. The capillary zone extends from the water table up to a limit where the capillary action in the geological formations can be effective.

Zone of Saturation: The groundwater fills all the interstices in the saturated zone. Hence, the porosity is a direct measure of the water contained per unit volume. The molecular and the surface tension forces always retain a portion of the water in the saturated zone, and therefore, the water contained in the saturated zone is not emptied totally by pumping or by natural drainage. The specific yield of groundwater in the saturation zone is a fraction of porosity of an aquifer. The specific yield values hinge on the grain size, shape and distribution of pores, and compaction of the strata that hold the groundwater. In the case of uniform sand being the aquifer material, the specific yield may be up to thirty percent, but most aquifers comprising unsorted alluvial sediments give specific yields in the range of ten to twenty percent. The specific yield of the aquifer materials can be determined in laboratories by determining the porosity, the water content and the mass of the materials with and without the saturated water in them, but all these laboratory methods have limitations. This is because, one should be sure that the samples collected are actually from the zone of saturation and below water table.

Geomorphic units as Aquifers

Well-defined aquifers consist of unconsolidated sedimentary rocks such as gravel and sand, which constitute beds of considerable thickness. These aquifers may occur along the watercourses, as stream channel fill sediments, in abandoned and buried valleys or stream channels, in plain terrains and in intermontane valleys. The watercourses consist of alluvium that forms and underlies stream channels, as well as form the nearby flood plain of the streams. Wells located in highly permeable strata bordering streams produce large quantities of water, as infiltration from the streams augment groundwater supplies. Abandoned or buried valleys/stream channels are those locations where the streams/rivers flowed in the geological past, and are no longer currently occupied by streams/rivers that carved them in the past. Although such valleys may resemble watercourses in terms of permeability and quantity of groundwater storage, the scope for recharge and capabilities for perennial yield are less in them, unless they are frequently recharged by rainfed process.

Extensive areas under erosional regime by a network of streams and tributaries, characterised by plain country covered by unconsolidated and unsorted sediments are present in almost all continents. Geomorphologically, such terrains are known as *pediplain*. These plains flank highlands or other geomorphic features that serve as the source for the sediments. In some places, gravel and sand beds form aquifers under these plains. In these plains, the groundwater reservoirs are recharged chiefly in areas accessible to downward percolation of water from precipitation and from occasional streams.

Intermontane valleys are stream-cut valleys developed in between two linear chain of hills or ridges. These valleys are underlain by huge volumes of unconsolidated rock materials interspersed with unconsolidated sediments (like sand, gravels and alluvial deposits), both derived by erosion of the bordering hills/ridges. Most of these intermontane valleys constitute large groundwater basins. The sedimentary deposits in these valleys are good quality aquifers, which are periodically replenished with surface runoff from the hill slopes and seepage from the streams that flow along these valleys.

Geological formations as Aquifers

Porous and permeable sediments deposited by river system, such as the alluvial deposits comprising unsorted sand and gravel, interspersed with boulders, cobbles and pebbles, are considered as the geological formation best suited for a good aquifer. Earthy material of similar composition is derived under non-alluvial conditions by the mechanical weathering combined with chemical weathering of a variety of rocks such as granite, gneiss, quartzite and sandstone. A major part of the central Indian terrain constituted by parts of Karnataka, Andhra Pradesh, Madhya Pradesh and Orissa is a terrain comprising granite and gneiss, which have produced sandy soil in this way. Further, the granite and gneiss have fractures and joints serving directly as voids for groundwater or filled with the sandy soil derived *in situ*. This sandy soil cover on, or fracture fill in, granite and gneiss too serve well as a good aquifer.

Sandstone and conglomerate are the sand and gravel that have undergone induration with their grains held together with the aid of cement. Therefore, the cement material reduces their porosity and permeability. In order that they become permeable to percolating water, they need to develop fractures and joints by deformation. Similarly, crystalline and metamorphic rocks are relatively impermeable and are poor aquifers. Where such rocks are fractured and are in decayed condition due to weathering (as described above), they can give rise to aquifers. Clay, mudstone, shale and siltstone are generally porous to varying measure, but their pores are so small that they become practically impermeable, unless they are jointed and fractured. Volcanic rocks, such as the basalt flows of Deccan Trap occurring in the Deccan plateau of India, can form somewhat good aquifers, as they are permeable. Their permeability is attributed to the flow breccias, porous zones between the lava beds, lava tubes, lava shrinkage cracks and joints. The laterite, derived by *in situ* chemical weathering of basalt, is a porous and permeable formation that can be a fairly good aquifer.

The density, porosity and permeability of limestone and other carbonate-rich sediments vary widely, depending upon the degree of consolidation and development of permeable zones after deposition and induration. The solution activity of water on limestone and carbonate rocks creates voids of varying dimensions from microscopic size to huge solution cavities and caverns. Frequently, subterranean channels sufficiently large to carry the entire flow of stream are formed in limestone terrain. Large springs are frequently found in limestone areas. Under a well-developed groundwater system, the limestone terrain eventually forms a *karst* region or topography where subterranean drainage through the limestone creates large ground water reservoirs. The drawback, however, in the quality of the groundwater in the limestone aquifers is that, the solution of calcium carbonate by the water imparts *hardness* to the groundwater.

The unique feature of the Indian terrain is the diversity of its rock formation. Added to the variety of rock types, the diverse weathers such as arid to semi-arid, and the hot and humid conditions prevailing over different parts of India have created a variety of soil types, which are mostly *in situ*, but are frequently transported to the present locations. These soil formations occur in an array of geomorphic units such as channel fills, intermontane valleys and as covers over vast pediplain. These rock and soil formations are good aquifers under a rainfed regime, provided sustainable practices for groundwater reservoir development are adopted.

Types of Aquifers

Aquifers are generally of large area and they can be visualised as large underground storage reservoirs. The groundwater flows out to the surface by the action of gravity or is drawn by the people for various human activities such as personal hygiene, agriculture and a host other purposes. The water percolates into the groundwater reservoir system and replenishes the aquifers by natural recharge fed by rainwater or as artificial recharge through human activity. The aquifers are of two types, namely (1) Unconfined aquifer and (2) Confined aquifer.

An *unconfined aquifer* is one in which a water table serves as the upper surface of the zone of saturation. It is also known as a free, phreatic or non-artesian aquifer. The water table of the unconfined aquifers is in undulating form and slope, depending upon the areas of recharge and discharge, pumpage from the wells, and permeability of the formations. Rise and fall in the water table correspond to changes in the volume of water in storage within an unconfined aquifer. Figure 9.3 is an ideal section through an unconfined aquifer. Contour maps and profiles of the water table can be prepared for conceptual illustration of the groundwater reservoir in an unconfined aquifer environment, from the elevations of water in wells tapping the aquifer, to determine the quantities of water available in the aquifer, its distribution and movement.

Isapach maps showing the zones of groundwater saturation having equal or almost equal thickness are also prepared based on the conceptual projection of the three-dimensional form of the aquifers on a map or plan. With the availability of appropriate computer-based satellite image-processing software, Geographic Information System (GIS) software and the remote-sensing tools in current times, Digital Elevation Models (DEM), which give us three-dimensional picture on the zone of groundwater saturation having equal thickness and isopach maps, as also on the groundwater reservoir as a whole, are prepared for the aquifers and reservoirs. Such studies are currently undertaken only in small isolated pockets in India. Studies employing these modern concepts and tools can be carried out in speedy manner with satisfactory accuracy even in the case of aquifers in arid regions such as Rajasthan. There is an urgent need to launch both sustained long term approaches as well as crash programmes for quick results for areas all over India using the modern developments in computer technology, besides using conventional methods of field geological and geophysical data in groundwater exploration.

Confined aquifiers, also known as *Artesian* or *Pressure Aquifers,* occur where ground water is confined under pressure greater than the atomspheric pressure by the overlying, relatively impermeable strata (Todd, 1959). Figure 9.2 illustrates the concept of a confined aquifer also. In a well penetrating the confined aquifer, the water level rises above the bottom of the confining bed due to the pressure, shown as Artesian or Flowing Wells in this figure.

Water enters a confined aquifer in an area where the confining bed rises to the surface or ends underground and the aquifer becomes unconfined. The region supplying water to a confined aquifer is known as a recharge area. Rises and falls of water in wells penetrating confined aquifers result primarily from changes in pressure rather than changes in storage volumes. Hence, confined aquifers have only small changes in storage and serve mainly as conduits for conveying water from recharge areas to locations natural or artificial discharge.

The imaginary surface coinciding with the hydrostatic pressure level of the water in the aquifer is called as the *piezometric surface* of the confined aquifer. The water level in a well penetrating a confined aquifer defines the elevation of the piezometric surface at that point. A flowing well results if the piezometric surface lies above the ground level, as shown in Figure 9.2. A confined aquifer becomes an unconfined aquifer when the piezometric surface falls below the bottom of the upper confining bed. Contour maps, profiles of the aquifer, isopach maps and digital elevation models can be prepared for the confined aquifers, just the same way as in the case of unconfined aquifers.

A special case of an unconfined aquifer is the *Perched Aquifer*, which is illustrated in Figure 9.3. This occurs wherever a groundwater body is separated from the main groundwater by an impermeable formation or

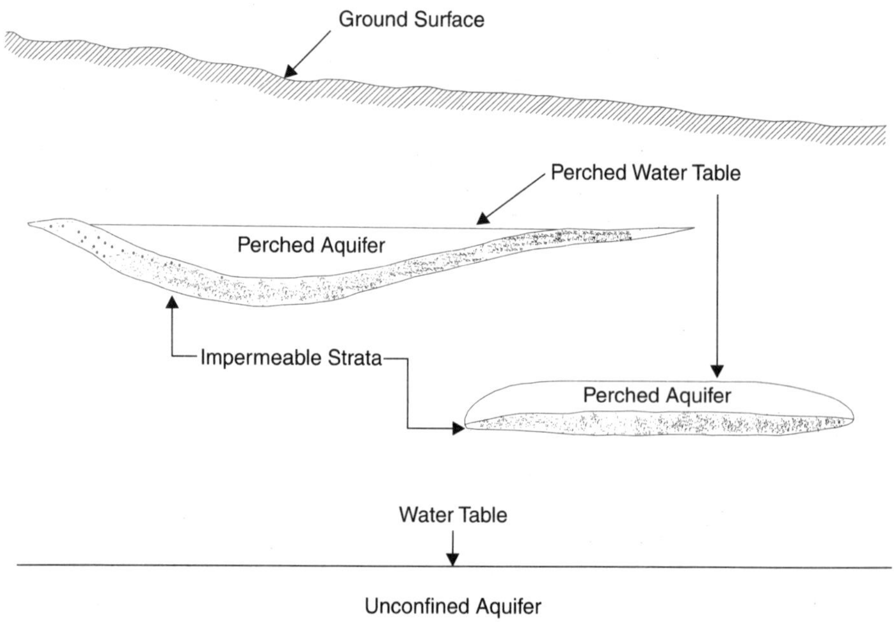

Figure 9.3. Illustration for perched aquifers. (Redrawn from Todd, 1959).

stratum of small area and by the zone of aeration above the main body of groundwater. Clay layers or beds in sedimentary deposits often have shallow perched aquifers overlying them. Interestingly, wells tapping perched aquifers yield temporary small quantities of water for short intermittent periods. Over a period of time, when the water that accumulated in the perched aquifer is totally tapped, those wells totally dry up. They remain dry till replenishment takes place in the subsequent season of rainfall.

Water recharged to, or discharged from, an aquifer represents a change in the storage volume within the aquifer. For unconfined aquifers, this is expressed by the product of the volume of aquifer lying between the water table at the beginning and at the end of a period of time, and the average specific yield of the formation. In confined aquifers, assuming the aquifer remains saturated, changes in pressure produce only small changes in storage volume. Thus, the hydrostatic pressure (the pressure exerted by the water column) within an aquifer partially supports the weight of the overburden while the solid structure of the aquifer provides the remaining support. When the hydrostatic pressure is reduced, such as by pumping water from a well penetrating the aquifer, the aquifer load increases. Compression of aquifer results, which forces out some water from the aquifer. In addition, lowering of pressure causes a small expansion and subsequent release of water.

A physiographic unit containing one large aquifer or several connected and inter-related aquifers is defined as *Groundwater Basin*. The concept of groundwater basin is important because of the need for basin-wise development of groundwater potential in a region. The approach to the development of a groundwater basin has intimate links to the management of water resources by watershed development. In fact, the efforts towards the approaches for both may overlap.

The water-yielding capacity of a confined aquifer can be expressed in terms of its storage coefficient, which is defined as the volume of water that an aquifer releases from, or takes into, storage per unit surface area of aquifer per unit change in the component of head normal to that surface. The storage coefficient for an unconfined aquifer is the same as its specific yield. Storage coefficients can be determined from pumping tests of wells. For a detailed account on the coefficient of permeability, coefficient of storage, pumping tests and several other field and laboratory practices relating to groundwater exploration, the readers may refer to Todd (1959).

Ingression of Seawater in the Aquifers

The groundwater basins or reservoirs or aquifers come in contact with sea along the coastline. The density of seawater is considerably higher than the density of the groundwater because of the total dissolved salts and minerals in the seawater. Owing to the difference in the densities of the seawater and the groundwater, under normal circumstances, the groundwater from the aquifers gradually flows into the sea. The overdraft of groundwater in the coastal areas results in the fall of the groundwater table levels in those areas. This results in the reduction of the seaward flow of groundwater, or even total reversal causing seawater to enter into the aquifers (Figure 9.4). Further, due to the lowering of water table in unconfined aquifers or the piezometric surface in confined aquifers, the natural gradient sloping downward toward the sea is reduced or reversed. Because of the difference in the densities of the two waters, an interface or boundary surface forms between the groundwater front and the seawater front (Figure 9.4).

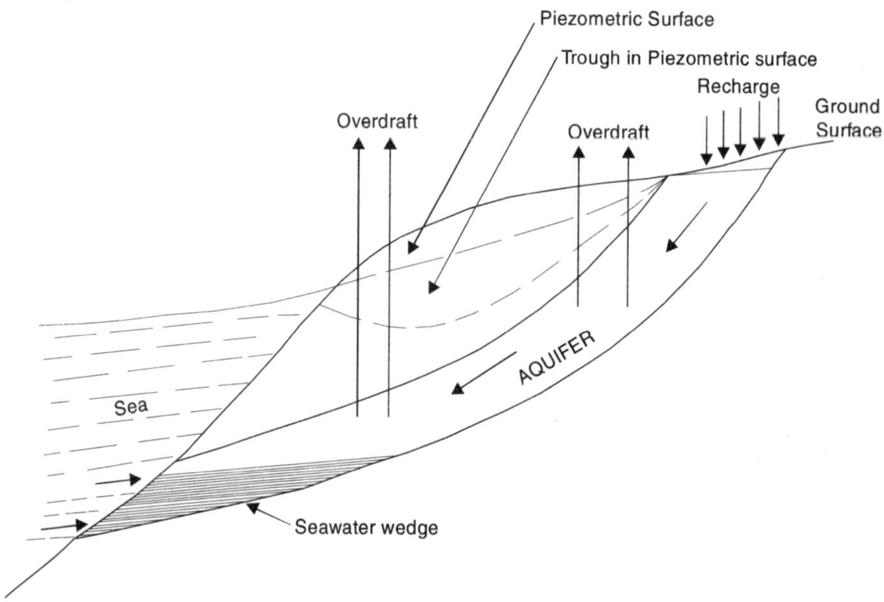

Figure 9.4. Seawater ingression in a coastal aquifer (Adopted from Todd, 1959).

The balance in the dynamics between the seawater and fresh groundwater governs the shape of this interface. When the interface moves landward due to the drop in the groundwater dynamics, the ingression of seawater into the aquifer takes place. The ingression of seawater into the groundwater system in coastal areas leads to contamination of groundwater with the salts contained in the seawater. Seawater ingression wrecks the groundwater basin or reservoir or aquifers situated near the coastal areas, and it takes several years despite copious recharge of fresh water into the groundwater system, to reverse the situation for better. Starting from the late part of the last century, almost all countries including U.S.A, U.K., Japan and Germany had to face the seawater ingression into the coastal aquifers due to water-stressed conditions prevailing in the coastal urban centres. India is no exception. The condition of groundwater in Chennai, Nellore, Ongole and Visakhapatnam in the east coast, all along the Gujarat coast and some places in the west coast, especially in the northern Kerala is affected this way.

The only way of prevention of seawater ingression into the aquifer is by gross reduction in the draft of water from a coastal aquifer. This helps in the rise of groundwater levels above sea level, thereby maintaining a seaward gradient of the groundwater table. Such a measure would entail proclamation of an unambiguous water policy by the government, followed by its strict implementation. Alternatively, in order to prevent the seawater ingression or nullify the effect of the ingression that has already taken place, large amounts of freshwater recharge into the affected aquifer has to be made. However, the moot question is, can we, in an already water-stressed status, afford to find such a large quantum of freshwater to recharge the depleted and salt-contaminated coastal aquifer. Finding an alternative sources of water to support the human activity and for augmentation of water resources is the only way to mitigate the stress on the groundwater source, especially on the coastal aquifers.

GROUNDWATER SCENARIO IN INDIA

The Central Ground Water Board (CGWB) takes care of the developing and assessing the groundwater potential of India. In addition, there are groundwater development departments in almost all the States. Together they carry out hydrogeological surveys, exploration, evaluation and monitoring of ground water regime on a continuous basis round the year. Their role in mitigating the water crisis has been pivotal to the successful tiding over of the crunchy summer months, and therefore crucial to both people and the governments alike. With the advent of remote-sensing technique, large areas are examined for groundwater potential in shorter time with more accuracy. But it should be borne in mind that the empirical field data painstakingly collected over a period of half-a-century has been the foundation for the exploration methods using modern techniques like remote-sensing. It should be further borne in mind that for all the field-based scientific studies, like the groundwater exploration, there is no substitute to the empirical data collected directly from the field. Therefore, the remote-sensing scientists dealing with the groundwater study would do well to add the strength of field inputs to their conclusions.

Several remote-sensing organizations patronized both by the State and Central Governments are currently engaged in locating potential groundwater basins and aquifers, as well as developing algorithms for processing the satellite data in digital form in relation to the groundwater studies suited to the Indian conditions. Several aquifers and groundwater basins/reservoirs have been delineated in the peninsular Indian terrain using remote-sensing technique. This has been possible partly because of the availability of sophisticated satellite imagery showing better spatial and spectral resolutions, as well as appropriate computer software for processing of the imagery in digital form. Several States of Indian Union have instituted Space Application Centres or Remote-Sensing Application Centres, which are making efforts to locate new potential zones for groundwater in their States by remote-sensing technique coupled with field data collection for validating the remote-sensing data. The continual and cumulative efforts of several scientific departments have yielded phenomenal success, as several aquifers have been identified in the recent past in the whole country, especially in the drought-prone areas of Gujarat, Rajasthan, Maharashtra, Andhra Pradesh, Madhya Pradesh and Uttar Pradesh. World Bank aided water resource development projects using remote-sensing data are underway in many States.

Remote-sensing using satellite imagery, with concurrent use of topographic base maps, is an effective technique for identifying and locating the abandoned and buried channels, valley-fills, channel-fills, paleo-channels, cutoff meanders and intermontane valleys, and tracing their whole length and width, in a short time with an accuracy of very high order. Typical examples of the aquifers of channel fills and valley-fills in intermontane valleys interpreted from the satellite imagery and corresponding geological maps are illustrated as Figures 8.3, 8.4, 8.5 and 8.6.

The Research, Reference and Training Division (2003) has reported that the total replenishable groundwater in India is estimated to be about 43.39 million hectare-metres per year. This can be expressed otherwise as 433.9 billion cubic-metres. About 7.13 million hectare-metres per year of groundwater are reserved for domestic and industrial uses and 36.26 million hectare-metres per year are assigned for irrigation. It is also estimated that '37.24 percent of the available groundwater resources have so far been developed'.

An information furnished by the Research, Reference and Training Division (2003) which has significance as far as groundwater resource is concerned, and water resource management specifically, is quoted at this juncture: 'Out of 5711 blocks/ talukas/ mandals/watersheds in the country, 310 ...have been categorized as "over-exploited", i.e., the stage of groundwater development exceeds the annual replenishable recharge and 160 ...are 'Dark', i.e., the stage of groundwater development is more than 85 percent'. Further, considering the complex nature of the groundwater-related affairs, a separate body known as the Central Ground Water Authority was set up in the year 1997 under the Environmental Protection Act of 1986. This Authority has been given the mandate of 'regulation and control of the development and management of groundwater resources in the country'. Indeed, so much has been done, and India has successfully tide over extremely difficult times in terms of the available water resources on the strength of the good work done singly and jointly by several organizations.

Paradoxically, however, all our efforts always fall short due to continuous stress on the water supply against the demand of a billion population and need for irrigating large areas of agricultural lands. The status of groundwater resource in several pockets of India, especially densely populated urban and semi-urban places, and intense agricultural belts, is extremely precarious. The whole system of groundwater has to function under a perfectly balanced mechanism of replenishment followed by exploitation. Any imbalance in this system would spell doom, and the humanity might have to pay a heavy penalty for creating such an imbalance. Those penalties include degradation of soil and land, creation of vast new wastelands bereft of any water resources, creation of arid zones followed in the next stage by desert-type biome, incursion of seawater in the coastal aquifers and so on. Therefore, we need to continually attempt to conserve the groundwater during dry spells instead of over-exploiting it, develop new groundwater basins by arresting or slowing down the surface runoff of rainwater, recharge the aquifers and enhance the water resources in the groundwater basins. Needless to mention at this state, that these measures would be possible only if we ease the stress on the groundwater resource by developing new alternative methods such as recycling of wastewater and so on (discussed in later chapters). A massive exercise combining the remote-sensing application in identifying the above-mentioned geomorphologic features, with groundwater recharge in watersheds shall yield tremendous results in this direction.

Some of the State Governments in Indian Union, notably Delhi, Tamil Nadu and Andhra Pradesh Governments, have launched vigorous campaigns educating the public about the advantages of groundwater recharge pits. Nevertheless, there have been conflicting press reports during the peak summer of the year 2003 that despite provision of the groundwater recharge pits in many building complexes all over Andhra Pradesh, the groundwater table in many parts of the State as also in and around Hyderbad city has *not* risen to the extent anticipated. That would mean that the method applied to the design of the groundwater recharge pits has been faulty or deficient. It would *not* mean that the concept of groundwater recharge pits, by itself, is faulty. It only underlines the necessity to closely supervise the method by which the recharge pits are dug, the depth up to which they are dug, and whether they are refilled with appropriate materials. That would also point to the need for the work-audit or close monitoring of the entire exercise carried out by the officials, contractors and labourers alike.

Recharge pits in residential complexes and colonies are bound to divert seasonal rainwater directly into the groundwater basin, and their efficacy in recharging the groundwater resource is phenomenal, and cannot be doubted. We cannot cast aspersions on a proven technology, if the technology does not yield desired results due to human manoeuvres. Resorting to groundwater recharge by digging pits in residential complexes and colonies is, indeed, a process by which surface runoff is arrested and rainwater, which would otherwise go waste, is allowed to infiltrate into the groundwater reservoir. Therefore, the Urban Development and Housing departments of the governments both in Centre and States have to concurrently promulgate ordinance or to make the groundwater recharge pits in residential complexes and colonies all over India as mandatory, as one of the clauses of the Water Policy for the country.

Groundwater Recharging

Though the concept of groundwater is known right from the Vedic times in India and also referred to in the Old Testament, it was only in the nineteenth Century the drilling for groundwater was started, and the tapping of groundwater resources along with the other surface resources commenced worldwide in the late nineteenth Century. It is because the water resources were once considered to be an eternal boon to the humanity and, therefore, the mankind kept tapping this natural resource to limitless extents. Added to that, the exponential population growth and the resultant need for foodgrain to the humanity further led to the overexploitation of surface water resources for personal and agricultural needs. The industrial revolution also saw a sudden spurt in the need for water. The result is that most of the countries, the affluent United States and several European countries like Germany inclusive, are now faced with water-stressed conditions. Gradually, groundwater has come as a mainstay in agricultural activity in many populous countries as in India, perforce to grow enormous foodgrain to feed the population. The result is that the levels of groundwater table in these countries have gone down to abysmal depths. In order to increase the yield of groundwater, man has been attempting to recharge the groundwater basins with natural supply of water during the peak rainy seasons of the year.

With the size of a subcontinent and good rainfall over a large part of the terrain for about four months in a year, the groundwater basins in India get automatically replenished to a large extent without any human effort. This is known as *natural recharge* of groundwater. However, as mentioned in an earlier context, the lack of adequate soil cover in many parts of the country coupled with extensive deforestation leading to the destruction of the soil-binding plant roots, has often resulted in quick transportation of the surface runoff, thus impeding the precipitation from percolating to the groundwater system. An orchestrated twin approach by effective surface runoff management and a planned watershed approach to rainwater harvest is bound enhance the capacity of natural recharge over vast segments of terrain.

In order to replenish and develop the groundwater resource, man has been trying to find out ingenious ways to artificially recharge groundwater basins. The *artificial recharge* may be defined as augmenting the natural infiltration of precipitation or surface water into the underground rock/soil formations by some method of construction, water spreading and altering the natural conditions.

Water spreading: This is the process of releasing the water over the ground surface in order to increase the quantity of water infiltrating into the ground, and allowing the water to percolate to the aquifer. Though many factors govern the rate at which water will enter the soil, the area of water spread and the length of time water is in contact with soil are most important in terms of quantity of recharge.

1. *Flooding method* is the least expensive and it is conducted by diverting excess water over a flat terrain. Irrigated farms are example for the flooding method.
2. In the *basin method,* water is released into basins, which are formed by sand and rock quarries. In still better methods, bunds are constructed to harness water for artificial recharging by infiltration in low-lying areas.
3. In the *Ditch or Furrow method,* water is distributed to a series of ditches or furrows, which are shallow, flat-bottomed, and closely spaced to obtain maximum area for contact with water. A collecting ditch is provided at the lower end of the area to convey excess water back into the main channel. This method can be adopted in undulating terrain with least expenses towards cutting ditches or furrows, and also can be resorted to in the wasteland areas.
4. In the *natural channel method,* the water for artificial recharge is diverted into natural stream/tributary channels, and the aquifers below these natural stream channels get recharged.
5. *Irrigation method* is yet another water spreading method in which

excess water is irrigated during agriculturally dormant season or in the wastelands; the aquifers below the land surface, which comes in contact with the irrigated water thus gets recharged.

A fundamental and justifiable debate arises at this juncture regarding the water source for artificial recharge. Under rainfed conditions, artificial recharge is exactly similar to the natural recharge of the groundwater in the aquifers and groundwater basins. However, in India where there is rainfall only for about four months in a year, the water source for the artificial recharge in dry months becomes a crunchy issue. The southwest monsoon and the tropical depressions are themselves so capricious for various reasons that when there are floods caused by excessive rains in one segment of the Indian peninsula, the other parts get only traces of rainfall resulting in scanty infiltration of rainwater in the aquifers. Paradoxically, many parts of peninsular India come under the grip of dry spells due to scanty rainfall right through the monsoon season, while on the other hand, much of the rainwater from floods is emptied through rivers into the sea, an annual feature witnessed in Mahanadi, Godavari and Krishna Rivers. This is where strategies need to be evolved through legislation or appropriate water policies to divert the excessive floodwaters to the rain-deprived terrains for effective artificial recharge by the water spreading methods discussed above.

Groundwater recharge through pits and shafts: None of the water spreading methods discussed above can be effective in the areas where subsurface strata is sealed from receiving downward passage of water. For example, in the urban areas, foundations of residential and commercial buildings, metalled roads and several other constructed structures render the earth's surface impervious to percolation and impede the rainwater and artificial recharge water from remaining in contact with the natural surface. Therefore, the rate of infiltration becomes almost nil over large areas of the urban sprawls. Similar is the case with areas where clay layers are deterrents to recharge by natural as well as artificial methods. In such areas, recharging can be effectively achieved by digging pits or shafts in such a way to penetrate the permeable substrata and directly percolate to the aquifer or groundwater reservoir. This method of recharging can be employed independently or in combination with the ditch or furrow method described earlier. Figure 9.5 is a cross section of a typical groundwater recharge pit or trench. The type of materials used and the thickness of each material used in building the recharge pit or trench, as well as the design for the civil work and the dimension of the built-in structure are also shown in this diagram.

In comparison with natural and water spreading method, the recharging by pits and shafts is certainly less effective since less area comes in contact with the water spread, as well as there are expenses involved in the construction

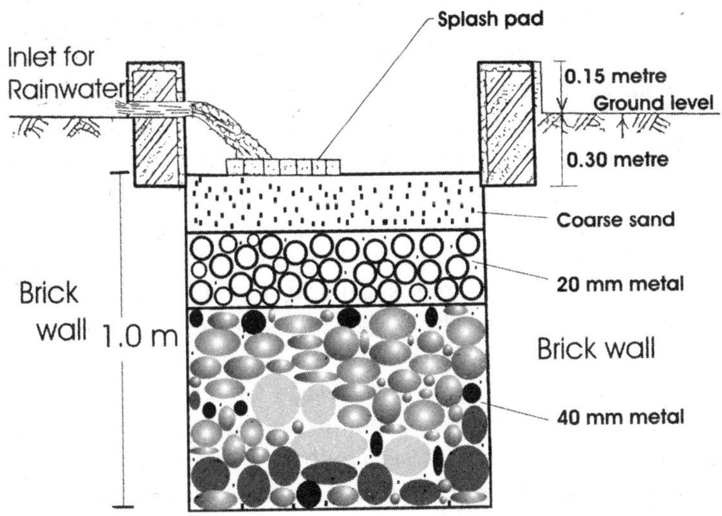

Figure 9.5. Cross section of typical groundwater recharge pit or trench.

of pit/shaft. Several State Governments in India have taken decisions to make the groundwater recharge pits in new buildings as mandatory and also encouraged the owners of the existing buildings to provide with these pits, but the implementation of these decisions is not done with seriousness. It has been reported in a section of the Press in the summer of 2003 that even after several years of decision taken by some of the State governments regarding provision of groundwater recharge pits in new constructions as mandatory, the levels of the groundwater table in several cities and towns have gone down alarmingly low. This indicates that the measures taken for construction of recharge pits by the builders are unscientific, the attempts in this endeavour are inadequate and the citizens flout the governments' decision on this with impunity.

In Indian climatic conditions of rainy season limited for about four months or so, only the recharge methods described will suit effectively, though there are a few more methods which will be effective only if and when there is rainfall all through the year though in varying quantities. Groundwater is an important source of water that the citizens of India rely on for their day-to-day life. In many areas in India where the Governments have not been able to provide protected water supply for drinking, the people are forced to use it directly as potable water and quench their thirst. Unfortunately, with repeated failures of monsoon resulting in low infiltration, indiscriminate use of groundwater for irrigation and excessive withdrawal for industrial use, the depths of groundwater tables all over India have gone down to alarming depths.

This fact can be exemplified by a frightening fact that has come to light in the summer of 2003. In ancient times, Palar River was a perennial river draining the far South region of India. Continuous overexploitation of Palar River over a period of time to unsustainable levels transformed it into an ephemeral river, now situated in Tamil Nadu. This river has a large catchment area as also several watersheds over small areas. The catchment and watersheds are fairly adequately rainfed for about four months in a year, at annual rainfall rates in the range of 80cm to 90cm. Townships such as Gudiyattam and Vellore in Tamil Nadu and several villages in the adjacent State of Andhra Pradesh are the beneficiaries of the Palar River waters. In the summer of 2003, the drilling operations right on the Palar River bed carried out by the authorities of Tamil Nadu Government's engineers had revealed that the groundwater levels just below the riverbed have gone down to depths ranging from 12 to 15 metres. Further, it is learnt that about ninety percent of the borewells drilled on the riverbed to tap water for distribution to the towns of Vellore, Katpadi and Gudiyattam have *not* been successful as they have not yielded water. This speaks volumes of our approach to the management of groundwater resource, and it starkly exposes our disregard to sustainable development of water resources.

If we do not take adequate and result-oriented measures to develop the groundwater reservoirs in a sustainable manner through rigorous movements of groundwater recharging, we may face disaster on the face of further droughts and monsoon failures. The sooner we realize this and act appropriately it is better for us.

Chapter 10

In this chapter, we explore the:

- Water needs, and desirable water quality in various human activities.
- Human activities, including industries, in which the currently used pristine water can be substituted by recycled water after treatment.
- Precision Farming technology as a means for sustainable agriculture.
- Status of Indian agriculture, livestock and wasteland development vis-à-vis water resources.

WATER RESOURCE MANAGEMENT FOR VARIOUS NEEDS

The total area of the planet Earth's surface is estimated as about 510 million square-kilometres. Of this, about 149 million square-kilometre area, making up about 29.22 percent, consists of land surface of five continents. Oceans occupy the rest of the 361 million square-kilometre area, which works out to about 70.78 percent of the Earth's surface (Arthur Holmes, 1944). There is an uneven but continuous cover of water around the Earth, saturating the rocks and soil of the continents, and over the enormous depressions of the ocean floor. The waters of all categories on the Earth's surface together constitute what is known as the Hydrosphere of the Earth. Oceans, seas, lakes and rivers together cover about three-quarters of the Earth's surface. Despite such a ubiquity of water in the oceans, the liquid freshwater available on the Earth's surface for use of the entire eco-system accounts for a mere one percent of Earth's total water resources. In addition, the distribution of this one percent freshwater on the land surface is not equal.

Below the surface of the ground, as we have reviewed in the previous chapter, for a few hundred of metres, at places even a few thousands of metres, the pore spaces, cracks and fissures of the rock strata are filled with enormous quantity of water. This water resource is known as groundwater. Though it is possible to estimate the yield of groundwater for specific periods (*for example*, for second or minute or hour), the exact quantity of the groundwater resource in various terrains is, at the most, a matter of conjecture, since there are no established or standard scientific methods to investigate the near-exact quantity of this resource. Further, the quantity of groundwater resource in an area fluctuates from one season to the other, as also the quantity of water trapped in the rocks depends on the properties of the rocks such as the pore spaces, cracks and fissures. Expectedly, the stock of groundwater during and just at the end of the monsoon season will be more than the average resource, and the stock would dwindle alarmingly in the summer seasons.

Until recently, it was felt that the water resources in the Earth are perennial and is inexhaustible. The depletion of water resources world over is noted only about three decades ago, as a result of rapid industrialisation and population explosion on a global scale. Since then, all the countries have been taking adequate steps to explore, budget, conserve and augment their water resources. Despite those efforts, it is estimated that over two billion of the world's population is affected by the scarcity of water, and in all likelihood this number would go up steeply in the years to come. The malady of water scarcity is acute in what are known as the 'Less Developed Countries', and the World Bank has forecast that this situation would turn more acute. This is, however, only partly true, for water scarcity is a problem that has affected many parts of the industrialised European and North American countries too. With a view to augment the water resources, heavy investment in plants, machinery and other installations for water treatment and resource augmentation are made in recent times in several countries. Extreme stress on the generation of clean potable water resource poses yet another biggest challenge for the governments of many countries today. The status of water resources in the African continent (notably Ethiopia, Sudan and other central African countries), Middle East, and to a large extent in the northwest and central parts of India has been causing extreme concern, due to the frequent failure of seasonal rains and the resultant lack of replenishment of the basic water resources.

Apart from Climate Change, Global Warming and El Nino phenomena, which might lead to droughts, over-populous countries like China and India have the additional responsibility to feed billions of their people, and droughts would undoubtedly upset the applecart of planning in terms of agriculture, economy, exports and commerce of these countries. Many of the European countries like Germany, Middle East countries like Israel and Saudi Arabia, and some parts of United States, such as California, face hardship due to insufficient natural water resources, and, therefore, have resorted to technological means to augment their water resources.

Some of the organizations affiliated to the United Nations Organization as well as special study groups continuously monitor the availability of basic facilities required by human beings. Depending upon the ease with which the basic facilities are available in different countries of the world, these organizations or study groups arrive at what is known as 'human development indices' or a parameter of that sort, which directly reflect the ease at which the citizens of the respective countries can get their basic needs. Easy availability of basic food at affordable price, fundamental education, medical aid, higher education and speedy justice, as well as prevailing crime rate, safety of women, character ethics of the people in public service, simplicity in implementation of the States' policies and geriatric care are some of the most important aspects that are considered as yardsticks in arriving at such parameters or indices. Nevertheless, at the top of the consideration is the ease with which the population, both urban and rural, of a country gets the most basic of the facilities for the day-to-day life, namely, *the water*.

The fundamental requirement of water for the masses of population is for the purposes of direct oral consumption, use in cooking, maintaining personal hygiene such as bathing and for washing of cloth, residential premises, toilets and automobile. It needs to be understood at this juncture that in view of the water-stressed condition in recent times, these basic needs of water for human beings can be distinctively classified into three types as follows:

1. Bacteria-free potable grade water for drinking and cooking.
2. Clean water for bathing and personal hygiene.
3. All other miscellaneous needs such as for toilet-flushing in which the quality of water can be compromised.

Supply of bacteria-free drinking water and sufficient quantities of clean water for personal hygiene to their citizens are the major challenges that the governments of several countries face today. Besides, providing water for irrigation of the arable lands and to the water-dependent industries amidst a drought-like situation in the arid and semi-arid terrains of various countries has become almost the whole-time work of the local governments. There has always been an increasing gap between the demands against the quantity of water available for supply in many counties, including India.

Way back in the early 1990s, the World Bank authorities have cautioned the Governments of Indian Union and its State Governments that judicious management and sustainable development of water resources for human use, agriculture and industry are crucial to the Indian economy, and that all our efforts should be directed towards the water resource development, without any more delay.

Though India has enormous amounts of water resources, and also the monsoon rains in varying quantities annually replenish the partly spent resources, the demands of water for various human activities always outsize the resources available as supply. There is tremendous pressure on the water resources due to the population growth, rapid development of the urban sprawls, growth of innumerable semi-urban townships in all the States of the country, a sudden curiosity of a major section of the urban population in water-sports, and a spurt in industrialization evenly almost in all the States. The annual rainfall due to the monsoons, a boon to the Indian subcontinent, is the only source for replenishment of the country's water resources. Though a large amount of the spent water resource is recouped by the glacial melt and snowpacks, such accretion is restricted to the Himalayan and the sub-Himalayan belts of India, and it does not reach the areas south of the Indo-Gangetic belt and the Yamuna River valley. There are strange instances of large segments of the area north of the Indo-Gangetic plains and the sub-Himalayan belts in Uttaranchal, Uttar Pradesh, Bihar, Haryana and Punjab coming under the grip of water scarcity in recent times. As a result, all parts of the northwest and central Indian terrain, as also the above-mentioned other parts of Indian terrain look for the only source, namely the monsoon rains, for periodic replenishments of the spent water resources. Except the monsoon seasons, several parts of the country are perennially under the grip of water scarcity, apart from the drought-like situation witnessed in the summer times. The one billion population of India, which accounts for one-sixth of the world population, depends mainly on the rainwater for all of its activities such as agriculture, industries, personal needs and power generation. Relying entirely on the capricious Nature, on which we have no control, is not good for a country that has a vision to build and develop itself as a strong and developed country by the year 2020. This vision warrants, in the first place, effective water resource management practices across the entire country. This is applicable not only in India, but to any part of the world, which depends solely on rain for the water needs as also for the purpose of replenishment of water resources.

Water Needs for Various Human Activities

The water-dependent human activities can be broadly identified as the *need of water for the:* (i) Day-to-day consumption for the population; (ii) Building construction; (iii) Livestock; (iv) Agriculture; (v) Industries; (vi) Balancing the Ecology; and (vii) Recreation and Entertainment. With ever-increasing population of the country, coupled with the need for massive agricultural production and industrial growth, there is always tremendous pressure to augment the water resources for all the above activities. The need of water in all these activities is bound to increase manifold in the near future, in view of the growth in all the segments of these human activities.

It is equally pertinent to note that in all the above-mentioned activities, India at present depends only on a single source—the monsoon rains. The water that is precipitated as rain is clean, pure and pristine. Therefore, the rainwater that is accreted as natural resource needs to be judiciously conserved and totally reserved as a source to provide potable water to the masses of the country. On the contrary, this source is now indiscriminately put to use (or misuse) in all the above-mentioned segments of human activity, besides several other activities in which the quality of water is *not* critical. In other words, there are several human activities, which can go on undisturbed when potable grade water that is currently used is substituted by alternative source of water, such as the *recycled water*. This conservation and substitution would, to a great extent, lessen the stress on the potable grade water, and therefore, adequate quantities of potable water can be supplied to the masses of the country's population.

The first step in a holistic approach to the water resource management, therefore, should begin with identifying those areas of human activity that require only potable grade water, and such areas where quality of water used is not important or critical and therefore the use of potable-grade water can be substituted by recyclable treated water. This also calls for strict classification of various water-related activities of the people and assigning appropriate quantities of water to those activities. This should be followed by diligent planning for formulating the water-related policies and conscious implementation of those policies for an effective management of the available water resource.

There are several areas of human activity where the quality of the water is *not* important. The pristine rainwater source can be largely allocated for human consumption where the water quality *is,* indeed, critical. The answer lies in augmenting the water resources by various known technological means, allocating requisite quantities of water of desirable quality (both potable grade and the augmented source using the technological means) for various human activities. That in a nutshell sums up the water resource management, where we collectively identify different sources of water, their quality, usability in various sectors of human activity, classify these sources based on their quality and quantity available, create a total resource, and budget this resource and allocate it to different purposes as per the needs and the water quality desirable by each segment. It needs to be visualised that this exercise is as complex, but as important, as the preparation of a financial budget for a fiscal year. When we plan for the country's future with a vision for about two decades later, a stringent exercise on the water resource management becomes mandatory, without which an overall planning like fiscal budget based on economic fundamentals of the country becomes, to some extent, meaningless. Let us analyse this aspect in the light of the current situation *vis-à-vis* in a near future of about a decade later.

(*i*) Water needed by the people for day-to-day consumption

The amount of water that the human beings consumes daily is highly variable and it depends mainly on the geographic location of his stay, overall climatic pattern of that location, and his lifestyle. In an arid to semi-arid or humid subtropical climate of India, a man orally consumes about three litres of water in the months of March to July and about one-and half to two litres in other months. It is further estimated that a man requires at least twenty litres of water per day for personal hygiene in a modest life. With the change in urban lifestyle, this basic need doubles. Further, in summer when the water supply in most of the Indian cities is curtailed, inversely, the need of water for human consumption and hygiene escalates to about fifty litres. Therefore, if we employ a simple arithmetic, for the purposes of oral consumption and personal hygiene alone, a billion-strong population of India would require fifty billion litres of water per day in summer and about twenty billion litres per day in winter and rainy periods. This does not include the necessity of water for the common activity of the family such as cloth- and dish- washing for which, an estimated quantity of one hundred litres of water is used additionally. Thus, a four-member family uses a barest minimum of approximately three hundred litres of water daily in summer and about two hundred and sixty litres in winter. With the ever-growing population and migration of the population to urban centres in search of jobs and livelihood, the need of water for personal consumption and hygiene in the urban centres is expected to increase steadily and exponentially in the years to come.

The quantity of water used by the people for the other household chores varies widely. With affluence and a better lifestyle, the urban citizens start acquiring automobiles as also maintaining domestic gardens, lawns and plants in pots, all of which too require water for maintenance. In most of the Indian cities and towns, the supply of potable water is organised by the corporation or municipality authorities. Drawing a limit for the use of this potable water in direct oral consumption and for the other purposes is difficult. Often, we can see that the water supplied through pipelines by the authorities is put to use in watering the garden, automobile washing and for a host of other purposes where the water quality is not an issue.

Organized urban greenery and landscapes in houses, parks and social forestry need plenty of water for maintenance, especially in the water-starved summer months. The water-supply departments require bulk of water for this purpose too. Otherwise, bereft of greenery, the skyline of the metropolises and towns would appear strewn only with high-rise structures. The quality of water required for this purpose is not an issue. However, for the purposes of maintaining domestic gardens and automobiles, the citizens liberally use either the protected potable-grade water distributed by the municipal water works department or from the groundwater source. Either way, good quality water, which can be put to use for potable purposes and personal hygiene is misused for purposes in which water quality is not a serious matter. Evidently, this mounts a stress on the potable-grade water for direct oral consumption and other purposes where hygiene, indeed, matters.

Water dissolves several substances with which it comes in contact. Further, different agents, the most important being the human beings, let many substances into the water bodies such as lakes, rivers and ponds. The result is that the water's potable nature deteriorates, and improvement of its quality would require treatment to varying degrees. The auto-immunity in the human beings guards them to a large extent when the water contaminated by tolerable limits of bacteria and impurities is orally consumed. However, water-borne diseases frequently break out as epidemics over large areas, especially in the monsoon periods, when impure waters contaminate the potable-grade water. Whereas micro-organisms and bacteria form dominant of the pestilence-causing impurities, often a few deleterious ionic substances such as elements and salts too are present in variable quantities in the potable-grade water. Often, removal of several of these elements becomes necessary to convert the raw water into potable-grade water. The quality, properties and the permissible limits of dissolved substances in the potable-grade water have always been matters of debate and subjectivity. Table 10. I lists out the concentration of different substances in standard of potable-grade water.

The standard stipulated as above is for the water meant only for the oral consumption of the human beings. For the other functions of the day-to-day life, such stringent standard *need not* be imposed, though abundant care and caution should be exercised to ensure that lower grade water does not mix up with the potable-grade water, and also it is not administered or supplied wilfully or by negligence as potable water. The consequence of such acts of negligence is likely to inflict disastrous results as breakout of epidemics such as enteric diseases in short term, and carcinogenic disease in long term.

With a marked improvement in the life-style of the people as a whole in the country, their expectations from the government and the authorities responsible for providing basic amenity like water supply go tremendously high. So much so, the electorate in some parts of the country boycotts the general elections as a silent protest against the failure of the governments to

TABLE 10. I: STANDARDS RECOMMENDED FOR POTABLE-GRADE WATER.

Property / Substance milligram/litre (mg/l)	*Upper limit in potable water*	*Maximum concentration permitted by World Health Organisation (WHO)*
Iron	0.1	1.0
Arsenic	0.05	0.05
Manganese	0.02	0.5
Cyanide	0.1	0.2
Copper	0.2	1.5
Zinc	1	15
Lead	0.05	0.05
Calcium	not stipulated	200
Magnesium	not stipulated	150
Sulphate	not stipulated	400
Chloride	100	600
Mg & Na sulphates	200	1000
Phenol	0	0.002
Organic pollutants	0.05	0.2
Hardness ($CaCO_3$)	80	300
Total solids	300	1500
Selenium	—	0.01
Cadmium	—	0.005
Chromium	—	0.05
Fluoride	—	1.5
Mercury	—	0.001
Nickel	—	0.1
Nitrate & Nitrite Nitrogen	—	10
Lindane	—	3.0 microgram/litre
Monochlorobenzene	—	3.0 microgram/litre
Property:		
Taste and Odour	Tatesless, Odourless	Unobjectionable
pH	7.0—8.3	6.5—9.2
Turbidity unit (Jackson TU)	0.2	25
Coliforms	—	absent in 100 ml.

Source: This information is extracted from Smethurst, (1988), Table 2.2 and Tebbut (1992) Table 2.6.

provide safe drinking water. Often, many of the existing governments in the State lose the general elections due to their failure to provide water to their subjects. Whereas planning for buildings and roads in the urban and metropolitan development is one aspect, comprehensive approach to provide adequate good quality water to the residents takes the top priority. Otherwise the aspirations of the citizens for a comfortable living would suffer setback and the frustrated masses take to courses such as agitation and mass movement. It is expected that the water supply authorities gear up to the rising expectations of the masses and provide sufficient quantities of potable water, the lifeline. It would be not wise on the part of the water supply authorities to depend only on monsoon rain as water supply source even to meet the daily consumption of the growing population in India. We have to take cognisance of the innovative technologies (discussed later) in the development of water sources and adopt them to augment our water resources.

(*ii*) Need of water for building construction

The building construction activities in most of the Indian cities and towns reach a peak in summer for the reason that the builders desire to complete the roof casting, construction of the super-structure and exteriors of the buildings before the onset of monsoon. During the four-month monsoon period the builders concentrate mostly in the interiors of the structure. As the government has recognized housing as a priority sector, and due to the easy availability of house finance, owning of a house by an average Indian has become a reality as well as easier than in the past. This aspect in India is in sharp contrast to the housing need of citizens living in the metropolises of other countries such as United States, Japan and Korea, where owning a house remains pipe-dream for many during the entire or till the very late stage of their life.

In almost all metropolises and towns in India, most of the new constructions are undertaken after drilling bore-wells to tap the groundwater for the construction activity. The quality of water for construction activities is not an issue at all. Potable grade water from rainwater and groundwater sources is not required for the building construction purpose. If the use of potable-grade water from the pristine rainwater and groundwater sources were not stopped forthwith for building construction, it would lead to drop in the levels of groundwater table over vast terrains. Consequently, more and more open wells and borewells, which currently yield potable-grade water to the masses of the population, are likely to go permanently dry. Building construction being an endless process in our cities and towns, as also a water-intensive activity, there will be pressure on the water for building activities in the years to come. We have to identify enormous quantities of construction-grade water source, other than rainwater and groundwater, to meet the demands of the building construction industry. Otherwise, there would be extreme stress on the potable-grade water in the cities, towns and metropolises, because a considerable part of the potable-grade water would be weaned away by the building construction industries.

Loss of aquifers beneath the constructed structures

Though the urban development authorities and town planners in India consider several vital aspects in their attempts for planned development of the metropolises and towns, one aspect that is given scant regard is the geological consideration. As one of the consequences of this, several paleo-channels, channel-fills and valley-fills, which are aquifers that otherwise could be used for drawing water to augment the community water supply in the urban areas, are permanently lost beneath the concrete. Metal or pitch top on roads and the reinforced cement-concrete or stone-slab flooring of the constructed structures become impervious layers that prevent percolation of water into the groundwater system in the urban area. Therefore, seepage of water into the groundwater system during monsoon time is considerably reduced due to density of constructed area. Though the urban authorities in many States have urged the builders to provide groundwater recharge pits in new buildings, the steps taken in this direction is far from adequate. This prohibits seepage of rainwater during monsoon into the groundwater system beneath the Indian cities, towns and metropolises, resulting in many open- and bore-wells going dry at a fast rate. For this reason, geologic and geomorphologic considerations are necessary inputs for the purpose of urban planning. It should be ensured prior to the planning and developing housing complexes with metal or RCC topped roads that no aquifers or geomorphologic features having potential for groundwater is concealed by the built-in structures.

(*iii*) Need of water for the livestock

India can take credit to the fact that the world's largest population of cattle and livestock lives here. The Research, Reference and Training Division (2003) says (in page number 413) that India accounts for 57% of the world's buffalo population and 15% of the cattle population. This report, quoting the Livestock Census of 1992, adds that the country has 20.5 crore cattle and 8.4 crore buffaloes; and that the gross value of output for livestock sector during 2000-01 was a whopping Rs.167,205 Crores. These statistics point out that the country's sizable population owes its livelihood to the livestock. It is also heartening that since 1999 India has been an exporter of milk and dairy products, thanks mainly to the ongoing successful co-operative movement in the dairy industry of the country. In the year 2000-01, India has produced 81 million tonnes of milk.

Despite these bright spots, if we take into account the strength of the cattle in our country, India should have been the leader in the production of milk and export of the dairy products. But, in fact, we are not. As per the figures available for the year 1994, the 'annual milk yield per head of cattle' in India is just 500 kilograms, as against 9200 kilograms in Israel and 3000 kilograms in New Zealand (Abdul Kalam, 1998). The reasons are too obvious—the malnutrition and frail nature of our cattle. The health of our cattle needs to be improved with the aid of veterinary scientific means, and also the quality and the quantity of the fodder available for them have to be considerably enriched. It is a common knowledge that in India, no separate cultivable lands are earmarked for growing the fodder, except rarely in some States. Consequently, the cattle have to depend for their fodder mainly on the naturally grown grasslands and forestland. Exceptions in the organized cattle farms apart, the other cattle, especially a vast majority of the rural cattle in India are perennially undernourished. In summer, when most parts of the country experience water scarcity or drought-like situation, the quantity of fodder available to the cattle hits the nadir. The economically backward owners of the cattle just cannot afford to buy the fodder to feed their milch-cattle at a premium price from the local market. One comes across the news in media that in summer countless cattle perish across the country due to the non-availability of water and fodder.

The acute water scarcity in summer too takes a heavy toll on the cattle population. Barring the cases of organized cattle farms, no owner of the cattle takes adequate care in the quality of water given to the cattle. The cattle in rural India consume stagnant water from pools that is infested with micro-organisms, toxic weeds, and the eggs of intestinal worms and the parasites. Mostly, the water in these pools has very high levels of biological oxygen demand (BOD). Added to this, during grazing in weird places such as the outskirts of industrial belts that manufacture fertilizers, pesticides, dyes and basic chemicals, petrochemicals, rubber, textiles and so on, the cattle feed on the grass which carry the traces of the chemical substances both as direct deposits over the grass blades, and as substances absorbed by the grass from the contaminated soil. To add further to their miseries, the cattle consume the water contaminated with toxic chemicals coming out of these industrial units. Most of the time, the chemical oxygen demand (COD) and the bio-chemical oxygen demand in these waters are of high to very high levels, which cause a host of ill health to the cattle that consume this water. All these have a direct bearing on the health of the cattle as well as their milk yield and the quality of the milk.

This author has come across several cattle-owners across the country reacting bluntly out of frustration that when the human beings are hit by water scarcity repeatedly, how the less beings like the cattle can be provided with the care of good quality water and nutritious fodder. Strange and unacceptable, but this sort of curt statements too have hidden meanings. In an integrated approach to improve the health of the cattle and enhance the milk yield from the masses of the milch-cattle wealth that India possesses, the availability of copious amounts of potable-grade water holds the key.

It is important that just like we budget the potable water for the direct oral consumption of the human beings and distribute it as the protected water through pipelines laid by organized agencies such as metropolitan/urban water supply departments, adequate quantities of the same kind of protected potable water has to be arranged for the cattle masses too. In our water resource management efforts, we have to give due importance to the quantity and quality of the water we provide for the milch-cattle because the quality and the nutrient value of the milk that the people consume directly depend on the health of our cattle.

In our water resource management, a significant quantity of pristine and pure water (*not* recyclable treated water) of potable grade from the rainfall or from groundwater source has to be allocated for the cattle, for their oral consumption. Since the cattle wealth can help us generate new wealth that can bolster the country's economy, there cannot be compromise in the quality and quantity of water we provide to the cattle. As this is going to cost us considerably in terms of the available water source, we have to manage the water resources in a way that good quality water is not misused for subsidiary activities like servicing of automobiles and railway wagons, gardening, toilet flushing and a host of water-intensive industrial units where the water quality is not an issue.

(*iv*) Need of water for the Agriculture Sector

India has a total land area of about 328 million hectares. Out of this, the arable land is estimated as about 161 million hectares. It is also estimated that about 75% of the total cropped area is not irrigated but rain-fed, and it accounts for about 42% of the country's total food-grain production. Further, nearly 78% of cotton, 82% of oilseeds, 96% of sorghum, 98% of millet, about 62% of rice and 38% of wheat that are produced in the country are grown under rain-fed condition (Dhruva Narayana, 2002).

In India, there are three main crop seasons, namely, Kharif, Rabi and Summer Crop Seasons. The Kharif crop comprises rice, jowar, bajra, maize, cotton, sugarcane, soybean and groundnut. Wheat, barley, gram, linseed, rapeseed and mustard constitute Rabi crops, and the Summer crops mainly yield rice, groundnut and maize. The gross value of output from the agriculture sector in the year 2000-01 was Rs.561,717 Crores (Research, Reference and Training Division, 2003). This mind-boggling amount paints a picture on the vastness of India's agriculture sector and its major contribution to the Gross Domestic Product (GDP).

The short-lived food shortage in the early 1960s was a lesson to everyone connected to agriculture to enhance food production year after year and create buffer stocks of foodgrain. Aggressive researches in agriculture were launched and our scientists had pioneered in generation of high yield variety of seeds under Indian conditions. Nevertheless, the population pressure has also been steadily mounting during all these years, and it has gone so high that the steps taken by all the concerned in the agriculture sector are overtaken by further needs for the land, fertile soil and copious water, to meet the need of foodgrain to our population. Besides, excessive sapping of the fertility of land in most parts of the country in an effort to produce more foodgrain has shown its ill effects, as the liberal use of fertilizers and pesticides to augment the agricultural production has led to degradation of vast tracts of agricultural land in many parts of the country, and also affected the fertility of the soil. Degradation of soil and agricultural land, concomitant with our efforts to augment the agricultural production, has necessitated bringing several hectares of wastelands and non-agricultural lands into the fold of agriculture. Development of wastelands and non-agricultural lands for agricultural activity was possible by pumping considerable amounts of groundwater to the elevated and upland farms. These measures taken in the agriculture sector *alone* have taken a heavy toll in our surface water and groundwater resources. An immeasurable quantity of groundwater was drawn using high-powered pumps for agriculture. These unsustainable practices in agricultural sector would continue and remain unabated in the years to come.

India is envisioned to have a population of 1.2 billion by the year 2010 and 1.4 billion in 2020 (Abdul Kalam, 1998) at a growth rate of 1.5 percent at the approach of 2020, though currently it is 1.8 percent. Indian economy is always hinged to agriculture. Our annual budgets over the decades have shown major lean towards agriculture by allocation of major funds to agriculture and agriculture-related industry. Agriculture will ever remain the country's thrust area for the reason that, the Indian government considers starvation as its permanent enemy and relentlessly fights it. Every government not only identifies that providing food to its every citizen as its primary responsibility, but also strives in this mission. As long as agriculture remains as the topmost priority in our economy, water is also going to remain as the one single priority in our requirement.

The quantum of agricultural products produced in a season, as also the very success of agriculture in India are directly linked to the amount of rainfall over the agriculturally active sectors of Indian terrain. Improvement in the productivity of the crops holds key to the success of our agriculture, and the productivity under rain-fed condition hinges on the intensity of the monsoon and the quantum of rain received by the area during the monsoon season. There are always risks involved in choosing our route to success in agriculture through a single natural phenomenon such as the monsoon on which we have no control. It would always be prudent to have one or two alternatives as the sources of water, so that if one source fails or gets depleted, the other source comes to our assistance. If the nature's bounty such as the seasonal rain happens to be inadequate, our progress and prosperity are hampered, as demonstrated by the drought-like conditions during the summers of 2002 and 2003 in many parts of the country due to the weak monsoons preceding those summers. The recent report brought out by the Government of India sums up that '*the performance of crop husbandry greatly depends on the performance of monsoon as about 62 per cent of area sown is dependent on rainfall*' (Research, Reference and Training Division, 2003, p.395). The fact that the Indian economy is, to a large extent, dependent on monsoon is candidly emphasized by one of our former Union Finance Ministers thus: "*The Union Budget is largely a gamble on rain*". The State and Union governments have taken umpteen measures to offset the difficult conditions arising out of unfavourable results in agriculture in a particular season, by way of offering meaningful sops to the farmers, such as writing off loans, providing 'Kissan (farmer) credit cards' through banks, standardizing and enhancing the procurement prices for the agricultural produces and so on. However, such sops given by the governments to the farmers result in additional burden on the expenditure from the state exchequer and lead to cascading effects such as additional tax burden on the other segments of the society. Dependence only on the monsoon rain for achieving success in agriculture is not only a gamble, but also amounts to playing a risky game with the economy of the country as well as the prosperity and aspirations of the citizens. The researches of seed hybridization for high yielding agricultural products have indeed bolstered the agricultural production, but the role of water in agriculture cannot be substituted.

The scientific community in India, being second to none in the world and also quick to accelerate its work keeping in pace with the global technological developments, has been continually at work and research on cost-effective agriculture and has initiated a revolution of sorts in the farm sector. Conscious about the depletion of our water resources due to its use in the farm sector, they have infused innovative ideas in agriculture through a concept known as *'Sustainable Agriculture'*. It is an integration of various technologies, with one single motive: *Increase the agricultural output to ably support the society, by the optimum use of natural resources, simultaneously keeping watch on the overall environment and maintaining the ecological balance.* Fulfillment of this objective can be possible not by the services of one single organization or a group of individuals alone, but by the intervention of the governments, interaction between the whole scientific community in the educational institutions and professional organizations, and active involvement of the Indian farmers in the sustainable agricultural practice in close association with the scientists and technologists.

Conservation of Water in Agriculture sector

Agriculture is, perhaps, the largest water-consuming sectors of the nation's activities. The agriculture, despite being a water-intensive activity, cannot come to a halt under any circumstances including on account of weak or absence of monsoon rains, lest the country's population will be hit by food shortage. Alternatively, the country will be forced to spend considerably to import foodgrain to provide for its population. Due credit needs to be given to the scientists and the scientific organizations involved in the agricultural research in the country for their relentless efforts in the fields of soil conservation, nutrient enrichment to the soil, seed hybridization, and a host of other ancillary works related to the agricultural technology. 'The pressure of increasing population neutralizes all efforts to raise the standard of living and nutrition, whereas loss of efficiency in the soil itself nullifies the value of any improvements made' (Dhruva Narayana, 2002).

In a significant study in the last quarter of the 20th Century, it was found out that even in industrialized countries, on an average, only about 30 percent of the water budgeted for irrigation is effectively utilized by the crops. In the 'developing' countries, however, because of old methods of cultivation and irrigation, this percentage is still lower. The problem is compounded in tropical and equatorial countries where the water loss from arable lands is more due to transpiration and evaporation. The Food and Agriculture Organization (FAO) has estimated that only 40 percent of the water fed into the distribution system actually reaches the irrigated areas (Garbrecht, 1976). Therefore, with 30-40 percent as the level of irrigation efficiency, total levels of efficiency or effective utilization of water is only about 10-20 percent. Sustainable practices in agriculture are therefore extremely necessary for conservation and optimum utilization of water resources.

The industrialized nations have been able to improve the conservation of their water resources in agriculture with the aid of their economic strength, which helped them to introduce mechanization in farm sector, technical facilities and training to the farmers. The most important of the reforms in the agriculture sector is replacing the traditional gravity type of irrigation with efficient mechanical spray and drip irrigation (Garbrecht, 1976). In Israel, a country located in the desert biome, for instance, gravity irrigation has now virtually disappeared and irrigation in more than 90 percent of the lands is by using sprinklers or by drip irrigation. The philosophy of water budgeting in agriculture sector should be that the increased demand for water should be entirely met by improved efficiency in water utilization by new innovative irrigation methods. India should not lag far behind in adopting the innovative irrigation methods such as sprinkler technology and drip irrigation for effective conservation and management of existing and replenishable water resources. Though these methods are adopted in many parts of central and southern Indian terrain, as a matter of Water Policy, more and more arable lands need to be brought under the agricultural reform so that water loss in agriculture is minimized.

The main causes of water loss and thus the beginning for any substantial improvement in efficiency may be identified as:

(i) seepage from the distribution canals;

(ii) losses caused by improper regulation;

(iii) non-existent or inaccurate measurement of the volume of water in the irrigation system; and

(iv) irrigation losses such as seepage and surface runoff.

A critical analysis of these causes would reveal that prevention of these losses could easily be controlled with resolve if we have proper regulatory Water Policies as well as strict implementation of those policies.

The researches carried out by the Intergovernmental Panel on Climate Change (IPCC) have forecast certain scenarios in the field of agriculture all over the world too. These researches have indicated that where the crops are near their maximum temperature tolerance and where dryland non-irrigated agriculture predominates, the agricultural yields are likely to decrease with even small climatic changes (described in the context of Global Warming). In regions where there is a likelihood of decreased rainfall, agriculture could be substantially affected regardless of latitude of that region. Besides, the climatic changes seem likely to lead to decrease in yield potential in the tropics and sub-tropics. These researches have indicated that South Asian and African countries are likely to be affected by the small but definite climate change scenarios (Parry and Livermore, 2002).

Considering the current levels of micro-level changes in food production related to the change in weather pattern in many parts of India (leading to the development of arid and semi-arid conditions), deficit rainfall and the resultant transformation of the erstwhile water-saturated soil zones into semi-arid to arid environment, the forecast seems to be real. Until and unless adequate or excess amount of rainfall occurs to saturate the water-deficient tracts of the country, the current trend of intensive agricultural activity by irrigation method in many parts of the country appears difficult in future. On the other hand, we may have to resort to sprinkler or drip irrigation techniques to conserve water resources.

It is in this context that the adoption of the practices of sustainable agriculture would be an answer for the conservation the surface runoff and groundwater, as well as management of water in the agriculture sector. A host of scientific inputs is necessary for the implementation of the concept of sustainable agriculture in the entire country. Scientists involved in researches in the fields of agriculture, soil, water resource, remote-sensing, information technology and industrial and fertilizer chemicals, among others, need to integrate their scientific wherewithal, and evolve strategies to get directly involved in the agricultural production by keeping in constant touch with the government functionaries and the farmers directly.

PRECISION FARMING

One of the remarkable recent developments in Sustainable Agriculture is an innovative concept and practice known as 'Precision Farming'. Precision Farming is sometimes referred to as 'Precision Agriculture', 'Prescription Farming', 'Site-Specific Farming' and 'Variable Rate Agriculture', but the methodology and technology adopted in all these practices are same or similar. The Precision Farming is, indeed, a technology gifted to the world by, what Friedman (2000) terms as *'the democratization of technology'* in the era of Globalization. Though the Indian scientists are quick enough to appreciate the overall potentiality of this modern concept and apply it to limited areas of farms (Ray *et al*, 2001; Dwivedi *et al*, 2002), it is still in experimental stage in India. In this sophisticated agriculture, the Information Technology (IT) data obtained from the satellites launched under space programmes, Global Positioning System (GPS) and computer software known as Geographic Information System (GIS) are combined, and the desired results achieved with the support of local administrative authorities and co-operative farm movement activists. Precision Farming has virtually taken the agriculture to the Space Age and into the era of Information Technology. India being one of the frontline countries in both these modern technologies, ideally it is expected that we fall in the bandwagon of countries that have adopted

Precision Farming over wide arable lands and literally 'reaped' immense profits.

Precision Farming is essentially an agricultural management practice with a potential to increase the profits of the agriculturists by employing irrigation scheduling, water conservation and variable rate of application of farm nutrients and pesticides. Furthermore, under this management practice, vast amounts of data on farms can be gathered, viewed and analyzed, and later can be stored as agricultural data inputs in the form of reusable knowledge in Information Technology, to make better farm management. This technology is a means to create wealth and a strong knowledge base in the country's Information Technology pertaining to better farm management. This knowledge base can facilitate sound decisions on agricultural, taxation and fiscal policies. In other words, as a technology, Precision Farming is a '*new system of wealth creation*' conceptualized by Toffler (1990) and it can generate vast amounts of knowledge and meta-data on agriculture that can be stored in computers as wealth of information.

Precision Farming uses the space satellite-linked Global Positioning System (GPS) mounted on a farm vehicle such as a tractor. The GPS gives the precise location of a farm or an even a small unit of agricultural land within a large agricultural field area connected by a computer network, which is also equipped with an appropriate Geographic Information System (GIS) software. All the necessary information relating to the farm, such as the availability and requirement of water and soil nutrients by the farms, is stored in the GIS. As the tractor moves through the fields, the GPS mounted on it sends signals on the contents of soil moisture and soil nutrients in various farms situated in the vast field to the GIS through network, along with the locations of the farms. In consultation with appropriate supporting software, the GIS decides the exact amount of fertilizer or pesticide required by the various farms in the field, and sends back signals to a computer-linked equipment called 'applicator' mounted on the same tractor. The applicator is simultaneously sensitized by the computer network to sprinkle desired quantity of fertilizer in the farm. Further, the farmer monitoring that particular farm is alerted whether or not to irrigate the farm, or, if necessary, how much water need to be sprinkled in that field. Thus, the Precision Farming technology is highly effective and economical as well, in co-operative farming and it facilitates the farmers to save considerable money by the optimum application of fertilizers and pesticides. Further, most importantly, this farm practice helps in conserving the groundwater or any other source of water available for agriculture, and maintaining the water table by optimizing the draw of groundwater for irrigation to the requisite quantities.

There are five major components of technology employed in the Precision Farming. These are: (i) Global Positioning System (GPS); (ii) Sensor Technology, including Remote-Sensing; (iii) Geographic Information Systems (GIS); (iv) Variable Rate Technology (VRT); and (v) Yield Monitoring and Yield Mapping (Glen C. Rains and Daniel L. Thomas: www.ces.uga/pubcd/b.1186.htm). While attempting to understand these components of the Precision Farming technology, one can simultaneously infer that these five components involve in-depth study, acquisition of knowledge on, and constant improvisation of, the tools that are used in these technologies. These tools range from precision equipments employed in Global Positioning System, software in Geographic Information Systems and yield mapping, electro-optical components in the case of sensors, satellite imagery in remote-sensing and mechanical tools in the case of Variable Rate Technology (VRT) and Applicators.

Adoption of Precision Farming in a big way in the country would benefit a large number of sensor technologists, computer professionals, software developers, remote-sensing scientists, and entrepreneurs in the mechanical engineering and GPS-linked agricultural machinery manufacturing segments, to keep innovating as well as pioneering new tools and facilities in the above-mentioned five components of the unique technology in agriculture. Needless to emphasize that a large number of youth qualified in the above-mentioned branches of technology would find employment opportunities within the country, if more and more agricultural lands were regularly brought under Precision Farming. Besides, the country can demonstrate its technological prowess to the entire world, and prove that we are a mighty force to be reckoned with in the field of Precision Farming as well as all the other sophisticated technologies that support it. We have comprehensively everything to gain and nothing to lose if we implement the Precision Farming technique over large areas of agricultural fields as one of the practices in sustainable agriculture. It would be beneficial, therefore, to take an overview of various components of Precision Farming.

1. Global Positioning System (GPS)

The GPS is a man-made constellation of satellites that are put into high-altitude orbits (about 20,000 km away from the Earth) under the space research programmes. Currently, a set of twenty-four satellites with twelve-hour orbital period in six orbital planes at 55° inclination to the equator, which are developed by the United States of America's Department of Defense, incessantly transmit radio signals that are picked up and deciphered by specially designed gadgets known as the GPS Receivers. These signals are put into a variety of uses such as navigation, ground surveys, defense, communication, earth-science (including glaciology, structure of the earth's crust, seismic studies and river course monitoring) and several other applications where spatial and temporal variations are required to be determined. The GPS Receivers require at least four satellites to determine their geographic positions (or the locations where they are held) on the Earth. These positions are picked up by the GPS Receivers as the point of intersection of a specific latitude and longitude. The accuracy of the calculated location is often or always about 30 metres away from the actual location, as a result of a deliberate offset in the signal introduced by the US Department of Defense, as also due to the atmospheric variations. Therefore, the raw GPS signals are not sufficiently accurate to determine the precise locations on the Earth. An additional signal from preferably a known land-based reference position, whose precise location is known in terms of latitude and longitude intersection, is required to provide the necessary accuracy.

The GPS that utilizes a reference signal to gain accurate positional data is known as the Differential GPS or DGPS. The DGPS Receivers are the type of receivers, which are employed in the agricultural tractors in Precision Farming technology. The DGPS Receivers operate with an accuracy level of about one to one-and-half metre error in tracking the field positions. These receivers are mounted on the top of the tractor or on some high point to gain a view of the sky. The need for an instrument such as GPS or DGPS is for an ability to revisit a particular location repeatedly so as to ascertain that the adequate need of the farm such as fertilizer or water or pesticide is taken care and in right amount.

2. Sensor Technology, including Remote-Sensing

Sensors are necessary in Precision Farming to detect the stress on the crops, soil properties including the soil moisture- and nutrient-contents, and pest incidence. The simplest way of sensor technique is manual sample collection for determining said parameters and subjecting them for an array of prescribed analyses. There are sensors, which can be manually carried to the field for instantaneous checking of the crop health, insect- and weed-infestations, fertility deficiencies, soil properties and nutrient content. They use light reflectance on the leaves to find out the chlorophyll- and nitrogen-contents in the leaves. Several varieties of soil chemistry analytical kits are designed by various professional bodies to suit the needs of the individual farm practitioners. These kits directly determine the soil pH, and the contents of nitrogen, potassium and phosphorus in the field. However, it needs emphasis that this method is a long drawn and less effective in a technology, which expects high degree of accuracy in a quickest possible time, such as Precision Farming.

On the other hand, sensor techniques that give desired results in 'real time' (meaning almost instantaneous) are needed in Precision Farming technology. Generation of real time information is possible when the sensors are mounted in aeroplane or satellite, and they pass over the fields covered under Precision Farming *repeatedly* as also *frequently* in short gaps of time.

Remote-sensing is an integral part of Precision Farming technology. The remote-sensing data for Precision Farming can be collected from the sensors that are provided in the airplanes or remote-sensing satellites. These sensors provide instant maps displaying field or farm characteristics. Aerial photographs are optical sensors that show variations in field colour that correspond to changes in soil type, crop development, farm boundaries, road, soil moisture/water content, crop type, vegetative indices of different farms and several other parameters. The satellite-mounted sensors too provide all these data as multi-spectral imagery (Figure 10.1). Therefore, these imagery are more versatile than the data obtained from optical sensors. Repetitive passes of the satellites over the field under Precision Farming would be of great use since data can be collected more frequently for intensive studies. Temporal resolution (for repetitive data collection on agricultural farms) of about two weeks would be ideal in Precision Farming. Remote-sensing technology in Precision Farming hinges on the interpretation of the temporal variations in *reflectance* observed in the multi-spectral imagery. Repetitive observation of the multi-spectral imagery may bring out anomalies in some of the farms. Such anomalies may denote endemic pest or weed attack, which would require immediate management. Likewise, crop stress on account of the inadequacy or lack of moisture and nutrient deficiency is also imaged by the anomalies. The standing crops in the field and farms offer spectral response as a result of the vegetation index, and this response is highly variable depending on the type of crop and the health of the crops. Since subtle spectral variations result in the imagery as a result of variations in *reflectance,* those variations are most likely to be the functions of variability in soil moisture content, pests, weeds, humus- and chemical nutrient-contents in the soil from time to time. Study of these spectral variations, vegetation index and other parameters relating to the farms having standing crops is possible by computer-aided image processing of satellite-based data in digital form (see for example Figure 10.1). Numerous digital image processing software have been already developed for application to processing of the space-based digital data obtained from the sensors of the satellite platforms. Besides, the computer professionals engaged in digital image processing both in India and abroad continually strive to arrive at algorithms to facilitate the satellite data to display sensitive spectral patterns on account of factors such as soil stress, crop stress, moisture or nutrient deficiency, endemic crop diseases, pest and weed infestations, and so on. These advanced studies on agriculture based on space data and computer software help in arriving at proper farm management decisions (such as application of fertilizer, pesticide and water) is facilitated without delay under trying conditions in the farms.

Effective application of remote-sensing practices in Precision Farming technology in India, or for that matter in any country, depends on the availability of the space data (imagery) having the following characteristics:

1. High spectral resolution (from visual range of band through infra-red band to thermal band).
2. High spatial resolution (of about 5 m, to start with, and 1 m in due course).
3. High temporal resolution (with a data frequency of about two weeks).

India's space programmes in near future propose to generate multi-spectral and panchromatic (black and white) satellite imagery data ranging in spatial resolution from one metre to six metres through missions known as 'Resourcesat' and 'Cartosat'. Currently, USA's satellite imagery called as 'Ikonos' data that have one metre resolution are mostly put to use in experimental Precision Farming techniques in India. The launch of Resourcesat-1 on 17th October 2003 comes as a shot in the arm for those engaged in the algorithm development and remote-sensing with a slant towards Precision Farming, as also those are directly involved in the researches and farm activities with a thrust on Precision Farming, to bring India into the global map of countries which are practitioners of the innovative technology, *i.e.*, Precision Farming.

India is one of the leaders of the world in space research programmes, satellite launch, sensor technology and space data use (remote-sensing) for a variety of applications. Our expertise should be channelized towards whole-hearted adoption of modern technologies such as remote-sensing in Precision Farming, apart from initial experimental levels in this technology. Currently, in India, remote-sensing is applied to map the type of the standing crops and as a tool for an inventory of the crops. A host of benefits accrue both individually to the farmers and collectively to the Indian masses if we apply remote-sensing in a combined manner in locating aquifers and Precision Farming on a large scale. The most important of these benefits is conservation of precious water resources of the country. The Precision Farming, apart from being a technology in which an array of the state-of-the-art components is put into practice, also serves as an important system of water resource management.

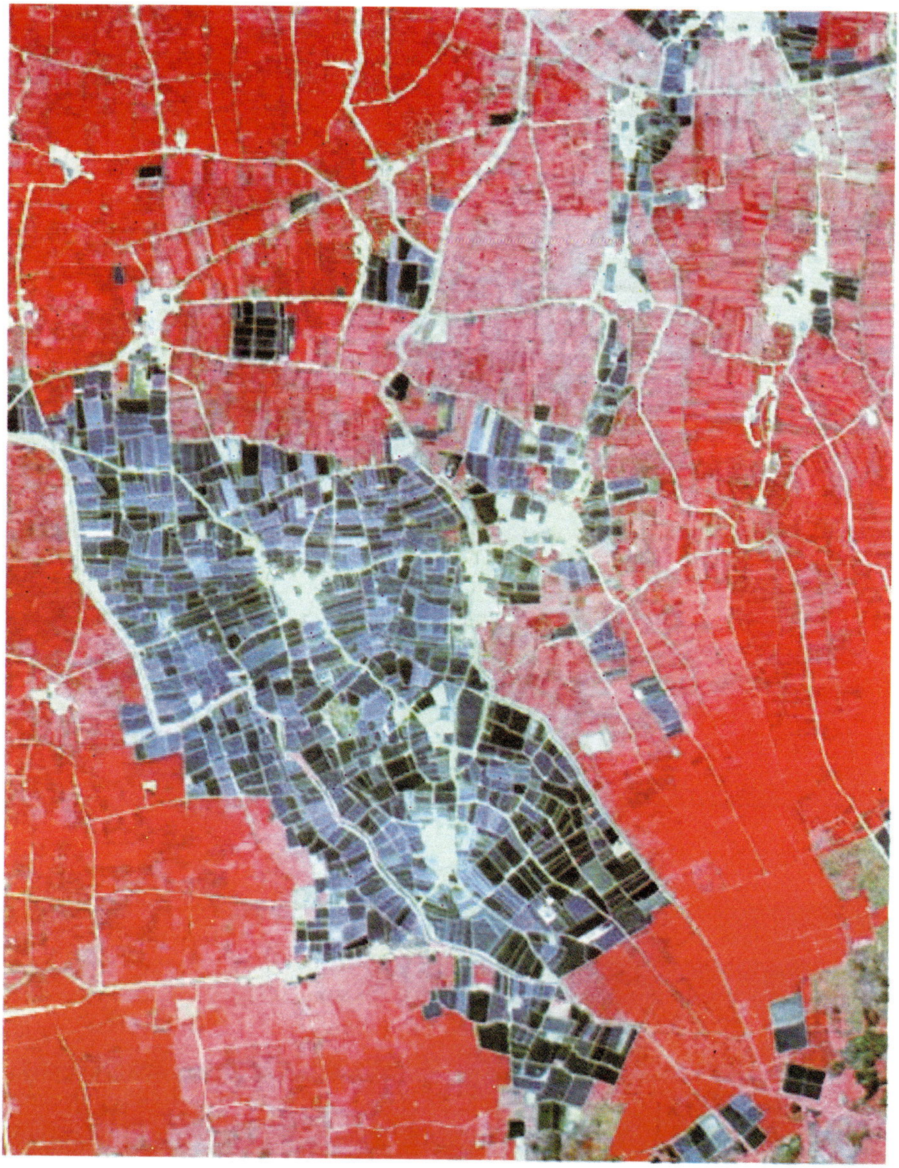

Figure 10.1. IRS 1D-LISS 3 (false colour compositeRGB--3, 2,1) imagery subset of a terrain of intense agriculture. Spectral variations in the imagery are due to variations in moisture content, vegetation index and types of crop. The farms are divided by clear boundaries. The area has a number of villages interconnected by roads and tracks. This type of farmland terrain having a network of service roads, tracks and individual farms with farm-boundaries is ideally suited for Precision Farming.

One of the major hindrances towards our efforts to adopt this technology may be the pattern of farm holding in India as small individual and independent units by the farmers. The farmers need to be educated on the advantages of the technology and encouraged to participate collectively in the Precision Farming. That would push India not only as a multi-faceted technological giant, but help in taking one more step forward towards the holistic approach in the country's water resource management.

3. Geographic Information Systems (GIS)

Broadly speaking, GIS is an extremely efficient and versatile tool for geographic analysis and decision support using a computer system. The GIS comprises software-linked exercises in which huge volumes and a variety of data describing the geographic locations on the Earth's surface can be stored, retrieved, integrated, analysed and altered in the event of subsequent changes, in shortest possible time. Therefore, GIS has the ability to pick up specific data in terms of geographic areas or locations and themes. It can merge one theme of data for a particular area with another theme and analyse their spatial characteristics. GIS is also capable of sifting, searching and sorting specific features in different areas, as well as update and model the data of various sorts in short time. The output data of GIS are in the form of maps, graphs, lists and statistics, all of which together churn out useful information to facilitate instant decision making.

The GIS comprises three components namely, the computer hardware, application software modules and a proper organizational context. These three components do not function in isolation, but they work in tandem. The GIS draws on concepts and ideas from several branches of science, and therefore, the term Geographic Information Science has been adopted to refer to the science behind the systems (Heywood et al, 2002). The geographic information science draws on widely diverse and intricate disciplines of science such as computer science, engineering, cartography, environmental science, geography, geodesy, landscape architecture, remote-sensing, photogrammetry, landuse, land cover and public policies. The GIS software is designed to handle and function with the aid of spatial (or geographical) reference data such as latitudes and longitudes, which are easily available for any geographic location. (This information is given by the GPS too.) The GIS can be visualized as a software package comprising various tools used for data entry, data processing, data analysis and obtain output information.

Thus, in order to achieve these ends, the following components are necessary:

(*i*) The computer system (hardware and operating system);

(*ii*) the software;

(*iii*) the spatial data;

(*iv*) data management;

(*v*) analysis procedures; and

(*vi*) experienced technical personnel to operate the software.

It can be seen that a number of elements are essential for an effective GIS management and obtain the requisite information as output data. These elements include:

1. Data input and output devices, such as digitizers, scanners, printers and plotters.
2. High-resolution monitors to facilitate the view of multi-spectral graphics.
3. Adequate memory in computer for storage of voluminous data in multiple themes.
4. High speed processor in computer for quick storage, retrieval, processing, analysis and yield outputs.

In Precision Farming technique, GIS stores the data on the farms, retrieves them as and when they are required, processes the data on soil type, nutrient and moisture levels, pesticide quantity and a host of other parameters based on the empirical data that are initially obtained from, and assigned to, the fields that are brought under Precision Farming management. GIS stores those data in different layers and assigns the information to that particular farm location in terms of latitudes and longitudes obtained from GPS. Some of the GIS software have a simulation model on ideal or recommended parameters for the farms or crops, and this simulation model integrates the actual parameters of the farm under study with those of the simulation models and processes them for a comparative study and recommend corrective steps in management of the farm. The simulation model parameters also pertain to the usual farm-related information such as soil type, soil health, crop stress, pesticide- and nutrient-requirement and so on.

In Precision Farming technology, the GIS can be made to function based on the whole array of computer systems ranging from laptops and portable personal computers to multi-user supercomputers. The computers can be programmed in a wide variety of software languages. The computer systems are available for the Precision Farming technology that use dedicated workstations with monitors and digitising tables built in. On the other hand, there are instances where this technology is applied using lower-end stand-alone high-speed personal computers. In a nutshell, the GIS in the Precision Farming technology can be tailored to suit the specific needs depending upon the size of the farms, size of the field area that is managed by this technology, and the financial capacity of the farmers for initial investment.

By integrating the GIS with the GPS, several maps can be generated showing the variability of nutrient levels, soil type and soil stress if present, topography of the farms and the whole field, pest incidence if present, moisture- and fertilizer-saturation, probable yield from the farms and a host of other themes. The GIS software are mostly designed in such a way to provide the effects of data smoothing and contouring the values of various above-mentioned parameters. In order to have a smooth and accurate map, it is necessary that the sample locations and the data pertaining those samples need to be significantly high and the sample intervals have to be small.

Precision Farming without the generation of maps with the aid of GIS is also at places resorted to, where farm management results rather than the critical study of the maps are considered as priority. For Indian conditions of agricultural practice under a complex pattern of climates, dry land and semi-arid tropical farming, agriculture under water-stressed condition as well as excessive water condition, the GIS software package that is required in Precision Farming is expectedly complex too. Indian software professionals engaged in developing appropriate GIS software for Precision Farming in Indian terrains need to take into account all these complexities while executing the software development. Further, in a condition where the majority of the Indian farmers not having adequate exposure to education, as also the educated farmers not being techno-savvy to grasp the jargons of computer speciality, it is expected that a sizable number of software and computer professionals in India need to work in rural atmosphere directly in tandem with those farmers as their consultants and advisors, if the Precision Farming were to be launched in India as an aggressive campaign.

The knowledge base the technology is capable of creating in the entire country and the power that the knowledge can give to the Indian society are enormous. More and more computer professionals should be encouraged to pay lot of attention to an attractive computer-based farm technique such as Precision Farming. That would not only create jobs to our youth who possess the zeal and knowledge to work with computer systems (both hardware and software), but also that would facilitate reverse migration of technically qualified youth from urban centres to rural bases to practise advanced technologies.

4. Variable Rate Technology (VRT)

In this component of the Precision Farming technology, an array of agricultural machines or what are known as 'applicators', pioneered by the engineers in association with computer professionals are put to use. The innovation of the computer-linked agricultural machines or applicators was, in fact, the triggering mechanism for Precision Farming technology. The applicators are otherwise known as 'precision applicators' or 'variable rate applicators'. The VRT comprises computer controllers and the associated hardware to suit variable output of fertilizer, pesticides and lime in different farms. The outputs have to be variable since the requirement of the fertilizer and pesticide by different farms are often variable too. The applicator is linked to the GPS so that the operator of the applicator can precisely locate the farm that needs care in the form of fertilizer- or pesticide-supply. Therefore, the hardware and machinery are also known as 'GPS-linked agriculture machines'. Likewise, the GIS software decides the exact quantity of fertilizer or pesticide, which is to be administered to the needy farms. The controller of the machinery then manipulates it in such a way to administer the correct amount of fertilizer or pesticide required by that particular farm. In most efficient stages of Precision Farming, the GIS, the GPS and the VRT machines are perfectly synchronised in such a way that in 'real time' (meaning instantaneously) the VRT applicators administer variable but exact amount of nutrients and pesticides required by different farms as the tractor moves in those farms.

If the Precision Farming technology is adopted over large areas of arable land, it would help the entrepreneurs in mechanical engineering to conduct researches to invent innovative GIS- and GPS-linked applicators and farm tools that are exclusively suited to Indian conditions of soils and environment, in aid of the Precision Farming technology.

5. Yield monitoring and Yield Mapping

Yield monitoring is the method by which the crop production in different farms is assessed and various protocols are arrived at to manage the crop production to its maximum level. It is, in a way, a crop inventory. The yield monitor measures the crop as and when it is harvested. As the yield is measured, the data are stored in computer along with the co-ordinates (latitude and longitude) of the point obtained from the GPS. The mapping software creates a map called the yield map based on the data generated by the yield monitor. The yield map displays two parameters namely yield variability and yield production, which are represented in different colours on the map, with each colour representing a range in yield.

Limitations of the Precision Farming

It is important to realize that the availability of technology by itself is not the panacea for the problems concerning the availability of water resources. Success of the practices such as Sustainable Agriculture and the technology like Precision Farming hinges mainly on the whole-hearted participation of the common man in the spirit of a mass movement. Large field areas as shown in Figure 10.1 are ideally suitable to start the practice of Precision Farming. In the course of discharge of duties, the scientists engaged in water- and agriculture-related research and development should not only sensitise the potential land owners and farmers in the movement of Sustainable Agriculture, but also function in tandem with the local administration to give a fillip to the movement.

Furthermore, the technology such as Precision Farming by itself cannot give solutions to tackle the water scarcity for agriculture. Even in the farms and fields that are covered by Precision Farming, water still remains a necessity. The technology just helps us to conserve our water resources and manage them in prudent manner so that they last for a longer period. It is emphasized that despite availability of intelligent farm technology such as Precision Farming to aid the sustainable agriculture practices, we need to develop water resources before, or simultaneous with, employing technologies in agriculture. As long as India depends solely on monsoon for activities ranging from day-to-day human activities to agriculture, and from recharge of groundwater to filling the multi-purpose dams, we are most likely to face scarcity of water perpetually. Instead of linking our progress and prospects to the vagaries of the Nature, we need to look for alternative sources of water to augment, develop and manage our water resources prudently. This should be our first step forward towards sustainable agriculture. Technology such as precision farming serves as supplementary tool to the sustainable agriculture practice.

WASTELANDS AND THEIR SUSTAINABLE DEVELOPMENT

With the pressure on land for residential and industrial purposes mounting day by day, almost in all parts of the country vast tracts of agricultural land are transformed into lands for human habitation, setting up industries and a host of other human activities. This is more so in the immediate vicinity of big villages, towns and suburban belts of metropolises. The last fifteen years or so have witnessed a virtual spurt in the industrial activities that required suburban agricultural lands. It is for the reason that those industrial establishments expect their highly qualified technocrats, engineers, management and administrative personnel working in their industrial units to live comfortably in cities within reasonable limits of commuting distance.

In particular, the industrial units dealing with information technology, software development, car and electronic manufacture and the export processing zones (EPZ) have proliferated at the expense of vast suburban agricultural lands. No city in this country was perhaps spared from this mutation of agricultural lands in its surroundings, but Delhi, Chennai (formerly Madras), Bangalore, Ahmedabad, Hyderabad and Pune had to face major changes in this respect. The growth of the industries cannot be resented for any reasons, as we have to welcome modern technologies to grow in our country at the same pace as the global level growth of these industrial segments, as also we have to provide jobs to the highly qualified and technically capable youth of the country. At the same time, there cannot be a dispute over the fact that the growth of these modern industries had its own fallout in the form of a decrease in the cultivable lands.

We have, thus, a dual responsibility of encouraging and facilitating the growth of modern knowledge-based and economy-oriented industrial sectors, even as we continue to nurture our age-old profession of agriculture. In order to achieve this, we have to necessarily foster the *'Lexus and Olive Tree'* concept of Friedman (2000). We have to not only manufacture Euro-II model high-speed cars, assemble light combat aircrafts, build remote-sensing satellites and blend strategic alloys to keep pace with the global technological advancements, but cling to our own olive trees (representing culture and tradition) too, by attaining our own sworn mission of eradicating hunger and poverty from our midst by increasing our agricultural production.

Ironically, while on one hand we face an acute pressure on the availability of the agricultural lands to grow foodgrain and cash crops, on the other hand we also have a dubious distinction of possessing vast tracts and a variety of wastelands in our country. Out of the 328 million hectares of total area of the Indian terrain, the wasteland accounts for about 63.85 million hectares, which works out to 20.17 % (Wasteland Atlas of India, 2000). Of this, about 15 million hectares of wasteland are estimated as reclaimable for agricultural purpose. In order to compensate the allotment or transformation of agricultural lands to the residential, industrial and industry-supporting activities, it would be prudent to reclaim these 15 million hectares of wasteland in a phased and sustainable manner and bring them under cultivation. Since this could be taken up as an entirely new activity by the concerned State governments, it is preferred that a lot of sustainable planning goes into the reclamation of the wastelands so that the country benefits fullest in terms of economic growth without making compromise on ecology and without adding too much of a burden on the water resources needed in this effort.

The wasteland is described in the Wasteland Atlas of India (2000) as the '*degraded land, which can be brought under vegetative cover with reasonable effort, and which is currently under-utilized and land which is deteriorating for lack of appropriate water and soil management or on account of natural causes*'.

There are thirteen recognized types of wasteland in the country. They are as follows:

1. Gullied land or ravinous land.
2. Upland with scrub.
3. Waterlogged or marshy lands.
4. Land affected by salinity or alkalinity, both in the coastal and inland parts.
5. Shifting cultivation area.
6. Degraded forestland.
7. Degraded pastures and grazing land.
8. Degraded land with plantation crop.
9. Sandy, coastal desert lands.
10. Lands under mining or quarrying.
11. Barren rocky or stony waste land or sheet-rock area.
12. Steep sloping area.
13. Snow-covered or glacial areas.

These are self-descriptive terms that do not require any more explanation.

Development of wastelands into usable agricultural lands has a potential to become a key factor in the sustainable agricultural practice. Barring a few categories of the above-mentioned wastelands (such as waterlogged marshy and glacial areas), many other types can be reclaimed into agricultural lands. The plans to reclaim the wastelands for useful purposes such as agriculture, horticulture, fodder cultivation and grasslands should take into consideration the parameters such as the slope of the area, the soil fertility, thickness of the soil and its depth to the bedrock, but the availability of water in those areas is the one most vital factor.

The wastelands can be reclaimed and brought under considerable vegetative cover with a little effort, provided certain amount of water is made available to these lands. Large areas that were once wastelands in the semi-arid areas of central India (Maharastra and Andhra Pradesh) and off-coast Tamil Nadu, have been reclaimed in recent times as horticultural lands with orchards of mango, guava and citrus fruits with the aid of drip and sprinkler irrigation. This has not only brought about economic development of the people who have ventured into the enterprising attempt to reclaim the wastelands, but in fact transformed the geography of the area too. Those formerly dry sandy and rocky belts are transformed into greenbelts with thick cover of grass that arrests the soil erosion and facilitates slow surface runoff and percolation to the groundwater system. Several square-kilometres of area surrounding such reclaimed wastelands have greatly benefited by the rising levels of groundwater table over a period of time (about three years) and turning entirely into lush green areas with thick canopy of green vegetation.

There are a few reports on the services rendered by the non-government organizations (NGOs) in transforming dry and bald lands without grass cover, which once stood as degraded wastelands in the drought-prone areas of Alwar district in Rajasthan and (in a small way) in Anantapur district in Andhra Pradesh into green agricultural fields. Water has been the key in their efforts to reclaim these wastelands and bring them under cultivation. Rainwater harvesting (described earlier) is one simple human effort, which can do wonders in the field of wasteland reclamation. These NGOs resorted to the practice of rainwater harvesting by construction of *johads* or bund structures to impound the surface runoff rainwater during the monsoon time. This exercise does not require much financial investment, but what is needed is the input of human labour for which there is no dearth in our country. With minimum financial investment and the locally available labour force and materials, construction of *johads* and bund structures to impound rainwater should not be a difficult task. Result-oriented missions such as construction of *johads* and bund structures would be the first step towards our efforts for sustainable wasteland reclamation.

Hillocks and hummocky ridges in the arid regions are most often bordered by smooth slopes, which extend downward to plain area without any slope. The slope part consists of eroded bedrock with veneers of alluvium derived partly by weathering of the *in situ* rocks and partly transported along the slopes. In terms of geomorphology, this slope is known as *pediment* (Figure 8.8). The lower parts of the pediment transit into aggradational (leveling up of the terrain by natural deposition of sediments) geomorphic feature called as *bajada*, with the sediment thickness more than that of the pediment. The upper parts of the pediment meet the steep hill front with a break in slope. Both pediment and bajada put together show slope angles in the range of about 3 to 10 degrees. Almost all the areas forming margin to the hills show geomorphologic setting comprising pediment and bajada, and ninety percent of the terrain consisting of pediment and bajada constitute unused wastelands in India. This type of geomorphologic setting and wasteland is present in most of the agriculture-intensive semi-arid and sub-tropical to tropical climatic regions of central Indian States, which receive restricted rainfall. Currently, the bajada parts of land arc to somc cxtcnt put into agricultural use in some States to grow seasonal crops. With the provision of small quantities of water and planned approach, several million hectares of area comprising pediment and bajada can be reclaimed for agricultural purpose and sustainable methods for water resource development.

Method of identification of the wastelands

All the thirteen categories of the wasteland listed earlier can be accurately identified by *remote-sensing* method. Satellite imagery being tools that bring out the empirical data on the Earth's surface from a space platform, the interpretation made by the able scientists well-versed in remote-sensing using these imagery is capable of providing quick and accurate results in demarcation and categorizing the wastelands. Geomorphic features such as pediments and bajadas are easily interpreted from imagery (Figure 8.8), and therefore, crash programmes to identify and map the wastelands formed by them and which are easy for reclamation in a speedier manner should be launched by the remote-sensing scientists. Such maps and data should be made available in most appropriate scales to those executives in the governments who are responsible for framing the country's policies on the management of water resources and wastelands. That would facilitate speedy implementation of the governments' decisions on the crucial matters related to the wastelands that require immediate reclamation and the infrastructure required to be provided in that direction.

The governments on their part would do well if they patronize remote-sensing technology simultaneous with field checks to generate a wealth of data in the form of maps on appropriate scales. These data would serve as an inventory and categorization of wastelands in different parts of the country, especially in the agriculture-intensive States and regions, which face urbanization at very fast rate, but would prefer to remain agrarian. Further, those States and regions which are in the semi-arid zones, but can afford to earmark limited amounts of water resources from the river basins to the farming community in those States/regions can also gct thc invcntory on the wastelands generated from the remote-sensing scientists of proven ability.

Water quality for the agriculture needs

Agriculture is pivotal to the economic growth of India, or for that matter of any country. In India, agriculture contributes about 25% of the Gross Domestic Product (GDP), and about 70% of the population is dependent on agriculture for their livelihood (Research, Reference and Training Division, 2003). Further, the growth under agriculture and allied sector is estimated at 5.7% during 2001-02 alone, from a lower level of –0.2% in the preceding year. The agricultural production in India scales new and new highs by every passing year. This can be exemplified by the fact that the agricultural production in the year was 2001-02 was 211.32 million tonnes compared to 195.92 tonnes in 2000-01.

Agricultural products being consumable materials as also items related to health of the masses and aesthetics, the water that is used in the agriculture sector needs to be pure in form from either rainwater or groundwater source. Recycled waters are likely to contain traces of biologically non-degradable chemicals and heavy metals. As the plant sap comprises mostly nutrients absorbed from the soil and the water, these nutrients reach the human system when the humans consume the plant parts. Some of these plant nutrients and heavy metals (such as chromium, arsenic and mercury) are likely to cause health hazards in humans, if they exceed in content limits. We cannot make compromise in the quality of water used in agriculture sector because of the health hazards that are likely to be posed by using the agricultural products cultivated using recycled waters, be it from industrial or municipal wastewaters. We need to allocate requisite amount of rainwater source in the form of surface runoff or harvested rainwater or groundwater to the agricultural needs of the country on a continual basis. The sustainable agriculture practices including precision farming and wasteland development are basically modern state-of-the-art methods using the information technology inputs, but water is a necessity in these methods of agriculture. These methods are different from the conventional methods of agriculture in that they conserve plenty of field inputs like water, fertilizers and pesticides, thus paving way for saving of money by the agriculturists. The water quality in these modern techniques is expected to be the same as in the normal agricultural practice, though the method of application of the water may vary as conventional or drip or sprinkler types of irrigation.

(*v*) Need of Water for Industries

The availability of water to the needs of the industries has a direct relation to the investments in the industrial sector in the country, specifically in the areas like food processing, beverages, dairy, sugar, breweries, textiles, leather, pharmaceuticals, chemicals, fertilizers, and so on. In the order of consumption of water, the largest industrial segments are: Oil refineries, Paper manufacture, Metal working plants, Chemical manufacture, Air-conditioning and Refrigerating units, Distilleries and Fabric dyeing. The industrial production in India, like the agricultural output, scales new and new heights by every passing year. Together, these two contribute considerably to the Gross Domestic Product (GDP). For example, the Monetary and Credit Policy of the Reserve Bank of India for the fiscal year 2003-04 has announced that 'better-than-expected monsoon and satisfactory industrial and export growth....are likely to have positive impacts on the Indian economy as a result of which the GDP in 2003-04 is projected to show a growth of 6.5 to 7 percent'.

There are undisputable indications that the growth of each of the above-mentioned industrial segments in India in near future will, in all likelihood, go up exponentially. Especially, the growth of the industries that manufacture petroleum products, fertilizers, polymers, fibres, organic chemicals, dyes and pigments, leather chemicals and speciality chemicals is projected to grow in the range of two to twenty times by the year 2020, and these industries have the potential to 'venture into the export market' (Abdul Kalam, 1998). It will be case also with beverages, dairy and food products, drugs and pharmaceutical products, cosmetics and animal proteins. The availability of requisite water for these industries is going to be a crucial factor when we have specific plans to give a fillip to the growth of these industrial segments. In fact, even in the industries, which are *not* water-intensive, such as steel, cement, automobile, electrical equipments, electronics, software and certain consumer goods, availability of water is a crucial factor, because these industrial units employ a large contingent of workers who need water in their workplace.

With the stringent implementation of rules concerning fire-fighting in the industrial hubs in recent times, the need for a large quantity of water in the vicinity of these hubs has considerably increased. The availability of water is an important consideration in setting up of the industrial units, even though the quantity of water required by the industrial units may be variable. With the Indian economy open to the global players, and foreign direct investments made simpler, India has already attracted global industrial units. This trend is likely to continue in the decades to follow. If the infrastructure such as availability of requisite water is developed further, the foreign direct investors will be ready to make huge investments in India, offer employment opportunities and share the profits. Failing monsoons, falling groundwater table levels, depletion of groundwater resource, permanent loss of aquifers due to construction of new habitats proximate to the industrial hubs and drying up or conversion of several water tanks are going to be major impediments in our efforts towards developing water source for industrial development. Water scarcity at a juncture when we have solicited foreign direct investment would certainly not augur well for the growth of these industries, which have the requisite potential to make India an industrial giant. We have to develop appropriate infrastructure congenial for the industrial development all over India, and water resource development needs to be the first priority in this direction.

Not all industries need pure or potable-grade water for the survival or growth. Only those industries that deal with aesthetic products such as food, vegetables, dairy, beverages, sugar, drugs and pharmaceuticals and food-grade dyes and chemicals require potable-grade water, and therefore, it is necessary that we allocate adequate quantity of it for those industries. For example, food processing for the domestic market has made rapid growth in our country in the last two decades. Abdul Kalam (1998) has indicated food processing as one of the major sectors in which India would not only compete with the 'developed' countries but also has a potential be a leader in the year 2020. The Indian entrepreneurs have the capability to establish large food chains comparable to McDonald, Domino and Kentucky Fried Chicken worldwide, given the right atmosphere and level playing ground. Availability of large quantities of potable grade water is crucial in this matter.

Gross production of patented tinned food and seafood for export from the Indian shores has significant potential, as the availability of raw materials for these is in plenty within India. But, the cardinal factor in food processing industry is hygiene, and water holds the key to the hygiene. In country like India where humid conditions prevail, decay of food materials sets in quickly if they are not processed with care. Use of bacteria-free fresh water is the crucial factor in food processing industry. Therefore, availability of bacteria-free water needs to be ensured to the prospective entrepreneurs of food processing industry. We should attempt world-class, in our efforts to compete with the global food processing industrial units. Since the standards set for materials, especially for processed food items, by the international markets are indeed stringent, we have to necessarily comply with those standards if we want to make an impact on the international market on processed foods. Needless to emphasise that production of bacteria-free potable water in industrial quantities should be the first step towards the promoting food processing industry in India. We have to look for bacteria-free value added water in large industrial quantity for food processing and dairy industry.

However, a majority of the other major industries are, in general, capable of sustaining by using recycled water. Currently, barring a few cases, all these industrial units, be it petrochemical or paper or refrigeration units, tap huge quantities of precious potable grade water from the rainwater or groundwater source for their sustenance. This undoubtedly mounts an avoidable stress on the availability of raw water to generate potable-grade water for the human consumption and livestock, besides creating a shortage of water for agriculture and other functions, which require good quality water.

In an effective water resource management practice, the industrial units, which do not need potable-grade water should be named and listed. Further, they should be insisted to set up water treatment plants to suit their own needs in terms of capacity and the wastewater they generate. As a part of the Water Policy, the major industries that can sustain on treated wastewater have to be insisted to recycle the wastewater they themselves generate and reuse it in their units. The huge quantities of rainwater or groundwater sources that is used by the petrochemical, textile and paper industries can be conserved for the needs of society and for the livestock, because, these industries are capable of functioning under treated recycled water, be it from the industrial or municipal source.

(*vi*) Need of Water for Balancing the Ecology

The Planet Earth being a wonderful place to live in, it is the duty of the Mankind who dominates to rule it, to give an equal chance for its subordinate beings too to live happily and in perfect harmony with him. The health and happiness of the mankind hinge to a great extent on the degree of harmony he maintains with the flora and fauna amidst which he lives, and the environment he shares with the others on this planet. It will not be out of context here to quote James and Jongeward (1971) from their classical work 'Born to Win': *"Human survival and continued development depend not only on how we transact with out fellow humans but also on how we relate to the rest of our environment. The inanimate world, which includes rocks, sunsets, water, and air, and the animate world of living plants and animals are at our mercy. We have the power to enjoy them, to enhance them, or to destroy them. When we misuse them by polluting air and waterways, by rendering land barren, by causing the extinction of a species, or by upsetting the ecological balance, it is our own existence and continuance as a race which are, in the long run, threatened. Our own exploitation of our environment unnecessarily can doom us to a tragic ending".*

In our efforts to live in harmony with a healthy ecosystem, we need to protect the existing forests and create new ones, develop urban forestry and grasslands, create environments that are suitable for the multiplication wildlife and wild trees in several places, and protect the soil cover on vast tracts of our country.

Needless to say, all these considerations would need good amounts of water. In order to take care of these natural beings, it would be fit to budget the water from the natural phenomenon that occurs annually—monsoon rains. As 'ethical persons', at least a quarter of the rainfall that occurs in a monsoon period has to be set aside by us for balancing the ecological system of which we also are a part. In the likely event of weak monsoon and the resultant scanty rainfall, treated recyclable water has to be assigned for this purpose. The pristine rainwater that is allocated by us to the eco-system would not be a liability for, a part of it percolates to the groundwater system and the other part undergoes evapo-transpiration. Either way, this water is received back by us manifold via Hydrological Cycles (discussed earlier). If we allocate a sizable quantity of water to the natural ecosystem, which is at our mercy, we need to look for alternative source of water for our own utility.

(*vii*) Need of Water for Recreation and Entertainment

Gone are days when people enjoyed their leisure time as also beat the summer heat by taking dips in the unpolluted waters of our rivers, lakes and ponds. The then lifestyle, such as crossing of turbulent rivers to reach places, necessitated most of our predecessors to know swimming. In contrast, the present generation learns swimming more as a recreation in the artificial swimming pools, which need to be replenished with large quantity of water very frequently. The participation of Indian competitors in national and international swimming events and their success rate are far from satisfactory. One reason for this is cited as the non-availability or non-affordability of the facility to learn swimming to those who have raw talents in swimming. There would not be such a handicap if the water bodies and swimming pools were available to those who can learn competitive swimming as a passion.

Zoological parks are the place for relaxation and entertainment to the young and old alike. Copious amount of water is required to maintain zoological parks wherein natural environment of forests with pools, ponds and dense vegetation, is simulated for wild life in captivity. Water theme parks are popular in land-locked metropolises, cities and towns, where the water needed to maintain those parks has to be purchased at a premium. Therefore, though popular, more number of such parks could not be developed for entertainment of the urban middle class, which dominates the Indian society. For the recreational facilities such as water theme parks to be developed evenly throughout this country, availability of large quantities of good quality water holds the key.

The non-availability of facilities to conduct water-related games in international standards may be one crucial factor in India not being a host to many international sports meets like Olympic Games. The European Community (EC) has fixed stringent standards for bathing water, which can be adopted as the water quality in swimming pools. This water quality is given in the following Table.

Parameter (Milligram per litre except where stated otherwise)	*Guide Limit 90 Percentile*	*Mandatory Limit 95 Percentile*
Cadmium	0.0025	0.0025
Mercury	0.003	0.0003
Dissolved Oxygen (DO)	80-120% Saturation	not stated
Faecal coli MPN/100 ml.	100	2000
Total coli MPN/100 ml.	500	10000
Salmonella MPN/litre	not stated	0
Faecal streptococci MPN/100 ml	100	not stated
Entero viruses MPN/10 litres	not stated	0
(MPN—Maximum permissible number)		

Source: Directive 76/110/EEC, as quoted in Tebbut (1992).

Likewise, though many people overseas find the cultural heritage and fine Indian traditions curious, the inflow of tourists is low considering the vastness of our country. Scarcity of quality water is cited, among others, as a main reason. In a well-known tourist guidebook on India published abroad, the tourists are cautioned about drinking water in Kerala, and quality of water for drinking and personal hygiene in Rajasthan, and several other towns. Though these facts are to some extent exaggerated, there certainly is an element of truth in these writings, as we ourselves can see and realise, the scarcity of water and the resultant lack of cleanliness of the environs in several parts of our country are, indeed, undisputable facts.

Availability of water is a key factor for the development and maintenance of urban parks, ocean- and water-theme parks, zoological parks, swimming facilities, tourist attractions, dancing fountains and gardens that project the rich Indian sub-tropical and rain-forest flora. What is more important, there has to be steady replenishment of good quality water for these recreational and entertainment activities in lean periods such as summer season, which is fairly long (February to June) in India. The quality of water that is required for these purposes has to be nothing short of potable-grade, as the applications involve aesthetic part of the human life. Moreover, this recreational segment of human activity needs steady and uninterrupted supply of large quantities of water. Since the monsoon rain is not a steady and uninterrupted source during lean summer seasons, we need to evolve strategies to create alternative sources of good quality water to meet this demand.

(*viii*) Need of Water for Miscellaneous Human Activities

Besides the human activities described above, there are several other day-to-day activities where enormous quantities of water are needed. These activities are:

1. Need of water by the railway authorities for washing the coaches as well as for providing it in the railway stations and the compartments of running trains for use of the passengers.
2. Servicing and washing centres for automobiles such as cars, two-wheelers, buses and trucks.
3. Uninterrupted supply of water in the airports and bus stations.
4. Provision of considerable quantities of water in large offices, commercial complexes, pilgrim centres, public toilets and conveniences, sports stadiums and large star hotels.

It is a fact that these activities cannot, and should not, come to a standstill; however, the fact that is to be realized is that enormous quantities of good quality raw water, which can be converted into potable grade water after adequate treatment, are put into use (or misuse) in these activities, which do not require water of that quality. Currently, the water required for these human functions is met from either the rainwater or the groundwater source. That is responsible for mounting a stress (especially in the summer months) on these two sources of water, which can otherwise be put into use in the areas of agriculture, human and livestock needs and a host of other activities where the water quality is indeed an issue. It would be an arduous and long-drawn task to make an approximate estimate of the water required in all these human activities individually and in a combined manner. It can, however, be stated with certainty that this quantity would indeed be in mind-boggling proportions. It needs to be realized further that in all these activities concerning the overall cleanliness of the environs of our life, recycled treated wastewater can be substituted for the potable grade water or its raw water.

1. Indian Railways organization makes use of a large quantity of water for washing the coaches and providing water in the coaches of running trains for use in the toilets. The network of Indian Railways is spread over the whole country and its activity is continual over all the 365 days of the year.

Indian Railways organization is one of the largest users of water. If the nationwide presence and uniform network of the railways, coupled with the number of trains that run across the length and breadth of the country and the millions of people who travel by these trains are taken into account, we can just infer the enormity of the water required by the Indian Railways. Barring the water required in the railway stations for supply as drinking water, the activities of the railway organization do not need raw potable grade water. Efforts should therefore be made by the railways organizations to install water treatment plants in large numbers wherever they function in such a way that the recycled water is used in washing the coaches and for providing in the toilets of running trains. That would spare the railway organization from depending on the local water supply authorities for its water needs, as well as ease stress on the groundwater that they consume in such huge volumes for the purposes where water quality is not critical.

2. Likewise, millions of cars, buses, trucks, three- and two-wheeler vehicles are serviced daily in workshops spread all over the country for which an inestimable quantity of water is used. Most often, the water for this activity is tapped from the groundwater source. The use of groundwater in enormous quantities in the vehicle service stations located in metropolises, cities and towns undoubtedly mounts a severe stress on the groundwater source in those places, sending the groundwater table to greater depths with every passing year. If this groundwater is conserved from abuse in the dry summer seasons, it can be used as raw water for upgrading to potable grades as also as water for use in kitchens, swimming pools and a host of functions in which the quality of water is critical.

Further, the effluent water emanating from the servicing stations of the railway coaches and vehicles has lubricant oils, grease and fine mud as contaminants. This effluent water is led into the nearby rivers/streams directly or through the drainage system, thus causing enormous pollution of the surface water bodies. Instead, if the effluent water were diverted to the water treatment plants, the recycled water would serve not only as a source for reuse for the same purpose, but also the pollution of the water bodies such as rivers, streams and urban lakes and tanks by the effluent waters can be totally averted.

3. People in large numbers visit the airports and bus stations all over India daily. The maintenance of airports and bus stations, as well as the common facilities such as toilets in them, requires uninterrupted supply of water for their upkeep. Allocation of large quantities of treated wastewater would be an ideal solution for the purposes of not only cleaning and upkeep of the public toilets in the bus stations and airports, but for conservation of the precious surface runoff water and groundwater source also.

4. Religious tourism, pilgrimage and holiday tourism are becoming more popular with the availability of better lodging facilities all over the country and attractive tourism packages offered or sponsored by the State Governments and the private tourism operators. The States such as Rajasthan, Kerala, Uttaranchal, Jammu and Kashmir, Himachal Pradesh and Sikkim, and a sizable population residing in the tourist towns such as Ujjain, Nashik, Udhagamandalam (Ooty), Kodaikanal, Tirupati-Tirumala, Mount Abu, Haridwar and Darjeeling have greatly benefited in terms of economic growth due to the religious and holiday tourism. The striking feature in the case of the tourist centres in India is that there is a quantum difference in the experience of the visiting tourists to the places having unrestricted water supply, and those tourist spots that have water shortage. In other words, the shrines and tourist spots, which are situated closer to the water bodies like rivers are much cleaner than the pilgrim and tourist centres that do not have a water body nearby to provide adequate supply of water to the visiting tourists. Whereas plenty of bottled potable water is nowadays available at a price in all the tourist centres in India, regarding the availability of water for personal hygiene much has to be still done. If the tourism departments of the States of Indian Union are very serious about promoting tourism in their respective States and attain tourism-linked economic growth, the most important aspect on which they need to concentrate is to generate plenty of water to transform the tourist centres clean and green. There is no need to emphasize at this juncture the reason why Goa is an extremely popular as a tourist destination.

Chapter 11

This chapter is devoted to discussions on:

- Test characteristics of Drinking Water.
- Water Treatment Processes
- Treatment to upgrade raw water to potable grade

MANAGEMENT OF POTABLE WATER

From the previous chapter, it could be inferred that the requirement of the water for the human activities can be broadly categorized into three slots, namely:

1. For direct oral consumption, personal hygiene, cooking and cleaning of cooking wares, food-related industries and for the cattle.
2. For agriculture and select industries in which the water quality is *not* critical.
3. All the other activities that are not covered by those purposes mentioned above.

The water of the category (1) as mentioned above forms the lifeline for the human activity in anywhere in the world. Currently, the degree of human development in a country is gauged by the ease with which the availability of, or access to, this category of water to the citizens of that country. In simple terms, this type of water is known as the *'potable water'*, which by definition means the *'water fit to drink'*.

Water from any source or any raw water can attain the status of potable water, if the protocol of treatment is followed, and removal of biological and chemical pollutants, heavy metals, and all the other contaminants that are likely to inflict diseases, both of long- and short-term, from the raw water are ensured, before it is supplied as drinking water. It is however important that the aesthetics related to the raw or original water source, and the processes involved in converting the original source to potable grade are also equally important. Proper selection and protection of water sources, both raw water and the treated potable grade water, are cardinal in the provision of safe drinking water to the masses of population.

World Health Organization (WHO, 1993) has recommended '*Guidelines for drinking water quality*' and mentions that 'if properly implemented, the guidelines will ensure the safety of drinking water supply through the elimination, or reduction to a minimum concentration, of constituents of water that are known to be hazardous to health'. WHO (1993) further emphasizes that '*the guideline values recommended are not mandatory limits. In order to define such limits, it is necessary to consider the guideline values in the context of local or national environmental, social, economic, and cultural conditions*'. Therefore, before any issue or action or movement on the matter related to the quality of the drinking water or water-based product is taken up, it is important to study the local conditions in the particular country that dictate the quality of potable water generated in a given location. Lest, such actions would lead to stirring up of passions and sentiments regarding the quality of water orally consumed by the society. It is, nevertheless, always better to protect the water from contamination by any substance, rather than to treat it after it is contaminated, as opined by WHO (1993).

Let us, for example, consider the scenario of a semi-urban residential setup as also an industrial belt nearby. In this sort of setup, the transport of the municipal wastewaters and the industrial effluents is not through lined sewers but by unlined canals or trenches, which open into the nearby water body like a stream, river or lake. Under this civic condition, as the wastewater and the industrial effluents undergo transport for ultimate disposal, significant quantities of them percolate to the groundwater system. Though the natural processes to some extent filter this wastewater as it percolates through the surface sand layers, cracks and fissures in the rocks, certain traces of micro-organisms such as *Escherichia coli* and amoeba, as well as several dissolved industrial wastes that cannot be filtered, and they thus join the groundwater system. When the groundwater from this setup is pumped out, it would *appear* colourless, odourless, clear as crystal and, therefore, fit to drink. However, when used as potable water, it is *likely* to cause a health hazard to the masses of population due to the dissolved chemicals and contamination by micro-organisms. This has prompted the water supply departments in several countries worldwide to adopt the concept of '*safe drinking water*', which is produced by means of multiple stages of treatment to neutralize the possible wastes and chemicals, and disinfect in a way that the micro-organisms are eliminated. It is again important to note the emphasis made by the WHO (1993) that the '*guideline values have a degree of flexibility and enable a judgment to be made regarding the provision of drinking water of acceptable quality*'.

The primary mission of the Government of India ever since we attained Independence has been to *provide safe drinking water* to the masses of Indian population. With this mission, the governments, both of the States and at the Centre, have brought major parts of the country's rural areas besides all cities and towns under the scheme of safe drinking water supply, through the Five Year Plans. The United Nations Organization (UNO) had observed the decade 1981-90 as the International Drinking Water Supply and Sanitation Decade. In this period, the UNO had urged the countries worldwide to supply safe drinking water to its citizens by adopting water purification under standard conditions. The UNO had also offered technical and economic assistance to several countries for this purpose.

In order that the water that is supplied for oral consumption is declared as the 'safe drinking water', different countries have adopted different standards with regard to the contents of various constituents in the water. The desirable limits of possible contaminants and ions in the potable-grade water are shown in a Table in a different chapter. WHO, Geneva, had set the International Standards for Drinking Water in 1984. Besides, the United States of America and the European countries follow their own standards, which are more stringent than the WHO's standards. WHO (1993) is of the opinion that '*a guideline value represents the concentration of a constituent that does not result in significant risk to the health of the consumer over a lifetime of consumption*'. After a careful examination of all the provisions and guidelines for the standards of drinking water, it would appear that the WHO's guidelines and provisions for relaxations to the quality of drinking water is the most practical way to adoption. In that respect it needs to be borne in mind that '*the greatest microbial risks are associated with ingestion of water contaminated with human and animal excreta*', and therefore, all precautions should be taken against such eventuality while generating drinking water.

In India, the Bureau of Indian Standards (BIS) has fixed the quality standard for the drinking water. The '*requirements for the essential and desirable characteristics required to be tested for ascertaining the suitability of water for drinking*' are given in BIS's publication No. IS 10500:1991 Edition 2.1 (1993-01) entitled "Indian Standard Drinking Water—Specification (First Revision)" (BIS, 2003). In the following Table, the substances or characteristics, their desirable limits or requirements and the permissible limits in the alternative source in the event of non-availability of the desired quality water, as recommended in BIS (2003), are shown. The readers may refer to the BIS's reference quoted above to know the *methods of test* suggested by the BIS in determining the limits of these substances and characteristics, as also the undesirable effects beyond the desirable limits.

TABLE – 11.I. TEST CHARACTERISTICS FOR DRINKING WATER

Substance or Characteristics		*Requirement (Desirable Limit)*	*Permissible Limit in the absence of Alternate Source*
	Essential Characteristics		
(*i*)	Colour, Hazen units, (Max.)	5	25
(*ii*)	Odour	Unobjectionable	----
(*iii*)	Taste	Agreeable	----
(*iv*)	Turbidity, NTU, (Max.)	5	10
(*v*)	pH value	6.5 to 8.5	No relaxation
(*vi*)	Total Hardness (as $CaCO_3$) mg/l, (Max.)	300	600
(*vii*)	Iron (as Fe) mg/l, (Max.)	0.3	1.0
(*viii*)	Chloride (as Cl) mg/l, (Max.)	250	1000
(*ix*)	Residual free chlorine, mg/l, Minimum	0.2	----
(*x*)	Fluoride (as F) mg/l, (Max.)	1.0	1.5
	Desirable characteristics		
(*xi*)	Dissolved solids mg/l, (Max.)	500	2000
(*xii*)	Calcium (as Ca) mg/l, (Max.)	75	200
(*xiii*)	Magnesium (as Mg) mg/l, (Max.)	30	100
(*xiv*)	Copper (as Cu) mg/l, (Max.)	0.05	1.5
(*xv*)	Manganese (as Mn) mg/l, (Max.)	0.1	0.3
(*xvi*)	Sulphate (as SO_4) mg/l, (Max.)	200	400
(*xvii*)	Nitrate (as NO_2) mg/l,	45	100
(*xviii*)	Phenolic compounds (as C_6H_5OH) mg/l, (Max.)	0.001	0.002
(*xix*)	Mercury (as Hg) mg/l, (Max.)	0.001	No relaxation
(*xx*)	Cadmium (as Cd) mg/l, (Max.)	0.01	No relaxation
(*xxi*)	Selenium (as Se) mg/l, (Max.)	0.01	No relaxation
(*xxii*)	Arsenic (as As) mg/l, (Max.)	0.05	No relaxation
(*xxiii*)	Cyanide (as CN) mg/l, (Max.)	0.05	No relaxation
(*xxiv*)	Lead (as Pb) mg/l, (Max.)	0.05	No relaxation

Contd ...

Substance or Characteristics	Requirement (Desirable Limit)	Permissible Limit in the absence of Alternate Source
(*xxv*) Zinc (as Zn) mg/l, (Max.)	5	15
(*xxvi*) Anionic detergents (as MBAS), mg/l, (Max.)	0.2	1.0
(*xxvii*) Chromium (as Cr^{6+}) mg/l, (Max.)	0.05	No relaxation
(xxviii) Polynuclear aromatic hydrocarbons (as PAH)	----	----
(xxix) Mineral oil, mg/l, (Max.)	0.01	0.03
(*xxx*) Pesticides, mg/l, (Max.)	Absent	0.001
(*xxxi*) Radioactive materials :		
(*a*) Alpha emitters, B1/l, (Max.)	----	0.01
(*b*) Beta emitters, pci/l, (Max.)	----	1
(*xxxii*) Alkalinity mg/l, (Max.)	200	600
(*xxxiii*) Aluminium (as Al) mg/l, (Max.)	0.03	0.2
(*xxxiv*) Boron (as B) mg/l, (Max.)	1	5

Source: Information given in this Table is extracted from IS 10500:1991 of BIS, 2003.

The quality of water used for direct human consumption is a matter of serious implication since it holds the key to the health and well being of the masses of population. Therefore, well-trained personnel should establish the above-mentioned test characteristics of the drinking water under stringent laboratory tests, employing a set of high-precision analytical instruments. A lapse in this respect may lead to the breakout of an epidemic of short term diseases like gastroenteritis and/or long term illness such as fluorosis or carcinoma, or may turn out to be even fatal, depending upon the pathogens that are involved in the contamination of the water. With this background, the Bureau of Indian Standards has stipulated the above-mentioned standard as test characteristics for the drinking or potable grade, with main objectives of:

(*i*) Assessing the quality of water resources.

(*ii*) To check the effectiveness of water treatment and supply by the concerned authorities.

The Bureau of Indian Standards reiterates that *'all essential characteristics should be examined in routine'*. Further, the Bureau stipulates also that *'all desirable characteristics should be examined either when a doubt arises or the potability of the water from a new source is established'*.

Similar to the WHO (1993), the Bureau of Indian Standards too recognizes *'that often it is necessary to relax the specifications, especially when no alternate resources are available and therefore, to enable the experts to exercise their discretion a maximum permissible limit has also been given'*. Under any circumstances, however, there *cannot* be a compromise in the case of the contents of major contaminants in the product water that is to be supplied as safe drinking water. These contaminants include all the ionic substances, and chemicals such as phenols and pesticides. Thus, the points of view of WHO and BIS on matters regarding the quality of drinking water are identical.

The specifications on the bacteria-contents and the bacteriological examination for the drinking water are extremely stringent anywhere in the world. The drinking water is not expected to contain any coliform organisms. However, even the IS 10500 of BIS (2003) mentions that such a standard is *'not always attainable'*. It recommends that:

(*i*) Throughout the year, 95 percent of the samples should not contain any coliform organisms in 100 millilitre of the sample.

(*ii*) No samples should contain coliform in excess of 10 per a 100-millilitre sample.

(*iii*) No samples should contain Escherichia coli in 100 millilitres.

(*iv*) No two consecutive 100-millilitre samples can contain coliform.

WATER TREATMENT PROCESSES

Water treatment processes used in any specific instance must take into account the quality and nature of the water supply source. The intensity of treatment depends on the degree of contamination of the source water. The groundwater extracted from deep, well-protected aquifers is usually free from pathogenic micro-organisms. On the other hand, for contaminated water sources, multiple treatment barriers to the spread of pathogenic organisms are particularly essential and should be used to give a high degree of protection and to reduce the reliance on any individual treatment step (WHO, 1993). The fundamental purpose of water treatment is to protect the consumer from pathogens and impurities in the water that may be offensive or injurious to human health. Urban water treatment is a *four-stage* multiple-barrier system for the removal of microbial contamination. These four stages consists of:

(1) Reservoir storage or pre-disinfection.

(2) Coagulation, flocculation and sedimentation (or floatation).

(3) Filtration.

(4) Disinfection (WHO, 1993).

Alternative or additional processes may be interposed to meet the local conditions. Disinfection is the final safeguard and also protects drinking water during distribution against external contamination. The whole treatment sequence may be regarded as conditioning the water for effective and reliable disinfection.

This multiple-barrier concept can be adapted for treating surface waters in rural and remote regions too, since the quality of drinking water obtained would be same or similar, but the consumers would be living in a larger area. A typical series of processes would include (1) storage of adequate water in a large reservoir, (2) sedimentation or screening, (3) gravel pre-filtration and slow-sand filtration, and (4) disinfection. Disinfection should be regarded as obligatory for all piped supplies using surface water, even those derived from high quality unpolluted source, as there should always be more than lone barrier against the transmission of infection in a water supply. In large properly run waterworks, the criteria for the absence of *Escherichia coli (E. coli)* and coliform bacteria can be met with high degree of probability. The current trend is to optimize the use of chemicals such as chlorine and coagulants in water treatment, and to develop physical or biological methods of treatment, in order to reduce the doses of chemicals required, thereby reducing the formation of disinfection by-products.

It is also of great importance to note that the water that is to be distributed as drinking water should not contain any faecal coliform organisms such *E. coli, Klebsiella pneumoniae, Vibrio cholerae* as well as pathogens such as *Salmonella* and *Shigell*. Equally harmful organisms whose presence is undesirable in the drinking water are the mesophilic bacteria, rotifers and protozoa. Therefore, it is clear that all the sources of the water meant to be 'safe drinking water' or 'potable water' have to necessarily undergo stringent laboratory tests for determination of both chemical and biological contaminants, before administered for the said purpose.

In addition to the above-mentioned test characteristics, the aesthetics of the water meant for oral consumption and use in the kitchens are also matters of serious concern. Practically, almost all (or any) types of raw water can be treated employing different mechanical techniques to obtain potable grade water, which strictly conforms to the standards stipulated in the above Table. But, it is doubtful whether the people at large would accept the 'potable-grade water' obtained by rigorous treatment of municipal or industrial wastewaters. Therefore, whenever such practice as generating the water from wastewater is resorted to, the general public needs to be adequately informed about it, so that they would avoid its use for oral consumption and in kitchen.

Any water that does not conform to, or not improved by treatment to, the quality bearing the 'test characteristics' as quoted above is *not* of potable quality water. Such water needs to be further treated for improving its quality conforming to the standards stipulated by various national and international authorities (BIS, UNO, WHO and so on) before it can be administered as *'safe drinking water'*. This water source, however, can be directly used without any treatment for the purposes mentioned in column (2) in the opening paragraph of this chapter, *i.e.*, for agricultural purposes and for the industries in which water quality is not critical. This water resource is a part of the Hydrological Cycle (Figure 9.1), and the water that accretes the terrain as seasonal rains. The rainwater is the bulk of the replenishable water resource available for various human activities, especially in India. This resource of water forms surface runoff and gets impounded in tanks and lakes, or becomes parts of the watersheds, or gets transported along the natural surface drainage channels such as streams, tributaries and major rivers before partly emptied into the sea. The raw water for the purpose of generating the potable water or 'safe drinking water' at present comes this source. A significant fraction of this resource enters the groundwater system. Effective management of this water resource by budgeting the right fraction of it for the right purposes would, in fact, reflect the prudence and ingenuity of any country or any water resource management body.

Due to the absence, in major parts of rural India, of the treatment systems (such as, for example, a basic facility of filtration plants) that is capable of transforming the non-potable raw water into the potable grade water, a significant fraction of the rural Indian population is forced to orally consume the untreated raw water and also use it for the purpose of cooking, tapping it directly from the source such as impounding tanks, rivers and the groundwater. Since many of the pathogenic micro-organisms can withstand high temperatures, simple heating before use does not remove these micro-organisms from this kind of raw water. This is the major cause for a host of health hazards such as gastro-intestinal disorders, bacterial and viral infections, and worm infestations that affect the people. The situation gets deteriorated and an array of water-borne diseases become epidemic in the summers when the humans living in sub-tropical, semi-arid to arid and coastal zones in the Indian sub-continent have to consume more water orally in order to balance the dehydration resulting out of the hot, dry or humid weather conditions. The situation turns even worse during the rainy seasons of monsoon during which the surface runoff carries the bacteria, viruses and other micro-organisms into the water bodies from where the people access the water for their drinking and cooking needs.

More often than not, it is estimated that the total coliform, faccal coliform, bacteria and many more pathogenic contaminants in the water from any terrestrial source during the monsoon multiplies manifold during the monsoon period. This, in part, explains why epidemics such as gastroenteritis and cholera break out and sicken the people in large numbers in the monsoon times. Even small quantities of raw untreated water contaminated with the pathogenic bacteria are sufficient enough to infect a large source of 'safe drinking water' and cause an epidemic, when the former gets mixed with the latter. That would give an impression that the drinking water under the distribution system might not be in conformity with the standards of 'test characteristics' and, therefore, responsible for spreading the epidemic, but in reality it is the small quantity of bacteria-infected water that contaminates the potable water and incites an epidemic.

A considerable part of the replenishable water resource that accretes to the terrain from the source of rainwater is diverted for the purposes outlined in the column (3) mentioned in the first paragraph of this chapter. This is the area where the water resource management in India goes by and large incorrect. It is in this area of water resources of the country that a lot of reforms are needed and policies for sustainable water resource management have to be brought about for a better use of the water. In the previous chapter, several segments of human activities where water quality is not critical are identified. It has also been stated then that the raw water that is up-gradable with minimum treatments as 'safe drinking water' or potable grade water is grossly misused in these segments, since the quality of water is not important in these segments of human activities. These human activities are widely varied. Some of them are the industrial units dealing with paper and pulp, petrochemicals, fertilizers and chemicals, plastics, ceramics and textile-dyeing; besides, building construction, and the automobile and railway wagon servicing stations too currently utilize a bulk of the raw water that is up-gradable to potable grade and for the agricultural purposes. No statistics are available on the approximate quantity of water needed by these segments of activities, but even a rough guess would indicate the enormity of water quantities that is required for these purposes. It is this segment that needs a bulk of alternative source of water generated by treatment of wastewater. Substitution of good quality raw water used (or misused) for these purposes with the recyclable treated wastewater would conserve the raw water and make it available for agricultural purpose and for upgrading to potable grade.

TREATMENT AND MANAGEMENT OF RAW WATER TO PRODUCE POTABLE GRADE WATER

Currently, in most parts of India, the protected drinking water source originally comes from pure and pristine rainwater source. As the rainwater moves as surface runoff before impounded in tanks and reservoirs for the supply of the general public, it picks up and carries along with it considerable amount of impurities, silt, contaminants and, both dissolved and undissolved substances. The treatment of water for supply as drinking water and potable-grade water for use in kitchen and personal hygiene is a fundamental necessity, which has a direct bearing on the health of the masses of population. Therefore, all these pollutants—namely, impurities, silt, contaminants and, both dissolved and undissolved substances—need to be isolated and removed from the raw water for purification before it is administered as piped drinking water or potable grade water and protected water for hygienic purposes. The generation of the drinking water, potable grade water and the protected water for hygienic purposes from the raw water impounded from rainwater source as surface runoff follows a specific protocol of methods for purification before the piped supply. Surface runoff water will usually require full treatment. The degrees of removal of micro-organisms by coagulation, flocculation, sedimentation and rapid filtration are, with proper design and operation, equivalent to those for slow sand filtration. Additional treatment, such as activated carbon treatment to remove organic carbon, reduces the potential for after-growth problems caused by nuisance bacteria in distribution networks. The ozonation stage may also have significant effect on reducing pathogens (WHO, 1993). Disinfection should be regarded as obligatory for all piped supplies using surface runoff water, even those derived from high quality, unpolluted sources, as there should always be more than one barrier against the transmission of infection in a water supply.

In *preliminary treatment* very coarse to coarse undissolved substances—such as plastics (bags and containers), papers, plant parts such as leaves and roots—which are picked up and carried by the raw water during its surface runoff are removed. This is mostly done before the raw water reaches in the reservoir, which impounds the water prior to the piped supply. This is achieved by installation of screens and rakes in the course of the channels through which the raw water flows into the reservoir. At times, manual removal of these is also resorted to.

The objective of the treatment in the second stage is to disinfect the water. Disinfection of the water involves elimination of the pathogens, which are responsible for the water-borne diseases. Besides pathogens, the raw water almost always contains contaminants (that are picked up during its surface runoff), which can impede the disinfection process and/or the substances not desirable in the treated water. These contaminants are inorganic, organic and particulate matters and micro-organisms.

Treatment of raw water in the second stage includes the processes of sedimentation, filtration, coagulation and floatation. Coagulation facilitates the aggregation of small particles into large clusters of particles, which can be later isolated and removed by sedimentation or filtration. Clay, silt and natural organic matters are some of the materials removed by coagulation and, therefore, the turbidity of the water is also removed. Alum (hydrous potassium-aluminium sulphate) is mostly used in this process, but at times, ferrous iron salts or synthetic polymers (organic substances) are also used in this process. This treatment reduces the pathogens as also removes the particles, which are capable of guarding the pathogens from their chemical or thermal destruction.

Sedimentation is employed in combination with coagulation and flocculation (described later). Sedimentation promotes the settling of solid particles to the bottom. Floatation introduces gas bubbles into the water, which attach to the solid particles and creates bubble-solid aggregates that float on the water surface; the floating materials can then be easily removed. Filtration (described in the later chapter) is the most commonly applied process in most water treatment plants. Filtration removes the suspended particulate matters such as clay and silt, colloidal, salt and organic matter precipitates, and micro-organisms. Therefore, the treatment employing filtration process disinfects the water to a large extent, since it isolates the micro-organisms and viruses.

The filtration technique applied in the case of up-grading raw water to potable grade water is more or less same as in the case of the filtration in the hard water treatment technology, which is dealt with later. Either Rapid Gravity Sand Filter or Slow Sand Filtration is used employing beds of sand, gravel, anthracite and activated charcoal. In the filtration process using the said materials, the pathogenic materials including micro-organisms as also the turbidity of the water are removed to a large extent. If the activated charcoal is used as a layer in the filter, some of the heavy metals and chemicals get themselves attached to the charcoal by adsorption process, the concentration of substances on the surface of a solid, which are removed later.

In the third and advanced stage of treatment for potable or drinking water, combinations of processes to remove all the contaminants of biological nature, other micro-organisms and pathogenic constituents, which are not removed in the second stage of treatment are employed. This treatment ensures total disinfection of water, and therefore, ideally it has to be performed after the piped supply of the water just a few hours before the actual consumption. Hence, advanced treatment is not generally resorted to by most of the drinking water supplying agencies. Pipelines supply the water generally after the second stage of treatment, followed by disinfection by chlorination. As chlorination alone does not guarantee the total removal of the biological contaminants, and also the water can be infected through the pipelines before it is made available for humans, it is necessary that further treatment at micro-level be conducted without fail to ensure that the potable water is free from organic contaminants.

The third and advanced treatment for disinfection of raw water to get potable water can be performed in any, and combination, of the following ways: By boiling, chlorination using free and combined chlorine or chlorine dioxide, ozone treatment and ultraviolet (UV) irradiation. Boiling the water and chlorination are the most common methods of household advance treatment, because of their convenience and satisfactory performance. Since free chlorine is likely to produce by-products in the water, the household bleaching power (sodium hypochlorite) can be used to purify the water. Calcium hypochlorite contains about 70% oxygen and therefore it can be used to chlorinate the water strongly to destroy the pathogens and micro-organisms. Hydrogen peroxide has to be added to remove the excess chlorine. Though chlorine dioxide is a powerful chlorination agent, it causes pungent odour. Similarly, ozone too is an effective disinfectant, but its unstable nature and expensive production cost makes it less worthy.

Treatment of water by ultra-violet rays is a sure way of killing the bacteria, viruses, micro-organisms and other biological impurities. There are ready-to-use electronic machines available in market nowadays, which treat the influent water in the 240-280 nano-metre (nm) region of electro-magnetic spectrum and generate potable grade water free from bacteria and other biological contaminants. Innovations are continually needed in the manufacture of cost effective electronic machines that produce ultra-violet rays so that these machines become affordable to the masses of population.

The governments, both local and central, have to consider offering subsidies to the manufactures of the UV-ray-based water disinfectants in the annual budgets and also exemption from sale-tax for these water disinfectants and encourage more and more masses of population to set up UV-disinfection machines in their houses and community water dispensers. That would be one genuine way of expression by the governments that they are indeed interested in the health of the masses of population in the entire country. That would be an indirect way to tell the general public that they need not buy drinking water at a cost comparable to that of milk, but by just one-time installation of an electronic instrument in the form of UV-ray disinfectant, they can get plenty of drinking water for a long period of time.

Fluoride removal

High fluoride levels, above 5 mg/litre, have been found in several countries, for example, in Algeria, China, Egypt and Thailand, besides India (WHO, 1993). High levels of fluoride in potable waters at times lead to dental and/or skeletal fluorosis. Fluoride removal techniques have been developed for both community water supplies and individual households. The most frequently employed fluoride removal technique uses ion exchange or adsorption with either charred bone-meal or activated alumina. Full scale activated alumina facilities and household defluoridators using charred bone-meal have been reported to reduce fluoride levels from 5 to 8 milligram per litre to less than 1 milligram per litre (WHO, 1993). Fluoride-spent bone-meal and activated alumina are usually regenerated for further use.

Chapter 12

Topics of discussion in this chapter are:

- Types and reasons for Hardness in water.
- Management of hard water.
- A review of Filtration Processes.
- Principles and function of Water Filters.

HARD WATER AND ITS MANAGEMENT

The mankind has always paid heavy penalties for its indulgence of overdrawing or excessive exploitation of the bounty that the Nature has bestowed on it, and the case of water resources is no exception to this. To be fair to the government authorities it has to be admitted that adequate steps have indeed been taken in recent years to augment the water resources and provide the necessary potable-grade water to the major chunk of the Indian society. But, the ever-increasing population in India, as well as the commensurate growth of the agricultural and industrial activities in the country, has offset the measures undertaken, and therefore, the positive steps taken are almost neutralized. As a result, water inadequacy remains at almost in 70% of the country's terrain. Through a participatory programme of self-help, the rural women of twenty villages in the Nagpur district of Maharashtra in central India have organized a water movement and achieved protected drinking water facility (Devasia, 2002).

The exploitation of groundwater in India during the last three decades for various purposes of use by the masses of the population has been so phenomenal that many of the Governments have taken note of stress on the groundwater resource in the country, due to its fast depletion at the shallow and medium depths. A few of these State governments are contemplating reforms in the policies on the tapping of groundwater through boreholes in residential colonies. Though this has been a long-overdue step, nothing has tangibly and uniformly been done in this respect in the whole country. The fallout of this inaction has been disastrous, especially in the coastal belts of India.

The groundwater basins and aquifers come in contact with sea all along the coasts. The density of the seawater is higher than that of the groundwater due to the total dissolved salts and minerals in the seawater. Under normal circumstances, the groundwater from the aquifers gradually flows into the sea, due to this difference in the densities of the seawater and the groundwater. The excessive exploitation of groundwater from the aquifers, which are proximate to the seacoasts results in considerable fall in the levels of groundwater table in those areas and reduction of the seaward flow of groundwater. Often, reversal of flow is caused, whereby seawater enters the aquifers and transforming the fresh water of the aquifers to hard saline water. This phenomenon is explained in the Figure 9.4.

In addition, frequently in the coastal areas, the water table drops low down in the unconfined aquifers or the piezometric surface in the confined aquifers. As a result, the gradient of the groundwater table, which has to slope normally downward toward the sea is considerably reduced or even reversed. On account of the difference in the densities of the seawater and the groundwater, an interface or boundary surface forms between the aquifer and the seawater plane, as illustrated in Figure 9.4. When this interface moves towards the land because of the drop in the groundwater hydraulics, it causes the ingression of seawater into the aquifer. Such ingression leads to the contamination of groundwater with hard saline water of the sea, a problem that is faced by the coastal towns and cities such as Ongole, Visakhapatnam and Chennai in the east coast of India. Gradually, this sort of problem has afflicted the towns in the west coast too, such as significant parts of the urban parts of Ernakulam district in Kerala.

Besides, as a result of excessive withdrawal of groundwater from the shallow and moderate depths, in the off-coast areas too, the level of groundwater at several places has gone down below 200 metres. The groundwater in such deep zones more often tends to be hard with dissolved salts of calcium, magnesium and sodium, depending upon the rock type that occurs in those areas. Most often, the surface water and groundwater that passes through extensive areas covered by limestone, marble and calc-silicate rocks dissolve the minerals constituting those rocks, namely the carbonates of calcium and magnesium, rendering it unsuitable for application to the human activities.

The above-mentioned types of water, which are dissolved with salts of different compositions, are in general, known as the *hard water*. The hard water in its raw form is not suitable for the purposes of main activities of human life, such as for potable purpose, and personal hygiene. However, with some treatment to remove the dissolved salts, the treated water can be put to use in a variety of purposes where the water quality is not critical. The hard water, if treated to isolate a certain amount of the dissolved salts, can be a potential source of water for subsidiary activities of the masses of population, including personal hygiene. Tackling of the hard water to convert it into a usable source of water has to be one of the main objectives in the management of water resources. Therefore, it would be of great importance to review some aspects of the hard water and its management to make it worthy of use by the masses.

HARD WATER

The *Hardness* of water can be defined as the effect of water on the soap. The less of soap the water takes to produce the lather, the softer is the water. The salts of calcium and magnesium react with the soap, and until they are neutralized lather cannot form. The greater the concentration of the calcium and magnesium in the solution, the more soap is needed before the lather forms, and the harder the water is said to be.

Hardness is measured by adding the measured quantities of soap solution to the measured samples of water to be tested, and increasing the amount of soap solution until the lather appears on shaking.

There are two kinds of hardness, the temporary hardness and the permanent hardness. The temporary hardness precipitates salts on boiling the hard water, and it is the hardness due to the bicarbonates of calcium and magnesium. It causes the white deposit in vessels. The permanent hardness deposits more slowly and is due to sulphates, nitrates and chlorides of calcium and magnesium. Hardness is expressed in degrees or parts per million (*ppm*) expressed in terms of milligrams of $CaCO_3$ (calcium carbonate), per litre of water. About 200 milligrams per litre is normally accepted as upper limit for potable water, but for subsidiary activities and certain industries, this upper limit can exceed provided the hardness does not cause any damage to the machinery and installation. Hardness is more of a nuisance nature to the consumers rather than a cause for health concern. Presence of fluorides as impurity, on the other hand, causes an extremely hazardous health setback called as 'fluorosis' and therefore isolation of fluoride (described in the previous chapter) from the potable water source is mandatory.

Detergents

The use of detergents for cloth washing considerably reduces the effect of hardness of water. The detergents are a wide range powders and liquids used nowadays to replace soap for household uses. The detergents do not react with the hard water to produce the scum and the thick sticky precipitate as in the case of soap. However, the increasing use of detergents tends to interfere with the sewage work, as the detergents cause foaming in the sewage lines and the rivers to which the detergent-laden wastewaters are directed. The biological water treatment of municipal wastewaters is also likely to be affected by the use of excessive detergents, as some of the detergents contain chemicals that hamper the life of micro-organisms, which aid the treatment of the wastewater. Concentration of anionic detergents in excess of 0.2 milligram per litre is considered undesirable, and about 1 milligram per litre is the maximum permissible. The detergents are difficult to remove, but filtration helps to a large extent. Altering the chemical composition of the detergents can reduce the problem posed by the detergents. Chemicals that do not interfere with the micro-organisms can be used in the manufacture of detergents.

There are two methods of softening the water:

1. *Precipitation.*
2. *Base Exchange.*

In precipitation method, the hardness-producing salts in water, such as calcium bicarbonate, calcium sulphate, magnesium bicarbonate and magnesium sulphate and minor quantities of nitrates and chlorides of calcium and magnesium are settled by precipitation by adding lime and soda ash. Sodium-aluminate is frequently used at high pH to precipitate magnesium hydroxide. As hardness is frequently due to the presence of carbonates, the use of saturated limewater is common practice to soften the water. The precipitation method causes the formation of carbonate sludge, which needs to be disposed frequently in an appropriate manner to avoid the environmental concerns.

The base-exchange method is much simpler and less cumbersome, but it is a little more expensive than the precipitation method. In this method, the hardness-producing salts of calcium and magnesium are replaced by salts of sodium, which do not cause hardness, by the use of zeolites, the hydrated silicates of calcium and aluminium. The calcium in the hard water replaces sodium in the zeolite. The hardness-producing salts of calcium and magnesium are replaced by salts of sodium, which do not cause hardness. The following reactions occur:

$$Ca(HCO_3)_2 + Na_2(zeolite) \rightarrow Ca(zeolite) + 2NaHCO_3$$

$$CaSO_4 + Na_2(zeolite) \rightarrow Ca(zeolite) + Na_2SO_4$$

$$Mg(HCO_3)_2 + Na_2(zeolite) \rightarrow Mg(zeolite) + 2NaHCO_3$$

$$MgSO_4 + Na_2(zeolite) \rightarrow Mg(zeolite) + Na_2SO_4$$

These processes are reversible, and the sodium in the zeolite is restored by washing with brine (common salt solution), and the restored zeolite is reusable. The following reactions take place:

$$Ca(zeolite) + 2NaCl \rightarrow CaCl_2 + Na_2(zeolite)$$

$$Mg(zeolite) + 2NaCl \rightarrow MgCl_2 + Na_2(zeolite)$$

During this process the calcium and magnesium chlorides are discharged as wastes. It is important to note that the zero-softened water is corrosive because of an unfavourable pH/alkalinity ratio. Therefore, it is preferable to soften only a portion of raw water and mix it with the water that is not softened to attain a raw water to treated water blend at a ratio, which will be non-corrosive, reasonably soft and cheaper than a fully treated supply.

This method of water softening is a little more expensive than lime softening, and it does not reduce the total dissolved salts. It works well only on clear water having low iron content (< 0.5 milligram per litre). Chlorination should not precede softening. Base exchange process will not reduce high concentrations of sulphates or chlorides, and all treated water should be checked to see that these remain at acceptable levels. The base-exchange method for softening the water is widely used because of its simplicity and the reason that it does not produce any solid sludge.

The plant design for water-softening looks very similar to the filters, but instead of sand, an about 80 centimetres-thick zeolite bed rests on about 50 centimetre thick bed of graded gravel. The salt is mixed as a 25 percent saturated brine solution, but is let down to about 5 percent before being used in the regeneration process.

The process of *filtration* is useful for removing the undesired impurities after the treatment of hard water by the precipitation and the base-exchange methods to make it soft. Therefore, it would be beneficial to take a brief review of the process of filtration.

FILTRATION

Settlement of precipitates, impurities and other undesired materials from the water meant for treatment has to be followed by filtration process. Filtration is the simplest form of getting rid of medium- to fine-sized suspended particles from any water meant for cleaning. The process of filtration consists of passing the water through a bed of sand at a low speed. The sand retains suspended matter while permitting the water to pass, and the filtrate becomes clear and transparent in appearance. To achieve an efficient purity, it is expected that the incoming water by itself is low in impurities and in turbidity. The suspended solids cause the turbidity, but as such, it is an optical effect caused due to the dispersion of, and interference with, the light rays. It is measured in turbidity units (mostly Jackson Turbidity

Units—JTU) in various kinds of turbidimeter or by comparing a tube filled with the water on test with tubes containing standard suspensions of formazin or Fuller's earth. The World Health Organization's (WHO) international standard for highest desirable impurity levels requires turbidity not exceeding 5 JTU, and that for Maximum Permissible impurity levels requires a turbidity of less than 25 JTU (Smethurst, 1988). The process for the filtration of water is carried out in mechanical facilities known as *'Filters'*. The efficiency of the filters depends mainly on the very design of the filters. The mechanical engineers all over the globe strive continually to make innovations and improvisation in the existing filter designs in such a way that the efficiency and performance of the filters improve. The idea is to produce *large quantities of purest possible water at shortest possible time by least expensive means*. The mechanical engineers involved in the exercise of designing filter plants are expected to have thorough grip on the concepts on *'hydraulics'*—the science of the study of hydrodynamics (the dynamics or the force and energy of the water which is in motion) in general, and its practical application to the water-pipelines.

The treatment in the filters is often preceded by the treatment in the settling basins. A substance known as the *coagulant*, usually a salt of aluminium or iron, is used often to precipitate suspended solid particles in water. The precipitate appears gelatinous and it comprises hydroxides of aluminium and iron, and is referred to as *'floc'*. The floc accelerates the settlement of the suspended particles by adhering to the suspended solid particles in water and neutralizing the negative charges that may be present. The process known as *'flocculation'*, wherein the water is gently stirred so as to induce the particles to grow and coalesce as floc is frequently resorted to in water purification for the suspended solids.

Often, an appropriate substance known *'coagulant aid'* is added at the time of flocculation; it intensifies the settling action and makes the floc denser. The bigger and denser the floc particles, the quicker is the settlement of the solid suspended matter, and the filtration becomes easier in the filters.

FILTERS

The process of water filtration is carried out in mechanical plants known as Filters. The filters are of various types, but basically, there are two main categories known as (1) Rapid Sand Filters; and (2) Slow Sand Filters.

1. Rapid Sand Filters

There are a few designs of rapid sand filters, but most of them are improvisation to suit the local needs. All of them operate on the basic principle of passing the raw water through beds of sand and other media so that these beds retain suspended matter.

Rapid Gravity Sand Filters

The most common filters are of the Rapid Gravity Sand-type, which are operated with coagulants. They give good result when the incoming water has less than 10 JTU of turbidity; if the filters are operated for a short time to produce an acceptable filtrate when the settled water has a turbidity of up to 50 JTU. In this, the flow of water is through sand. The column of water above the filter bed (water-head) is about 2 metre thick. The flow of water through the sand is streamline flow (meaning that the water flows along a line followed by a liquid, along least resistance) and the loss of head is proportional to the velocity. The reduction of the water-head due to filtration is also a function of the viscosity of water and the porosity of the sand. Therefore, the temperature and the cleanliness of the filter sand too control the performance of these filters. The filter sand is of grain size of 0.4 to 1.0 millimetre. The rapid gravity sand filter can only operate effectively as intended when coagulants have been earlier used for the water in the settlement stage. Interception of the suspended matter by sand is a complex process, but most of the suspended matter is removed by adhesion to the surface of the sand grains. As the filter becomes dirty due to (and in the course of) filtration, the pores between the sand and gravel particles diminish in size; under such a condition, the velocity in the waterways increases and the suspended matter is carried deeper into the filter. The filtering action proceeds at depths and is only to a limited extent influenced by the formation of a film on the surface of the sand. If the filter works normally, the sticky floc particles do not pass through the intricate pattern of channels between the sand grains, and thus perfectly clear water emerges after filtration.

It is important to note that in the early part of the use of the filter, when the sand is clean and the floc particles are stopped at the higher levels of the sand bed, the pore space in the top layer of the sand gets clogged due to filling of the floc and dirt. This results in the increase of the hydrostatic pressure (pressure due to the water column above the sand bed). Under this condition, the floc particles penetrate to deeper levels of the sand bed. Consequently, the loss of water head due to filtration of water increases. If the head loss in the sand at any point exceeds the static head of water on the filter, a vacuum gets induced. Expulsion of the air from the pore spaces between the sand grains, and the resultant binding of the filter bed are caused. Washing of the sand bed and the filter removes the vacuum and thereafter the filter would start working normally. The use of coarser sand is preferable in the filter as the turbidity of the filtrate occurs at lower loss of head than if fine sand is used.

A schematic diagram of the Rapid Gravity Sand Filter is shown as Figure 12.1. The whole system of a rapid gravity filter is enclosed in an open-topped concrete chamber. The filter unit consists of a bed of sand supported on a bed of gravel. The sand grains need to be sharp, fresh, siliceous and free from weathering, and of uniform size of 0.5 to 1.5 millimetre. A set up of under-drains are embedded in these filters. The sand layer has a thickness varying between half and one metre. The supporting gravel bed may have a thickness of half metre, making the total thickness of the filter medium to 1 to 1.5 metre. The size of the gravel should be graded between 5 to 50 millimetre.

The bottom set up of the filters can be designed to suit the users' need. The components of the filter such as collectors, nozzles and the porous plate at the bottom can be of any locally available designs and can vary based on the filter design. Likewise, the numbers of these components as well as the valves to control the inlet, outlet, raw water, filtrate and the drain to waste can also be variable to suit the size of the filters and the requirement. The most important factor and requirement in the filters are the number of orifices that are uniformly arranged to collect the filtrate water, and deliver the air and the input water to the filter bed. The rapid gravity filter units can be constructed over an area of about one hundred square-metres. More than this area may result in complications such as raw water's higher velocity, scour of sand bed by the passing raw water and so on, which may affect the performance of the filters. Higher capacity filter system can be achieved by increasing the number of the filter units, and not by increasing the area of the filter unit. The velocity of the raw water over the sand bed may be manipulated to about 10 to 15 kilometre per hour. Washing and replenishment of the sand, and scouring of the sand bed to release the interstitial air as also other maintenance of the filter can be carried out by machines and appropriate designs can be given to such machines. The idea is to carry out these exercises in minimum time so that the filtration process goes on unhampered.

Multi-Layer Filter or Mixed Media Filter

This filter type is a refinement of the Rapid Gravity Sand-type filter. In this, instead of a bed of sand supported on gravel with particles of roughly similar density but greater in size as in the rapid gravity filter, layers consisting of media of varying densities are employed. A very coarse upper layer of lightweight material can provide increased void spaces to accommodate the larger impurities removed first from the incoming water. Usually anthracite coal or vesicular pumice, both of which are lighter than the sand, is placed as a layer over a sand layer in the filter design.

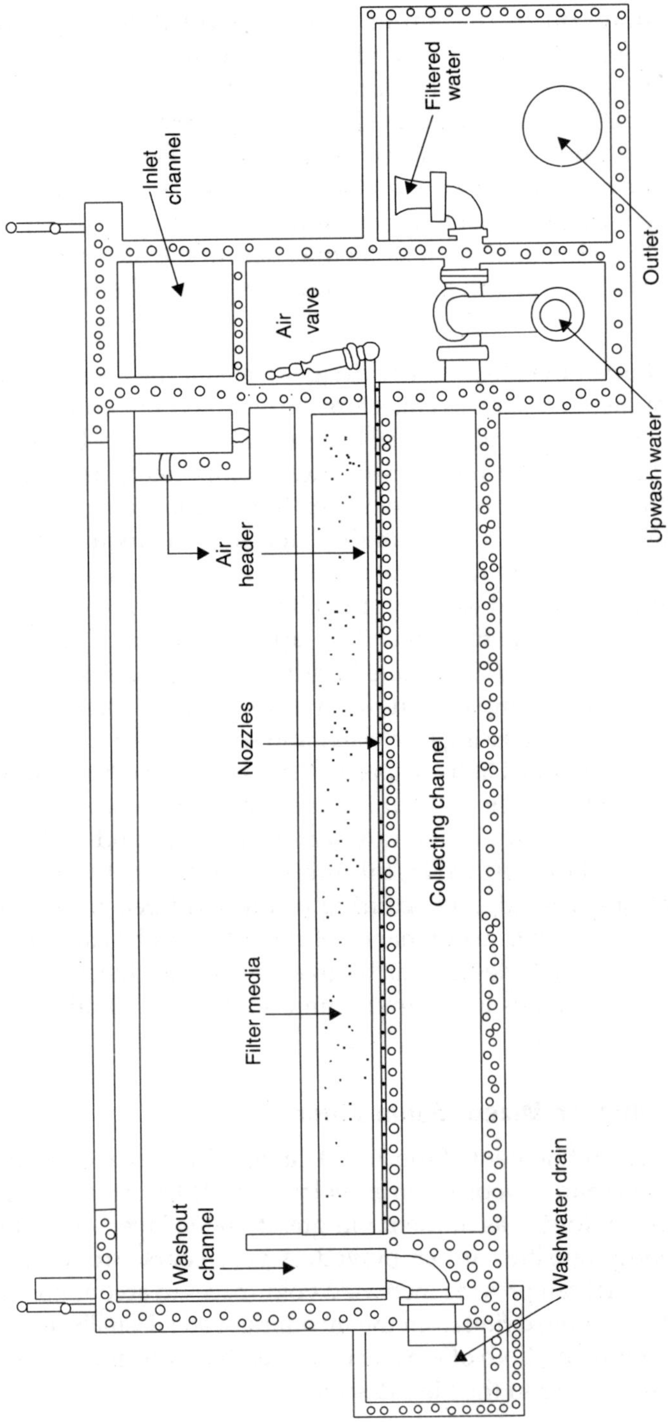

Figure 12.1. Rapid gravity filter (Adopted from Smethurst , 1988)

The void spaces between the sand grains of the sand bed in the rapid gravity filter being very small, there is limited or no capacity to retain the impurities. Therefore, the rapid gravity filter gets dry quickly and very frequently. In the multi-layer filter, this problem is overcome due to the improved design of the filter, which consists as mentioned above, the lighter but coarser layer on top over a bed of sand arranged over the gravel bed. This arrangement removes the floc at the top itself, before it reaches the sand layer that lies quite down. With the turbidity of input water less than 5 JTU, this filter gives good performance and the filter runs longer than the rapid gravity sand filter. The filtrate water that comes out of this filter is of superior quality.

Pressure Filters

The Pressure Filters are very similar to the Rapid Gravity type filters, except that the pressure filters are enclosed in steel pressure cylinders. The filter shell cylinders are of usually 2.4 metre diameter, and they can be installed either vertically or horizontally. The horizontal units have a length of 15 metres. Therefore, these units have 15 x 2.4 = 36 square-metre sand area. They are used where hydraulic conditions in the system make their adoption desirable and they can be installed at any point in a pressure pipeline without interfering with the hydraulic gradient. The rapid gravity filters always are preferable to the pressure filters because not only one can see and monitor them closely due to the direct visibility, but also it is possible to regulate the flow, mix the coagulants and perform pre-treatment of the input water in the rapid gravity filters. The pressure filters are common under such conditions where the input water from the reservoir is clear and it is transported in long and steeply inclined pipelines. In the pressure filters, water can be filtered without preliminary treatment in settling tanks and without disrupting the hydraulic gradient.

Iraq-type Filters

In keeping with the adage that 'the necessity is the mother of inventions', in the middle of the 20th Century, Iraq had indigenously designed simple form of treatment plants with the aid of inexpensive locally available materials, to cut down the delays due to import of the filter-plant equipments. These filters comprised three basic elements of functions (Figure 12.2), namely a settling basin, a rapid gravity filter and a high-level tank (Smethurst, 1988). The settling basins are of common horizontal-flow type, and the rapid gravity filter is the same as that described earlier. The high-level tanks are built on concrete platform structures or from bolted steel plates in such a way that they can take the desired load.

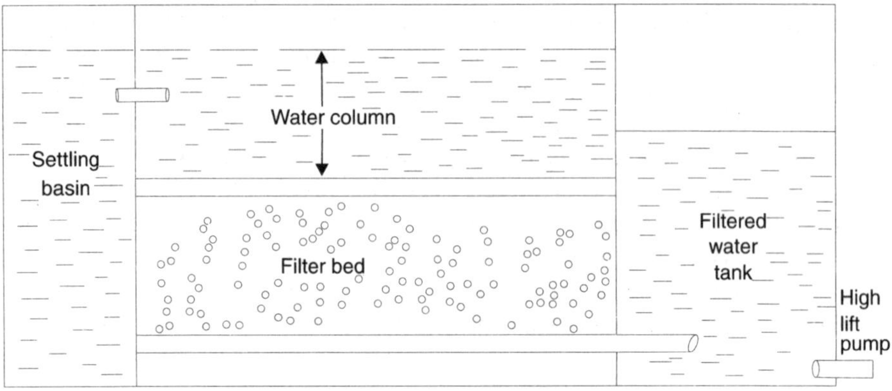

Figure 12.2. Iraq-type filter.

These filters have clear advantages. The flow control is automatically provided because the increased loss of water head due to the filter becoming dirty is balanced by a fall of water level in the pure water tank; this averts the need for the flow controlling devices. Secondly, constant output is ensured by the high lift pump, since whatever is the state of the filter, the speed of operation has to match the constant speed of the pump. Third, there is no artificial head loss in this type of filters and therefore there is significant gain in the efficiency. Finally, the low cost of Iraq-type filters and the use of locally available materials makes it an attractive initial proposition. However, the disadvantage of the Iraq-type filters is that they are small in size and therefore where bulk filtered water is needed, one has to go for more parallel similar units; this offsets the advantage of the basic low-cost of individual filter units.

There are a few other rapid sand filters such as Declining Rate Filters and Greeenleaf Filters (Smethurst, 1988), which show a few minor variations in design and addition of a few accessory parts such as control-valves, but the basic design of these filters is similar to that of the rapid gravity sand filters.

2. Slow Sand Filters

In these filters, a bed of about one metre thick sand is spread over graded gravel bed in which under-drains of open-jointed tiles are buried. The sand is spread to the full depth of the bed near the outer walls. The important difference of the slow sand filters from the rapid sand filters is that the sand in the slow sand filters is finer than the latter. Further, in the slow sand filters, no pre-treatment or chemicals are necessary. The raw water needs to be strained to remove coarse suspended materials and can be directly fed into the filter for filtration. Locally available sand can be used. The water column above the sand bed can be of about 1.5 metre thick. From top to bottom, the filter beds consist of sand, fine-size gravel, medium-size gravel and coarse gravel of about one metre, 5 centimetre, 5 centimetre and 15 centimetre thickness respectively. The size of the sand should be of 0.2 to 0.4 millimetre, of the fine-size gravel can be 5 to 10 millimetre, of medium-size gravel can be 10 to 25 millimetre and coarse gravel is generally of about 25 to 80 millimetre size. A conceptual illustration of the slow sand filter is shown in Figure 12.3.

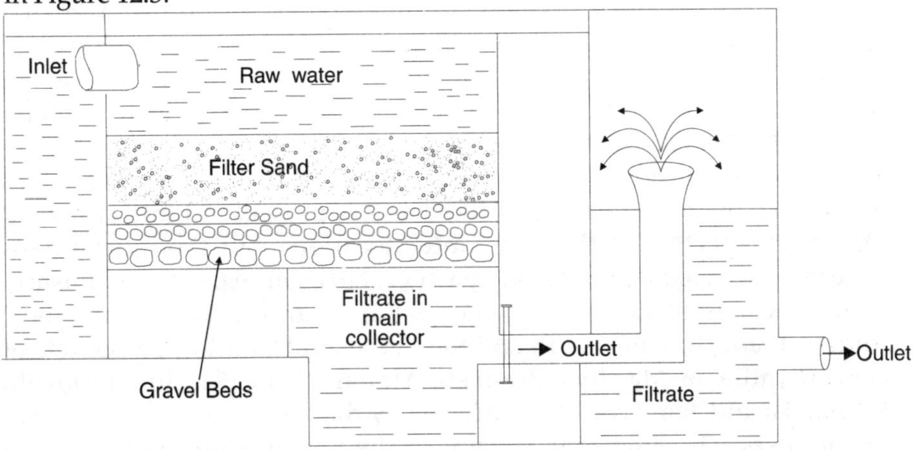

Figure 12.3. Slow sand filter (Redrawn from Smethurst , 1988).

The main advantage of slow sand filters is that the output water is of high quality without bacteria. Since no chemical treatment is required before filtration, there is no chance of chemical traces affecting the quality of the filtrate. However, they occupy large areas and the construction cost is higher. Highly turbid water gets stagnated in the slow sand filter, which is another disadvantage.

The main difference between the slow sand filters and the rapid sand filters is that the filtering action in the former takes place mainly at or near the surface of the sand. A thin veneer of impurities gets built up at the top of the sand bed and this requires frequent cleaning. The filtration is slow and the loss of water head is small depending upon the cleanliness of the sand bed's top. The higher water-head required over the sand bed is achieved by making adjustments using the outlets. When the filtration rate goes down, the top layer (about 10 centimetre) of the sand bed is scraped, taken out and washed to remove the veneer impurities. The process of filtration is continued after restoring the sand bed thickness to the original level. Similarly, the gravel beds would also require periodic servicing.

Hard Water Management

The water from the groundwater source is mostly clear, but it may be hard. So is the case with the aquifer water that occurs at shallow depths in many locations in the coastal areas, and the water that passes over, or associated with, the limestone, marble and calc-silicate rocks. The soluble salts of magnesium, calcium and to minor extent sodium, as well as iron and manganese, which can be present in the water obtained from any source can be converted into suspension by chemical reactions, and that suspension can be removed by settlement and filtration as described above. The water obtained thus, in sharp contrast to the treated recyclable industrial and municipal wastewater, is aesthetically acceptable because of the total absence, or presence only as traces in acceptable proportions, of the biological impurities in it. Hard water has become a bane to the people of India as it is present in India at several places, irrespective of the geographic locations, but the coastal belts have been affected with more vehemence. For example, in the Saurashtra region of Gujarat and adjoining parts of Rajasthan, coastal belts of Orissa, Andhra Pradesh, Tamil Nadu and Kerala, several land-locked pockets of central India in Madhya Pradesh, Maharashtra, Andhra Pradesh, Karnataka and Tamil Nadu are affected by the presence of hard water in the aquifers. The limestone terrains in the northeastern parts of India and some pockets of the Himalayan belt are also severely affected by hard water due to excessive dissolved calcium carbonate. Brackish water is present in several pockets near the coastal belts, both on the eastern and the western sides. Incidentally, these are some of the major locations, (apart from Punjab and Haryana) where maximum agricultural and industrial output comes from, indicating high degrees of water-intensive human activities in those places.

If the water-softening plants and filtration plants in large numbers are distributed in these locations, those would, to a large extent, ease the stress on the potable-grade water, especially in the water-starved months of summer, which lasts for fairly long durations of about nine months in these locations. Effective and meaningful water resource management should address this aspect and insist upon local governments and the departments that deal with the water resources and public water supply to install adequate number of the water softening plants. India is blessed with the requisite wherewithal, namely the materials and the expertise of a large number of civil and mechanical engineers, entrepreneurs and the large labour force of skilled and unskilled category, to install these plants. Further, the impact that would be made by installation of water softening and filter plants by way of employment generation is for sure to be unimaginably tremendous. The filtrate waters obtained from the rapid gravity sand filters are ideal for personal hygiene and a few food industries in which water quality is critical. Therefore, it is expected that over a period of time, we tide over the water crisis and meet the demands of water by setting up the softening-cum-filtration plants in adequate numbers evenly distributed in all the areas that have hard water resource, with the aid of locally available materials and workforce.

Chapter 13

In this chapter:

- Current situation in India justifying augmentation of water resources by alternative technologies is reviewed.
- Alternative methods using technologies for water augmentation are identified.

AUGMENTATION OF WATER RESOURCES

The Capital city of Aizawl in Mizoram, one of the northeastern States of India receives an average of 250 centimetres of rainfall in a year. Rainfall takes place almost round the year in this place. The paradox, however, is that the city reels under water scarcity and the water needs are met through water tankers (Dunglena, 2001). The situation is not very different in Guwahati, the Capital city of Assam, another State in the northeastern India. What is more embarrassing is that Guwahati is situated right on the bank of Brahmaputra River, one of the mightiest perennial rivers in the world. Blame it on the unsustainable methods in the urban development or inefficiency of the authorities concerned in foreseeing human needs, one thing is certain—we have not done enough in the area of harnessing the rainwater the moment it is precipitated on the land of our country.

It would be apt to quote Agarwal (2001), an exponent in the water harvest movement in the country, at this juncture: *"In reality, there is no village in India which cannot meet its drinking water needs through rainwater harvesting. Why is Cherrapunji today short of drinking water when it gets more than 11 metres of rainwater every year? Simply because it does not capture the rain that falls over it".* Prudent water resource management entails wise ways of utilizing the Nature's bounty that we receive, not squandering it away and then blame it on the very Nature that offers us its bounty.

The water resource management methods that have been discussed so far are already in vogue in all parts of our country in some form or other and at different degrees of efficiency. The results accrued by the measures adopted are also of different magnitudes in various parts of the country. The rainwater harvesting has been practised in many parts of Rajasthan and Saurashtra region of Gujarat by which large areas are brought under irrigation and a river, which had gone totally dry in the past have been revived as had been discussed earlier. Several paleo-channels of rivers and channel-fill deposits have been located by remote-sensing during recent droughts and mapped by the Central Government agencies and many State Government departments that deal with groundwater.

Rooftop rainwater harvesting (discussed in the next chapter) is yet to become a popular practice in India mainly for the reason of non-availability of sufficient space in the residential plots of the Indian people in general for the construction of the sump or storage tanks and due to unaffordable economic condition of the Indian masses at large. Furthermore, the author of this book has found out during discussions with a cross section of people (urban, rural, farmers, middle level employees, computer and medical professionals, and so on) residing all over the country that there is a lack of awareness in a majority of the Indian population on the concept and methodology involved in the rainwater harvests. A vast majority of these people feel that the rainwater harvest is the same as providing groundwater recharge pits in their buildings. However, this situation of awareness is changing fast in recent times, thanks to the vigorous campaign organised by several local governments, notably Tamil Nadu, Andhra Pradesh and Delhi. It is hoped that the national scenario with respect to the rainwater harvesting, including the rooftop rainwater harvest and groundwater recharging or integrated rainwater harvest-cum-groundwater recharge system will change for better in a near future, as the sustained campaign all over the country on the rooftop rainwater harvesting has attained a peak since the summer of 2003.

The State Government of Tamil Nadu has already introduced a new law making the rooftop rainwater harvesting structures as mandatory in the new buildings, failing which no approval for constructions would be accorded. Such legislative policies are extremely necessary in the present era of modern civilization, if the humanity has to survive without exploitation of each other and for equitable distribution of the Earth's natural resources. It is in this context that both the governmental and non-governmental organisations, and the entrepreneurs dealing with this technology conduct aggressive campaigns through various media highlighting the virtues of rooftop rainwater harvest, as also provide the storage tanks and installation charges at competitive levels so that more and more members of the general public resort to rooftop rainwater harvesting in the residential colonies. That would, to a large extent, relieve the public water supply organisations of the stress of water supply management to the residential colonies.

The methods of water resource management and augmentation discussed in the earlier section hinge directly on one single factor, namely the monsoon, coupled with the rainfall it brings. The logical questions, which arise at this stage, are—What would happen if the monsoon fails, or if it were a weak monsoon, or the rainfall that the monsoon brings were not widespread so as to saturate the parched lands of major parts of the country. The one billion population of India has witnessed the first near-drought-like situation of the millennium and one of the worst water scarcities in living memory, in the year 2003. The farmers in most part of the country had to skip one cropping season due to inadequate water for irrigation in that year. The buffer stock of foodgrain has considerably eroded. Added to that, the major industries that are water intensive have witnessed sluggishness in their growth. The economists kept their fingers crossed on the matter relating to the planning for the next fiscal year. When the matter related to Cauvery River water sharing, which had peaked in the year 2003 seem to have got abated, there were fresh disputes over Paragodu and Almatti reservoirs and over the water that is actually in short supply in the catchment areas of these reservoirs. There had been perceptible concern in the eyes of the common man every day during the long four-month period spanning March to June, as he had the lurking doubt whether he would be able to get his bucket-full of water daily to keep him clean. To compound his woes, the weathermen had forecast that the 2003 monsoon would be delayed, which indeed played truant in many parts of southern peninsular India, resulting in deficient rainfall.

The logical answer to this grim scenario of water resources is to augment the water resources by alternative methods. A clear vision of a strong and developed India is directly related to *inter alia* the availability of enormous and copious quantities of water for agriculture, industries, *per capita* consumption of the man, entertainment and improving our living environment by creating greenbelts everywhere. Besides harnessing and harvesting the seasonal monsoon rainwater for these purposes as is done now, almost an equal quantity of water resources needs to be generated and accrued to the total water resources available in the country so they can be put into use for various needs of the society. It would be prudent in this objective to divert the water that is excess in one part of the terrain to the water-deficient locations. Let us now consider such alternative sources of water, which could supplement the existing water resources.

Three important means by which augmentation of water can be achieved are as follows:

1. Rainwater Harvesting.
2. Generating usable water by recycling wastewater after treatment.
3. Desalination of seawater.

A pre-emptive statement is made at this stage that small quantities of water are, indeed, locally generated and accrued to the current total water resources, but this quantity is embarrassingly low considering the tremendous potential of water quantity that can be generated by these processes. Much smaller countries like Germany and Israel have pioneered these processes and techniques and benefited to a large extent by way of augmenting their water resources. If all those concerned in water resource management in India are serious about water resource management and mitigating the water scarcity and droughts, they have necessarily to divert their polarized attention from the monsoon rains to these two methods for their large-scale implementation. Fighting grim situations and pitched battles against the nature and also winning them require drastic measures and not soft and superficial attempts, which are also given up when the situation temporarily improves, like during the monsoon times. The following chapters are devoted to consider the above techniques as also to study the positive and negative points of these aspects in their implementation to build up our water resources.

Chapter 14

This chapter is devoted to an overview on Rainwater Harvesting. Main topics are:

- Rooftop Rainwater Harvesting: Principles, Method and Practice.
- Illustrated description of Rooftop Rainwater Harvest System.
- Scope of Rooftop Rainwater Harvesting as a Proviso in Water Policy.

RAINWATER HARVESTING

The rain, when precipitated, is the first form of water in the geo-hydrological cycle. Therefore, the rainwater is the least polluted and softest of all the water sources. The rainwater, besides, does not contain any salts or contaminants or pathogenic substances when it falls on a clean surface or storage system. Therefore, if collected the moment it is precipitated in its purest form, the rainwater is the least expensive potable water.

The Rainwater Harvesting is a practice in which the rainwater is harnessed as and when it reaches the land in a clean state, stored for a certain length of period and used for consumption with or without appropriate treatments. The concept of rainwater collection in some form or other for a later use is not new to India, because, from time immemorial, the rainwater is collected and stored in ponds and reservoirs as surface runoff, as discussed earlier and as illustrated in Figure 6.1. The rainwater thus collected is put to use for consumption by a community of population and for irrigation, even as a part of it percolates beneath the surface without any efforts on the part of humans and gets accumulated in the groundwater system. The practice of rainwater collection is about 3000 years old in the West Asian countries. In India, though many researchers in anthropology believe that the practice of rainwater harvesting is of recent age (for example Rajankar and Dolke [2001] believe that the tanks of Bhandara district, Maharashtra are about 250-300 years old), there are reasons to believe that in reality it seems to be much older as it used to be practised even during the Vedic times. A debate on this would be currently out of context.

The water-stressed condition in recent times has necessitated an aggressive campaign to resort to rainwater harvesting by individuals and small communities as well as at macro-levels. Development of technologies in the manufacture of low cost plastic cisterns and the state-of-the-art plumbing, has given a fillip to micro-level rainwater harvesting by the individuals belonging to large urban and rural communities. By far, the easiest and least expensive method of rainwater harvesting on a large scale for a small community use as that of a village or a small group of villages is the practice of collecting the surface runoff in ground surface-level tanks, reservoirs and bund structures as shown in Figure 6.1.

This practice does not involve installation of any particular structure except bund or barrier to impound the water in the tank or reservoir, and therefore, it is least expensive. This practice would involve only periodic desilting and deepening operations for the tanks and reservoirs, so that larger quantities of water can be collected and impounded, and larger communities of population can use the water thus collected. Except periodic costs towards labour for the desilting operation, this method does not involve recurring expenditure. The bulk of the water impounded can be used for agriculture and human consumption, though certain quantity is lost due to the evaporation and some quantity percolates to the groundwater system.

The rainwater harvesting by a family or a small group of families in the urban residential complexes, such as apartments or clusters of group housing, would involve a different approach, where rainwater is collected directly from the rooftops of the houses (where they first fall) and diverted for storage in the artificial tanks or sumps or any storage system built in exclusively for the purpose of water storage. In turn, water is drawn from these storage tanks for day-to-day consumption. This approach would involve installation of cisterns or storage tanks and proper pumping and plumbing facility. In order to differentiate the rainwater harvesting method by impounding the surface runoff in reservoirs and tanks from the practice of collecting the rainwater in residential cisterns and storage tanks, the latter practice is called as *Rooftop Rainwater Harvesting.*

Rainwater harvesting by surface runoff harnessing and the watershed-linked rainwater harvest have already been dealt with in earlier chapters. The rainwater harvesting as mentioned in this section refers to the practice of rooftop rainwater harvest, and it is described here with the aid of a conceptual diagram (Figure 14.1) to give the readers an overview of the Rooftop Rainwater Harvesting. Jeyakumar (2001) has given additional insights into this technique with the aid of illustrations for the plan and cross-section, and details on the dimension and cost factors of 'rooftop water harvesting'. It is recommended that civil engineers and entrepreneurs interested in the technology of Rooftop Rainwater Harvesting may refer to that article.

In order to visualise the concept of rainwater harvesting at micro-level, to be precise at a domestic level, let us consider that a house has a rooftop catchment area of 100 square metres. If we assume that the average annual rainfall of the area is 80 centimetres or 800 millimetres, the total rainfall that would be received by this rooftop area will be 80,000 litres (= rooftop area multiplied by the average annual rainfall). If a deduction of 25 percent towards splash effect from the roof, spill over and other wastages is considered, the total water that can be collected works out to 60,000 litres. Assuming that the per capita consumption of water per day by an Indian family consisting of five members is about 200 litres, the 60,000 litres of water thus collected from the rooftop of the house will last for 300 days. By minimizing the wastages such as spill over, and increasing the rooftop area the rainwater collected can be as high as 75,000 litres, which would be sufficient almost for the whole year. The rainfall in India is spread over a four-month-long monsoon season. Hence, adequate replenishment during that four-month period is possible. If it were presumed that at the end of the monsoon season the total collection and storage of water from rooftop is 75,000 litres, this stored water would thus be available throughout the next dry season. This, in brief, is about the domestic-level rooftop rainwater harvest.

Collection of rainwater directly from the rooftop is much preferred to the collection from surface runoff since the contamination will be almost absent in the case of collection from the rooftop. The time and cost involved in the transportation of water for day-to-day can be avoided in the case of rooftop method of rainwater harvest. It would not be necessary to wait in anticipation of the piped water supply provided by the water supply authorities. With a one-time expenditure, which in fact is an investment, at the time of construction of the rooftop rainwater harvesting facility, the advantages that accrue are indeed tremendous and multiple. Besides collection of potable water from the pristine rainwater source, further advantages of the rooftop rainwater harvesting include participation by individuals at grass root level in movements like soil conservation (by preventing soil erosion due to floods), flood mitigation in the urban sprawls and groundwater conservation. In a recent work, Jeyakumar (2001) has highlighted that the *"builders in Chennai provide low cost solutions to augment water supply and reduce problems associated with urban waterlogging"* and also provided detailed illustrations for the rooftop rainwater harvest system

The rooftop rainwater harvest system consists of the following four aspects:

1. Catchment or Rooftop area for collection of the rainwater the moment it is precipitated.

2. Storage tank with provision of input piping from the catchment/ rooftop.

3. Treatment of water during storage.
4. Pump and piping system to the residence for distribution.

1. Catchment or Rooftop: Rainwater can be captured from any constructed structure that has a roof, including a small residence. The structures, which have large roof surfaces, have clear advantage in that they can catch more volume of water and divert it to the storage system. It is preferred that the roof has a gentle slope on either one side or both sides (as shown in Figure 14.1). In modern type construction it is customary to provide for terrace with parapet walls, in which case plastering of the floor of the terrace needs be provided with a gentle slope of about 5 degrees. The rainwater shall be collected on the down-slope side and directed into the drainpipes, which in turn, empty the water in the sump or other storage system.

It would be advantageous to keep the catchment surface area always clean, as the clean surface averts contamination of water. Alternatively, it would be beneficial to thoroughly wash the catchment surface just at the onset of rainy season. Screens or sieves have to be provided at the mouth of the drainpipe that drains the rainwater into the storage system, as well as the inlet point at which the rainwater enters the storage tank from the drainpipe. These screens and sieves have to be cleaned periodically to clear dirt, silt, fine sand and other sediments that are likely to choke the entry of the rainwater into the drainage pipe and the storage tank.

2. Storage tank: A storage tank or cistern of appropriate capacity ranging between 60,000 and 75,000 litres needs to be provided for collection of the rainwater from the rooftop. This is the expensive part of the domestic rooftop rainwater harvest system. Provision for a 60 to 75 cubic-metre storage tank or space for a cistern of similar capacity can to be made within the house site at the time of planning the house. Polyethylene or ferro-cement storage tanks, which are available in the market, shall suit the purpose. Several numbers of cistern or storage tank may be interconnected in a cascading or any other appropriate manner in such a way that the input-output process is suitably manipulated to fill all the cisterns or tanks. Since the rainwater is free from dissolved salts and mineral contents, its storage in tank as well as transport through pipes do not cause scaling or corrosion or deposition on the system as a whole.

A conceptual diagram of the integrated domestic rooftop rainwater harvest-cum- groundwater recharge system is shown as Figure 14.1. In this integrated system, just one tank is shown to collect the rainwater. The rainwater that is in excess after filling up the water storage part of the harvest system, spills over to the adjacent part of groundwater recharge system. The excess spill

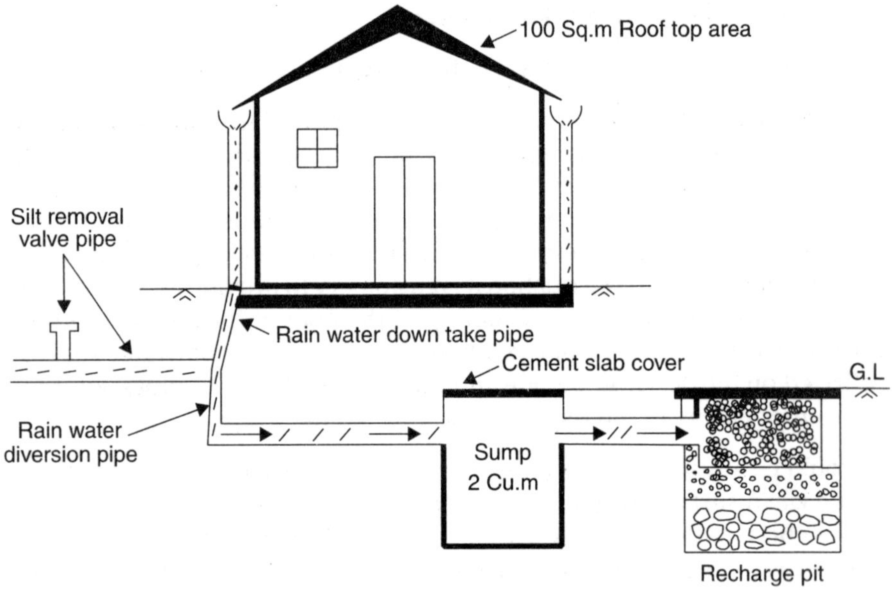

Figure 14.1. Cross-section of an Integrated Rooftop Rainwater Harvest-cum-Groundwater Recharge system

over water infiltrates through the sub-strata and reaches the groundwater system and get stored there. This built-in structure thus facilitates collection and storage of pure rainwater for direct oral consumption as well as to recharge the groundwater and raise the water table in the local area. If several such structures were built in close-by houses and apartments in urban areas, it would also facilitate substantial percolation of rainwater to the groundwater system to raise the water table on a regional scale. They would also help to a large extent in averting soil erosion, groundwater contamination through urban pollutants and hazards of urban floods.

The significant aspect of this integrated system if installed in large numbers in the coastal metropolises and cities is that a cluster of several such installations would recharge the coastal aquifers through the groundwater recharge component of the integrated system, even as the rainwater collected in the other compartment can be drawn daily for use. As a result of this recharging of the coastal aquifers, the ingression of seawater in to the aquifer would be arrested and the aquifer would remain saturated with fresh water.

The government authorities in the coastal States would do well if they mount aggressive campaigns on this integrated rooftop rainwater harvest-cum-groundwater recharge system. The governments should impress upon the citizens on the twin benefit of self-reliance for their drinking water needs and contributing to the cause of improving the health of the coastal aquifers, which have been over-exploited in the recent past resulting in seawater ingression into them. The stress on the potable grade water in coastal cities such as Chennai, Visakhapatnam, Nellore, Ongole, Ernakulam and Mangalore can be relieved to a great extent by this way, simultaneously paving way to recharge the aquifers that are situated nearby these cities. This is an ideal way to involve the common man at the grass root levels in the country in the participation of the movement concerning rainwater harvest and groundwater recharge at one single stroke.

3. Treatment of water: The water that is collected in the rainwater harvest storage tanks is retained for a certain period of time before it is used for drinking or cooking purposes. It is likely that during this storage period its quality might be affected due to the contact of external infectants such as bacteria and micro-organisms. Therefore, a few basic treatment would be required if it is to be used for direct oral consumption and for cooking purposes. For all the other activities, such as washing of cloths, use in toilets and for personal hygiene, only suspended particles and fine sediments need to be isolated from the stored water.

As has been already mentioned, the use of screens and sieves at a few places before the rainwater gets accumulated in the storage tank or sump acts as a coarse filtration process in the beginning stages itself. Further, filtration to upgrade the water to potable level by removal of the suspended particulate matters such as clay and silt, colloidal and precipitated matters and micro-organisms can be achieved to a large extent by use of domestic filters or by using granular media filters using materials like sand, crushed anthracite, coal or granular activated carbon (Pathak, 2002). Activated charcoal filters the water by adsorption, as chemical and heavy metal impurities are attracted to the surface of the charcoal and attached to it.

The basic treatment for the stored water needs to be conducted mainly in order to disinfect the water to ensure that the consumption of the stored water does not cause water-borne diseases. The disinfection of the water can be defined in simple terms as the *elimination of pathogens that are responsible for the water-borne diseases*. Disinfection at the domestic level stored rainwater can be effectively carried out by simple methods like *boiling, chlorination* and *ultra-violet irradiation*.

Simple boiling of the water for about five minutes ensures elimination of a large number and variety of pathogens and micro-organisms. Chlorination is a popular and easily performable disinfection process on the stored water. Use of free chlorine or calcium hypochlorite or bleaching power is some of the simple forms of chlorination method adopted for water disinfection. If the bleaching powder is used for disinfection, the frequency of disinfection by its use needs to be increased because of its unstable nature of the chlorine-content on storage for long time. The ultra-violet (UV) irradiation of the stored water in the 240 to 280 nanometre (nm) kills the bacteria most effectively. A few companies in India manufacture the UV irradiation-based water purifiers and their use ensures total elimination of bacteria and micro-organisms.

4. Pump and piping system: These two components are essentially of peripheral nature, but integral part of the rooftop rainwater water system and they can suit the size of the system or the number of the families that would use the facility. Currently, a virtual boom or revolution of sorts is taking place in the country in the field of innovating new types of pipe fitment as well as the pumps of different capacity in terms of horse power. The thrust should, however, be on the manufacture inexpensive power pumps to suit specific requirement by the users. Accordingly, continuous innovations in the techniques and researches to improve the plumbing materials, patterns in simple but effective plumbing and use of less expensive materials are essential to set up the rooftop rainwater harvest installations.

Rainwater Harvest as a Practice

China has been successful in large-scale rainwater harvesting as highlighted by Zhu *et al* (2001), who have reported that while there are floods in the south China, drought often occurs simultaneously in its northern parts. Areas receiving annual precipitation of less than 30 centimetres form one-third of China's total area. The local or provincial governments in the drought-prone areas in China have been conducting aggressive programmes of rainwater collection. Rainwater is collected in all possible ways—in water cellars, water kilns, tanks, jars and reservoirs. Water scarcity and wastewater disposal problem have forced the Germans to harvest rain, reports Buttner (2001). The maladies of water scarcity and pollution of water bodies faced by India, affect major parts Germany too, but adoption of rainwater harvest as a practice has changed the scenario to a large extent. Annually over 50,000 rainwater harvest systems are established and this practice is taken up as a marketing practice too (Buttner, 2001). Similar trends are seen in several places abroad where there is excessive rainfall but scarcity of drinking water prevails due to a host of reasons (see Agarwal *et al*, 2001 for a review).

Rainwater harvesting pilot projects were experimented on trial basis by dedicated groups of women in several countries around the globe, under the auspices of International Women's University, *ifu*. (Pathak 2002). The objectives of these pilot projects ranged from promotion and study of the feasibility of rainwater harvesting for water supply to the schools in the arid regions of Cameroon, Nigeria, Tanzania and Ibadan city and to an apartment in Tirana, Albania. In all these projects it was found that the objectives of rainwater harvesting for the purpose it was aimed at were successfully achieved and there had been tremendous positive results by way of accretion to the water resources for the use of the humanity. In India too, there have been a plethora success stories in terms of rainwater harvesting in many States of the country. Several State government organisations, NGOs as also private entrepreneurs are currently active in helping out the prospective individuals and organisations who are keen to have rainwater harvesting system in their premises.

There are reports in a website of Centre for Science and Environment (CSE), New Delhi that in November 1998, Mr. K. R. Narayanan, the then Honourable President of India had taken keen interest to install rainwater harvest system in Rashtrapati Bhavan (the official residence of the President of India), spread over 133 hectares, to augment the water resources in the premises, and for the use of a large contingent of workers in, and visitors to, the Rashtrapati Bhavan, and for maintenance of the gardens there. A rainwater storage tank of one hundred thousand litre capacity, recharge shafts for groundwater recharging, and diversion pipe lines for rainwater to a dry well were constructed under this project. Besides, a *johad* (a crescent shaped bund) within the premises of Rashtrapati Bhavan was also constructed to impound the surface runoff for seasonal use of the water accumulated, as well as for groundwater recharge in and around the premises. There have been press reports in the end of the year 2003 that the result of these measures have been encouraging and that the groundwater level within the precincts of the Rashtrapathi Bhavan and its environs has come up significantly high after the monsoon rains in the year 2003.

It is heartening that the First Citizen of the country, namely the President, has set an example to the countrymen to follow some of the modern water management practices and harvest the water that would have otherwise gone as wastewater. It is now the turn of the citizens of this country and all those other countries, which face water shortage to come over it by emulating the First Citizen of India. For a thorough overview of water harvesting practices and also the provisos that are needed in the policies of water harvesting, it is recommended that the readers may refer to the articles edited by, and projected in, Agarwal *et al* (2001).

Since cisterns, water pumps and plumbing materials are available in plenty all over India, those interested in installation of the rooftop water harvesting plants within their residential premises can give their own innovative designs to this technology with the assistance of the skilled workers like plumbers, duly taking into account the size of their plots and the rooftop area of their building. Likewise, the civil engineers, entrepreneurs and contractors can bring innovations to this design to suit the needs of the individual plot of the house, depending upon the shape and size of the residential building plots.

Rooftop rainwater harvesting is a micro-level water resource management adopted at a grass-root level by the common man. Encouraging more and more common people in the country to resort to rooftop rainwater harvesting practice would mean creation of awareness in the common man on the value of the water resource, which is becoming more precious and scarcer day by day, as well as involving the common man in the movement of the water resource management. There is an onus on the entrepreneurs and the engineers in the matter concerning the commodities of rooftop rainwater harvesting, as they need to find out ways to continually improvise the existing accessories of the technology so as to bring down their costs to make them affordable to the common man. That might be an easy and practical way to popularise the rooftop rainwater harvest technology and lure the masses into adopting it.

In spite of several efforts taken by various State governments and the Union Government, there has been inertia amidst the vast urban and suburban middle class population on the matter of adopting the rainwater harvesting. It is this segment of population, which predominates the Indian society, needs to be educated and brought into the fold of water resource management practitioners to give a thrust in the exercise of water resource management. This is because rainwater harvesting is a society-oriented concept to augment the water source by saving the rainwater, which would otherwise go as waste. What are needed are a one-time investment towards installation of the storage cisterns, and the cost towards pump and plumbing.

The urban development authorities of Indian cities, towns and metropolises need to make rainwater harvesting plants mandatory for independent houses and group houses of appropriate size groups, including the apartments. That would substantially reduce the stress faced by the municipalities and water supply authorities due to water supply that they need to provide to these houses and apartments. This rule in the case of apartments in prime locations in the urban centres may be difficult to implement due to the reason that the apartments provide housing to larger number of people distributed in several families that live in a limited area. Though collection of rooftop water may be easier under these circumstances, the difficulty would arise in providing space within the apartment complex for large sumps or cisterns, which can collect and store sufficiently large quantities of water for provision to the residents. For this, the water supply authorities need to intervene and install large rainwater storage tanks for the purpose of collection from the apartment rooftops at a conveniently central location and distribute the processed water at a cost that the consumers in those apartment complexes can afford.

If entrepreneurship in the sector of rainwater collection, storage, processing and distribution in the Indian urban centres is encouraged and nurtured by those in the authority in the government, it can tackle two problems, namely the unemployment of the qualified youth on one hand and the metropolitan water supply on the other, in one stroke. Rooftop rainwater harvesting and the integrated rainwater harvest-cum-groundwater recharge system not only help conserve water resource that would otherwise go waste, but also augments the recoverable natural water in its purest form for the fundamental use of a large human community.

The reforms and amendments in the country's Water Policy should make it mandatory on the part of those who propose to build new residential buildings, be it a single house or a group housing or a multi-storeyed apartment, to make provision for the rooftop rainwater harvest system in their proposed building. This can be done satisfactorily only if the ministries and organizations dealing with Urban Planning, Housing and Water Resources work in tandem in perfect harmony in an orchestrated manner.

What is more important is strict adherence to this orchestrated commitment in the provisos of the Water Policy by the prospective builders to actually provide for the rooftop rainwater harvesting facility in their buildings needs to be ensured by those in the authority in letter and spirit. That would be the one and only way by which the government would involve the common man at the grass root level in the movement of water resource management. Until and unless this is not done, the citizens of the country would continue to feel that the water supply is the sole responsibility of the government, without they themselves directly participating in the social responsibility and mass movement of conserving the country's water resource and using the precious water wealth with prudence and judiciousness.

Chapter 15

Devoted to wastewater recycling, this chapter contains following topics:

- Treatment of Municipal and Industrial Wastewaters.
- Illustration of municipal wastewater plant and description of treatment methods.
- Municipal wastewater treatment scenario and prospects in India.
- Status of Yamuna and Damodar Rivers and Hussain Sagar Lake (Hyderabad), highlighting the need for wastewater plants to salvage the pride of rivers/lakes.
- Treated wastewater as an alternative source for water resource augmentation for human activities and as influent into water bodies for environmental protection.

RECYCLING OF WASTEWATER

The concept of planned channels for the disposal of dirty wastewater from the human habitats is known from the Vedic times in India as well as in the ancient Indian civilizations. The 18th Century civilizations of Europe too had rudimentary forms of wastewater disposal. Despite such facilities, the wastewaters overflowed on the pavements and approach roads to the houses and the people came in direct physical contact with the wastewater. With the advent of the public water supply, the wastewater frequently polluted the water that was meant for human consumption and consequently epidemics broke out. Water-borne diseases were a bane for the whole world as they claimed in the past countless human lives including in, what are currently known as, the 'developed countries' (see Try and Price, 1995).

Globally, sizable human population started gathering and living together in communities for various reasons. The industrial revolution led many people who took up work in the industrial units to migrate to the towns and cities where they lived gregariously in overcrowded habitats. As the living conditions became more and more appalling under these circumstances, methods of treating and disposal of sewage water were thought of.

Wastewater is generated in large quantities in every walk of human life on a continuous basis round the year. The precise quantity of wastewater generated from various sources is currently not exactly known. However, on a very crude and naive calculation of wastewater generated as a result of a per capita use by a billion-strong population of India, for example, at the rate of an average thirty litres per head per day for personal hygiene, it can be estimated that a total wastewater to the tune of thirty billion litres is generated each day in India. This much quantity of wastewater, however, is not distributed uniformly over regions or evenly in the whole terrain of the country. This quantity varies from place to place depending upon personal habits of the people. Expectedly, the quantity is much higher in the metropolitan cities and other urban centres than in the rural places.

Added to this, wastewater emanates from innumerable water-intensive industries such as textile dying, petrochemical, paper and pulp, sugar and coal-washery industries, servicing centres of a large number of vehicles and railway wagons, and several other miscellaneous sources. Currently, these wastewaters are drained through the urban and suburban drainage systems and are led into the nearby streams, tributaries and rivulets that empty into the major rivers, which eventually join the seas. This leads to a twin malady such as:

(*i*) Pollution of the watersheds and subsequently the entire river basin.

(*ii*) Total wastage of a water resource that could otherwise be recycled and put into effective use in appropriate segments of human activity.

In the United Kingdom, a Royal Commission was set up at the end of the 19th Century to look into the matter of treating and disposing of sewage with the objectives of reducing the pollution of the water bodies and avert water-borne diseases arising out of accidental mix up of wastewater with the potable water. The Commission proposed treatment of municipal sewage water together with industrial wastewater (Try and Price, 1995) and recommended methods of treating these wastewaters, in their nine reports dating back from 1901 to 1915. It should, however, be borne in mind that the efforts of wastewater treatment in the 19th Century as well as the period up to the year 1965 were restricted to safe disposal of the wastewaters to avert pollution of the water bodies and clean up the environment in which the people lived. This was mostly on the premise that water resources were replenishable and they would last forever for the use of the humanity. Though there is a reality in this premise, the depleting stock of water all the world simultaneous with steep increase of world population and the need for plenty of water for human activities such as irrigation and industries have of late been exerting acute stress on the water resources.

Further, overindulgence, over-exploitation of water resources and excessively water-intensive activities by the people world over have created necessity to treat the wastewaters to usable levels and put them to use for recycling in such human activities where the quality of water is not a major issue. Frequently, the degree of 'development' of a country or 'human development index' is assessed based on the facilities available, and wastewater treatment plants that create a pollution-free environment are one such important facility. In United Kingdom, it is estimated that 83% of the population is served by sewage treatment, and therefore UK is considered as one of the most advanced countries on this count (Try and Price, 1995). Apart from providing recyclable water in appropriate activities and human use, many countries have employed the wastewater treatment to create clean environment in the rivers and other water bodies.

Supportive human activities such as washing of automobiles and railway carriages/wagons, maintenance of domestic gardens, parks, urban greenbelts and social forests, and so much so even flushing of domestic toilets do not require pure groundwater tapped from borewells or the treated water that is distributed to residences in metropolises and towns by the civic authorities as piped protected water. The same is true in respect of industrial sectors such as the petrochemicals, and the paper and pulp industries, which require enormous amounts of water. This means that we abuse good quality water from pristine sources like rainfall and groundwater for our marginal life-supporting activities, which do not require pure water. This is the first and foremost human activity in which we have to adopt a sensible water resources management. The quantity of potable grade water abused this way all over the country in each day would work out to a mind-boggling proportion.

We have to employ our judiciousness to put the recycled water into use and relieve the stress on the raw water up-gradable to potable levels for the main activities of human life such as irrigation of agricultural lands, industries related to food and food-processing, and for personal hygiene, cooking and cleaning kitchenware. In an efficacious water resource management, to start with, conservation of good quality water needs to be achieved by extensively substituting its use with the use of recycled or treated wastewater in the human activities in which the quality of water would not matter in any way. This would involve installation of a large number of wastewater treatment plants in urban, semi-urban and suburban localities.

Further, the ways and means for collection of wastewater in adequately large quantities to feed the treatment plants have to be planned. With the availability of a large bank of knowledge workers such as mechanical engineers who can design and install these plants, the civil engineers to plan and construct the wastewater collection tanks, and the semi-skilled and unskilled labour force for plumbing and maintenance of pipelines, establishing a large number of wastewater treatment plants all over the country would not be a difficult task. The governments both in Centre and the States would do well in promoting entrepreneurship in this segment, thus easing the stress on potable quality water, concurrently paving way for providing long-term employment to a multitude of educated youth as well as semi-skilled workers.

Wastewater treatment can be carried out both at micro-level for use in household and small community, as well as in large-scale quantities for the use of bigger population as that of metropolises, towns and groups of villages. At present, wastewater treatment plants are installed by a few large business houses, manufacturing units, municipal corporations and five-star hotels in Indian metropolises, to recycle water in such a way that the output can cater to their own needs. The current number of these treatment plants in India is measly, considering the immense potential the process and also the enormous quantity of water that this process is capable of generating for the purposes other than direct human consumption and use.

Incidentally, the general awareness among the moderate to less educated sections of the Indian society on the utility of the recycled or treated water is assessed as extremely low or nearly zero. Many people scoff at the idea of treatment of wastewater as they consider it dirty and a taboo in its application in domestic use of any kind. It is equally important to realize that under the prevailing conditions of economy of the Indian families, it would be extremely difficult to enthuse the people by and large to install micro-level wastewater treatment plants to cater to the family-level need of water for their own supportive activities.

Therefore, the emphasis should now lay on establishing wastewater treatment units focusing on the supply of the treated recycled water through separate pipelines to small local population in cities and towns in the beginning, till the concept of the usefulness of recycled water gains acceptability by a sizeable number of citizens. Subsequently, the wastewater treatment units may be proliferated to more and more locations evenly spread all over the country by which time, the average Indian citizen would get convinced on the utility as well as the acceptability of the treated recycled water. This two-stage approach would aid the fruits of technology to reach all the needy sections of the society.

The wastewater treatment and generation of usable water by recycling of wastewater should, of course, be a sustainable process, in terms of prevention of environmental degradation around human habitation, disposal of huge quantities of biologically non-degradable waste materials like plastics and polythene as well as the sludge that come out of the treatment plants. The generation of recyclable water from wastewater has to address the aesthetic issue such as laying separate pipelines for the recycled water so that it does not mix with the potable water source. In addition, among ethical issues, there should be proper policing also, lest crooks and conmen may sell or put the recycled water into use as potable water or good quality water, which may lead to health hazards as also spread epidemic diseases. Addition of an inert substance like a chemical dye, which can impart a particular tinge to the recycled treated water, can be employed as a remedy to overcome this problem. The exercise of wastewater treatment and recycling, and large-scale application of recycled treated water in the supporting human activities should, therefore, be undertaken with utmost care, only if the authorities concerned can plan the whole process as a sustainable project taking into consideration all the issues highlighted above. As a matter of rule, wastewater treatment for recycling should be taken up as an integrated project along with the other aforesaid aspects concerning waste disposal, pollution check and environmental protection.

Creation of a conscious awareness in water deficient countries in South Asia and Africa on the technology of wastewater recycling and the utility of the technology in augmentation and management of the water resources is one of the current objectives. Several developed countries in Europe and North America have successfully exploited this technology and reaped the benefit in the form of growing water wealth. Availability of a large think tank along with a large number of qualified mechanical engineers, entrepreneurs, semi-skilled workers like plumbers and an army of labour force is a boon to India—a boon with which we can not only take up challenges like augmenting our water resources, but also demonstrate that we are indeed serious in technology transfer to the benefit of the common man.

Abdul Kalam (1998) has given us a vision for the year 2020 to view India as a developed nation; development of technology in the country and availability of large water resources for a host of human activities especially for agriculture and industries are two crucial factors in this vision. Taking up challenges such as wastewater treatment is one among the several segments that can bestow the twin benefit of catapulting our country as a technological force to be reckoned with, and helping to build up or augment the country's water resources. The wastewater treatment is discussed here with a view to create an overall awareness in the readers about the technology and its capacity to augment the recycled but usable quality of water.

The wastewater treatment has focused on the control of the overall effluent parameters such as suspended solids, biological oxygen demand (BOD) and chemical oxygen demand (COD), but seldom on specific pollutants with the exception of some metals. The development and application of modern treatment technologies has been on a slow pace on account of the low priority attached to the wastewater treatment. This does not augur well for countries, which have specific plans and visions for the year 2020, especially the new economies including India. Instances and epidemics such as severe acute respiratory syndrome (SARS) that broke out in 2003 in the North Atlantic and some of the Asian countries left indelible blot in the standards of living conditions and health care in those countries, which drum up great levels of sophistication and advancement. Further, those instances made a deep impact on the economic conditions of those countries by diversion of investments, degradation of credit rating and slump in the stock market during that period. Though these instances and epidemics may not be directly attributable to the lack of wastewater treatment, they reflect the falling standards in general living conditions and the human development indices, lack of hygiene and decadence of environmental health amidst which the human beings live in the modern world.

No country in the world can currently deny the falling standards in living conditions of their citizens, including in the Europe, not to speak of the land-locked central African countries and many Asian and South-East Asian countries. The wastewater treatment would to a large extent wipe out the appalling living conditions and break out of epidemics by facilitating improvement of living conditions, which largely hinge on the availability of plenty of water for personal hygiene and the flushing and ultimate disposal of the municipal and domestic wastes.

Over the years till recently, the Paris Commission, which is concerned with marine pollution from the land-based sources, has set up lists of dangerous substances based on toxicity, persistence and bio-accumulation potential. This list includes organo-halogen, organo-phosphorus and organo-tin compounds, carcinogenic substances, mercury and cadmium and their compounds, mineral oils and hydrocarbons of petroleum origin, and synthetic substances (Johnston, 1992). Most, if not all, of these substances are contributed to the land-based sources from the wastewater—both municipal and industrial—hence, the importance of treatment of these wastewaters before letting into the streams and rivers, and in turn, into the sea need not be emphasized.

Furthermore, most of the waters from different sources need to be treated for purification before they can be supplied as potable-grade water. The raw water quality is infinitely variable and therefore, the treatment required for the raw water is also widely variable. The purpose to which the water is to be used decides the quality and aesthetic considerations. Wastewater, if treated to permissible limits of quality, can be a best substitute for any human activity that does not require potable-grade water, barring oral consumption and in kitchen. Wastewater is an important source of water that can be managed for up-gradation to usable standards. The treated recycled water would undoubtedly augment the country's water resource. In considering the wastewater or effluent water processing, three interdependent issues should be addressed. These are:

1. Generation of wastewater streams from primary manufacturing processes.
2. Processing of individual or bulk wastewater streams.
3. Disposal of the products of wastewater treatment; for example, recycling or other use of the treated water, and safe disposal of wastewater as the waste effluent.

These aspects are discussed a little later. The wastewater can be broadly categorized into two categories, namely:

I. **Municipal Wastewater.**

II. **Industrial Wastewater.**

Besides these two, there are wastewater due to infiltration, which is the extraneous water that enters the sewer system from the ground through various means, and storm water that is discharged from source such as roof, foundation drains, storm sewers and the water resulting from precipitation runoff. Almost all kinds of water can be purified to reusable standards, but some of the raw waters are unworthy because of the health risks and the prohibitive cost involved in recycling them. For example, though the wastewater generating from the hospitals may be of large quantity, on account of a variety of pathogens and infectants originating from the patients that find their way into the wastewater, such a wastewater would become unsuitable for recycling. In general, the economy and the requirement of water dictate need for the treatment of wastewater. Let us now discuss the broad strategies and methods employed in the processing of treatment of these wastewaters to convert them into reusable water by recycling.

I. TREATMENT OF MUNICIPAL WASTEWATER

The municipal wastewater contains undissolved and dissolved pollutants. In addition, a wide variety of micro-organisms, which are mostly pathogenic (disease-causing organisms) such as bacteria, protozoa, rotifers and fungi are present in the wastewater. But, the most important micro-organisms, in both treated and untreated waters, are a group of micro-organisms called as coliforms, belonging to the family *Enterbacteriaceae*, which include *Escherichia, Salmonella, Proteus, Citrobacter, Enterobacter, Klebsiella* and several other genera (WHO, 1993). The coliform organisms are aerobic, facultatively anaerobic, Gram Negative*, non-sporulating and rod-shaped bacteria occurring in large numbers in the faeces of humans and other animals. Therefore, they are referred to as Faecal Coliform. [* A method of staining and classifying the bacteria in which Gentian Violet dye is used to stain a bacterial smear is called as Gram's Method. If the bacteria retain the violet dye after washing with a solution of iodine and potassium iodide in water (Gram's solution) and counter-staining with safranine, they are Gram Positive. If they do not retain the dye, they are Gram Negative.] There are other coliform organisms having the same traits that flourish outside the human system. Thus, the coliform organisms with reference to their contents in the water are expressed as Faecal Coliform (FC) and Total Coliform (TC).

Besides, there are likelihood of the presence of 'faecal streptococci' in the faeces of human beings and animals. They are highly resistant to drying and their presence in water source is useful in detecting pollution by surface runoff to the potable water reservoir or even in the groundwater system.

Coliform organisms are capable of growing in the presence of bile (therefore in the human and animal intestines). Though many of the coliform organisms are considered to be non-pathogenic, quite a few of them (for example, *Salmonella typhi* and *Escherichia coli*, commonly known as E. Coli) are indeed pathogenic to the humans. These pathogens cause a host of ailments, such as diarrhoea, hepatitis, dysentery, enteric disorders accompanied with high fever, worm infestation in digestive system and several other health disorders. The water quality is a major issue concerning health and aesthetics, and hence, the municipal wastewater treatment has to tackle successful removal of the micro-organisms as well as the undissolved and dissolved components.

The municipal wastewater is currently disposed of in India (as also in many other countries) most often by draining it into the inland rivers through a network of subsurface or surface drainage. Treatment of wastewater, both at micro- and macro-level, before letting it into the rivers provides a twofold advantage of—(1) controlling the pollution of rivers, tanks, estuaries and groundwater system, and (2) making the treated water available for the use of supportive human activities.

Micro-level wastewater treatment

Globally, small communities of people at village levels depend on a single source of water for all their needs, which can be broadly divided into two—one for direct oral consumption and the other for all other supportive human activities. The use of unprotected drinking water by village-level people has been leading to a host of health problems all over the world, but this situation is somewhat acute in the African, South Asian, South-east Asian and South American countries. The high standard recommended by the World Health Organization (WHO) for potable water has been stringent and almost difficult to achieve in the event of large nation-wide plans and programmes. Therefore, the focus on (and the beginning of) the water supply management has necessarily to be shifted to providing water to the micro-levels of small local communities.

Since the quantity of water required for drinking water supply at micro-levels is small, logically it should not be difficult to locally generate that small quantity of potable grade water. The measures to be adopted towards the goal of providing water at micro-levels have to be sharply and unambiguously divided into two types of water. These two types are, one for direct oral consumption (drinking and use in cooking), and the other for miscellaneous life supporting activities. Such a division of the water needs is fundamentally the only logical approach towards an effective and efficient way for providing water to the masses of population, as well as conserving the good quality water resource for up-gradation to potable levels in future. This precisely needs to be our approach to the micro-level water resource management. In view of the dwindling potable grade water resources, the national Water Policy has to seriously consider making such a distinct division of water needs of the population. This micro-level management for water supply can be achieved in two distinct methods—1. Recycling of wastewater to meet the need of water for supportive human activities. 2. Treatment of raw water to potable levels.

Wastewater treatment has to always start from grass-root level organized for the benefit of small communities such as the residents of hamlets and villages, so that full advantage of water recycling can be gained. Besides, in the urban centres and metropolises too, those who live in independent houses with a compound around them, as also group-housing communities and apartment-dwellers, can ideally resort to micro-level wastewater treatment for full advantages in terms of sharing the cost of installation of the treatment plant as also sharing of the recycled water. The national Water Policy has to seriously consider issue of building permission only to those who provide for wastewater treatment facility in their proposed groups of houses and apartments, so that the residents can use the recycled or treated wastewater for flushing the toilets, watering the greenery and automobile-washing. That would ease the stress on the groundwater and potable grade water, which is currently used for the said purposes, and also conserve it to a great extent at micro-levels.

Single or interconnected pipelines connected to large tanks, similar to the septic tanks, can collect the wastewater meant for treatment. The treatment of wastewater comprises *pre-treatment* (with or without chemicals), *mixing, coagulation, flocculation, settlement and filtration*. These processes have been described earlier. These processes, individually and collectively, are important in some way or other, but the pre-treatment and settlement alone get rid of about ninety percent of the suspended solids and influence the performance of the wholesome treatment process for more than any other single process. Treatment of small quantities of wastewater (at micro-level) in order to avert pollution of the nearby water bodies such as rivers and lakes, and for the use by household or small community, is a common practice, at times mandatory, in some of the European countries and in Japan. This practice involves simple techniques such as removal of coarse undissolved materials using grits and sieves, followed by filtration with the use of gravel and coarse sand layers (similar to sand filtering earlier described) to remove the suspended particles and some dissolved substances. This recycled water can be put to direct use for supportive facility for day-to-day life such as house and cloth washing, toilet flushing, gardening and vehicle washing. Experimental studies in these lines have been conducted under the auspices of International Women's University *ifu* as pilot projects and found to be tenable in different parts of the world (Pathak, 2002).

The up-gradation of raw water to potable standards at micro-level too involves collection of the impure raw water in large tanks from sources such as surface runoff. It has been experienced almost always that the raw water thus collected is as impure as the wastewater. This collection is somewhat similar to the rooftop rainwater harvesting described earlier, but differs from it because in this process the water collected is from surface runoff, and the treated water is meant for use of a local community as a micro-level resource management. This raw water can be subjected to chemical treatment using calcium hypochlorite to kill the biological pathogens. In the course of micro-level treatment of raw water to up-grade to potable grade, biological treatment for isolating the other biological pollutants such as Total Coliforms

(TC) by using simple materials such as the seeds of *Moringa oleifera* (drumstick), a somewhat popular method, has been experimented by Pathak (2002). However, the use of *Moringa oleifera* seeds has not at times yielded the desired results and therefore not considered as a safe and sure way of removing biological pollutants in potable water.

The most dependable way of achieving potable grade for water at micro-level is by boiling the raw water after settling the suspended particles and treatment with calcium hypochlorite as earlier mentioned, followed by ultraviolet irradiation in the 240 to 280 nano-metre region, which gets rid of all pathogens, including the total coliform. Micro-level water treatment coupled with rooftop rainwater harvesting should go a long way in involving small-sized local communities of population to a large extent in the management of locally available water resources for their own use. Any national-level movement needs the active involvement of the people at grass root levels and the water resource management is no exception. Therefore, the national Water Policy has to seriously consider the involvement of local population in micro-level wastewater treatment and also up-gradation of raw water to potable grade water micro-levels.

Macro-level wastewater treatment

For a larger population base and larger industrial use such as automobile servicing centres, railway wagon washing and maintenance of extensively large urban greenbelts, simple techniques as described under micro-level wastewater treatment would not be sufficient, because the quantity of treated water output needed would be moderate to high. Further, this source is aimed to considerably ease out the stress on the supply of good quality water for the personal hygiene and drinking to larger groups of population in the urban and semi-urban environs, large towns and groups of villages. Therefore, macro-level treatment of wastewaters requires larger mechanical or technological means to facilitate attainment of objectives.

Methods of macro-level treatment in which the removal of contaminants in the raw water is brought about by chemical and biological reactions are known as Unit Processes (Metcalf and Eddy, 1979). Besides, the methods of treatment in which the application of physical forces predominate are called as Unit Operations. The unit operations and unit processes are grouped together to provide what is known as the *Primary, Secondary and Tertiary Treatments*. The tertiary treatment is otherwise known as *Advanced Treatment*. Together, these three processes are capable of generating large quantities of treated water that can be effectively put for large-scale use of the masses of population. Let us now take a look at these aspects, which are presented in brief to offer an overview of the technology rather than give every fine detail of the technology and machinery. While adopting these technologies, the concerned technocrats need to go into extremely fine details of the machinery and civil constructions involved in the wastewater treatment technology. A thorough reference to Metcalf & Eddy, Inc. (1979) would be beneficial in the effort to plan and implement the wastewater treatment plants. However, cost-effective innovations and improvisation in the currently known designs of the treatment plants to ideally suit the Indian or any other local conditions need to be attempted at continuous basis, which is a challenge to the civil and mechanical engineering technocrats involved in this technology.

In *Primary Treatment*, physical methods such as screening and sedimentation are employed to remove the floating and settleable solids found in the raw water. In *Secondary Treatment*, biological and chemical processes are used to remove bulk of the organic matters. In *Tertiary* or *Advanced Treatment*, additional combinations of unit operations and processes are employed to isolate and remove all the other constituents, such as nitrogen and phosphorus, which are not removed in the previous two treatments. Processes like nitrification, denitrification, and physical and chemical processes to remove nitrogen, phosphorus and dissolved inorganic substances are resorted to in the advanced treatment.

Primary Treatment

For over a century, community-level wastewater treatment for multi-purpose involved removal of coarse dissolved substances, undissolved substances and dissolved components using biological methods. Primary treatment is a sort of preliminary treatment, which protects subsequent processes from blockage or overloading and prevents damage to the mechanical equipments. In primary treatment, most of the large solids are removed from the flow of the wastewater by screens and grits, or thoroughly macerated. Inorganic grit is settled out and the flow is controlled. Major fraction of the pollutants in the wastewater including municipal wastewater, barring in the industrial wastewater, is undissolved components such as paper, objects of plastic and polythene (carry-home bags), cellulose and human wastes. Primary treatment of this wastewater is achieved by means of mechanical processes in which step-by-step removal of the pollutants is resorted to. Coarse and fibrous undissolved substances are first removed by the use of rakes and strainers, followed by removal of sand using sand catchers. In the third stage, removal of the substances that settle to the floor of the canal is achieved.

Undissolved substances: Use of a series of mechanical rakes (toothed bars attached to handles) or rake grid is an effective low-cost method to remove coarse suspended materials such as paper, plastic and polythene from the influent wastewater. Installation of a series of rakes or rake grid across the direction of wastewater flow in channels removes most of these undissolved components. Since the rakes have a tendency to clog if the fibrous or polythene materials get stuck to the rake bars or prongs, mechanical strainers are also installed along with rakes frequently to remove undissolved pollutants. There are appropriate mechanical devices that remove the materials clogged to the rake bars and strainers. Much care is to be devoted to the safe and convenient disposal of the bulky, water-soaked rake material. This part of the process is one, perhaps the only, ticklish issue concerning the environmental degradation due to the disposal of the waste materials recovered by the wastewater treatment. In most of the cases in municipal wastewater treatment plants, this rake and strainer waste material is subjected to application of pressure to extract the soaked water, and later on set on fire an exercise known as *incineration.*

Sand and Grit removal: The sand, grit, silt and the other coarse suspended particles are isolated from the wastewater in the coarse of treatment by the use of sand catchers. Sand catchers are the channel-cum-tank constructions through which the waster is allowed to flow for treatment. These channels separate out the sand particles and coarse suspended particles whose density is comparable to that of the sand particles suspended in the wastewater. Most of the wastewater treatment plants are designed in such a way that the sand catcher tanks follow the rake and strainer facility, so that undissolved materials of both large size as well as suspended sand-sized materials are removed at one stretch. If the sand and suspended materials are not removed, they interfere with treatment and transport on a later stage. Further, these particles erode the treatment plant parts such as pumps and pipelines and also clog the pipelines.

There are two common types, namely un-aerated and aerated types of sand catchers. Depending upon the purpose for which the wastewater treatment plants are provided other mechanical devices for sand and suspended particle removal, such as Hydro-cyclones, hydraulically revolved sand catchers and deep sand catchers are added to the municipal wastewater treatment plants.

Un-aerated sand catchers are channels with rectangular or trapezoidal flowing cross-section and consist of two parallel channels to be operated alternatively with intermittent time gaps for dewatering and cleaning of the non-operative channel by removal of the sand using mechanical scrapers. A constant horizontal flow speed of about 30 cm per second is maintained during the whole operation so that the sand is totally removed.

In the case of aerated sand catchers, the settling of sands and particles is not controlled by the horizontal flow speed of the water in the channel, but aeration of the water under treatment creates a revolving stream that restricts the settling of the particles. The speed of the flow is kept fixed irrespective of the employed capacity of the aerator. This sand catcher comprises a channel with aerator provided on the wall of the tank. Coarse bubble aerator pipes, laid 0.5 to 1 m, perform the aeration. The discharge apertures should be installed at approximately three-fourth of the height of the sand-catcher. The aerators should be designed in such a way that the amount of air can be reduced to about 0.1 cubic-metre. This avoids an air supply during treatment that is very high for the denitrification process. In the sand catching and removal exercise, ninety percent of the sand and grit particles having the grain size of about 0.3 millimetre and about fifty percent of sand and silt particles having diameters of about 0.1 millimetre is achieved. The bottom of the channel has to be semicircular in cross section with sand collecting channel in the centre. In addition, a few fat-collecting system such as small rake-cum-bag or –box, integrated with the treatment plant collect the floating substances such as fatty material.

The sand and other particles thus deposited in the sand catcher have to be scraped and removed periodically. This is done by a variety of automatic scraping devices or implements such as revolving scraper arms attached to the sand catcher in the case of circular sand catchers, or mobile scraping bridges with lengthwise scrapers in the case of long sand catchers, or mammoth pumps or screw scrapers installed lengthwise at the bottom. The sand, grit and silt thus collected may be dewatered to squeeze out the associated water and disposed of appropriately. In most of the plants, these are used for replenishing sand in the gardens, public parks, and social and urban forestry. The size of sand and silt that accumulates in the sand catchers is about 0.1 to 0.3 mm in quiet times and in the storm or flood times the size reaches up to 3mm.

The wastewater flows through the settling tank continuously at a slow speed. During its flow, the settling solids are separated from the clear water because of the density difference. The solids sink to the bottom of the tank, and therefore, they can be removed by appropriate attachments with the plant. In the preliminary treatment process, the settling tanks are designed to separate the suspended substances that are not of sandy and therefore passed the sand catcher, from the raw wastewater. As the preliminary treatment tanks are mostly followed by further treatment, a separation performance of about seventy percent of the filterable substances contained in the influent wastewater is considered adequate.

In the wastewater treatment technology, the settling tanks are mostly horizontal-flow tanks. These tanks may be circular or rectangular. With circular tanks the wastewater distribution in the tank influent is in the middle of the tank, and with rectangular tanks, it is on one of the narrow sides. The distribution device has to be designed in such a way that the inflowing wastewater can be distributed evenly over the entire cross-section of the tank. The clear water extraction is through the tank diameter for circular tanks, or at the side opposite the influent for rectangular tanks. Overflow weirs have to be installed for this purpose. Clear water with low solid contents can be ensured by extraction over the perimeter or the tank width sides. As the sludge settles over the entire tank bottom, it has to be collected mechanically and removed. Some of the suspended particles do not settle in the settling tanks, but float on the clear water surface. Appropriate devices such as strainers and grids have to be provided to remove these floating substances. The depths of various tanks in the wastewater treatment plants may be designed to vary from about 2.5 to 5 metres, depending upon the capacity if the plant. In the case of the rectangular tanks, the length of the tank can be about 75 metres, and the width about 10 metres. A tank floor inclination of about 1:100 for rectangular tanks may be provided. For round tanks of about 50 to 60 metre diameter, an inclination of 1:15 can be provided.

On a rough estimation, the sludge of the higher order of about 60 kilogram per cubic metre of water can be expected in Indian conditions in the case of

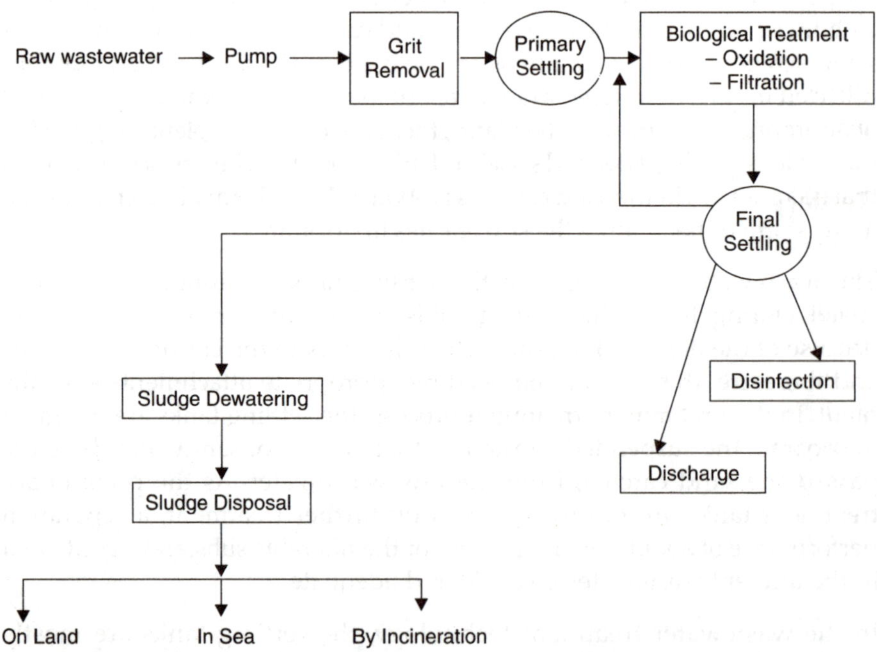

Figure 15.1. Stepwise processes of municipal wastewater treatment.

metropolitan, urban and suburban municipal wastewaters. Therefore, as has been said earlier, the municipal authorities have to devise ways for safe disposal of the effluents coming out of these wastewater treatment plants. Figure 15.1 is a diagrammatic representation to bring out the stepwise approach to the municipal wastewater treatment.

Secondary Treatment

The secondary treatment is meant for removal of the biological pollutants. The principle in the secondary treatment revolves around oxidization of settled sewage by micro-organisms (Try and Price, 1995). The pollution levels in the wastewater are determined using the parameters such as Biological (also known as Biochemical) Oxygen Demand (BOD), Chemical Oxygen Demand (COD), and the nitrogen and phosphorus contents. The COD and BOD represent the sum parameters that are employed to measure the dissolved and non-deposable organic pollutants in the wastewater. These parameters consist of the quantity of oxygen necessary for the microbiological or chemical oxidation of the organic pollutants up to the inorganic final product.

The BOD is a measure of the oxygen required by the micro-organisms while breaking down organic matter (Tebbut, 1992). The BOD is determined by measuring the amount of oxygen consumed by the microbial conversion during the degradation of pollutants for five days under aerobic milieu.

The COD is determined by chemical oxidation using boiling potassium dichromate and concentrated sulphuric acid (Tebbut, 1992). *The COD is a measure of the chemical oxidation by the pollutants.*

Tertiary Treatment

In the case of some of the wastewaters, further treatment may be necessary to enhance the effluent quality. The treatment includes the removal of residual solids, ammonia, nitrate and phosphorus, to reduce the nutrient enrichment of the water, or disinfection to reduce bacterial levels. The other major substances in the wastewater that require treatment are the plant nutrients like nitrogen and phosphorus. Nitrogen is an important element in the wastewater since biological reactions can proceed only in the presence of sufficient nitrogen. Nitrogen exists in four main forms in wastewater. They are:

(*i*) Organic nitrogen—nitrogen in the form of proteins, amino acids and urea.

(*ii*) Ammonia nitrogen—nitrogen as ammonium salts or as free ammonia.

(*iii*) Nitrite nitrogen—an intermediate oxidation stage not normally present in large amounts.

(*iv*) Nitrate nitrogen—final oxidation product of nitrogen.

In the parlance of wastewater treatment, the *oxidation of nitrogen compounds* or, in other words, the *biological conversion of ammonia nitrogen to nitrate-nitrogen* is called as *Nitrification*. Reduction of nitrogen from the wastewater is known as *Denitrification* (Tebbut, 1992). According to Kunst *et al* (2002), denitrification is the *utilization of carbonaceous organic matter by heterotrophic bacteria, nitrate-nitrogen being used as the terminal electron acceptor (i.e. the "oxygen source")*. The organic pollutants of the wastewater decrease the *dissolved oxygen* (DO) levels, and thereby affect the oxygen balance, in the wastewater. Phosphorus and nitrogen, on the other hand, facilitate the growth of algae and plants. Algae are minute organisms, which are classified as plants. Impounded water encourages growth of algae.

Eutrophication

Eutrophication *is a condition in the aquatic eco-system where high levels of plant nutrient concentrations stimulate blooms of algae including phytoplankton*. Although eutrophication is a natural process in the aging of lakes and estuaries, human activities accelerate the process to a large extent by increasing the rate at which the nutrients and organic substances enter into the aquatic eco-systems from the surrounding watersheds. The anthropogenic eutrophication can be visualized as water pollution caused by excessive plant nutrients (phosphorus, nitrogen and carbon) flowing into the water bodies such as lakes and estuaries. In urban environs, plenty of phosphates contained in the household detergents promote the algal bloom, when the wastewater laden with the detergents finds it way into the water bodies. Besides that, the agricultural runoff, urban runoff, leaking septic systems and sewage discharges can increase the flow of nutrients and organic substances into aquatic eco-systems. These substances can stimulate profuse bloom of algae, which create nuisance and interfere with the utility of water bodies for recreational purposes. Further, the algal bloom makes a negative impact on the health and diversity of indigenous fish, plant and animal population that make use of the water in which it is laden.

The algal bloom affects the aquatic eco-system in the following two ways:

1. It clouds the top surface of water and blocks the sunlight from penetrating the water, which causes the underwater grasses to perish. Since these grasses form nutrition to the aquatic creatures, the latter are starved of nutrition if these grasses die off and, in turn, they too perish.

2. When the algae die and decompose, the DO in the water is used up. The DO in water is essential to most organisms living in water, and therefore, danger is posed to the aquatic life.

Administration of chlorine in adequate doses stops the proliferation of algae. Because chlorination to such a large extent would be an expensive proposition, the effective and inexpensive way to avoid the occurrence of eutrophication in the aquatic eco-system is to subject the effluents and/or the inflows into the water bodies such as lakes and estuaries to tertiary or advanced treatment. This treatment ensures that the plant nutrients like nitrogen and phosphorus are removed from the influent water to the water bodies such that these nutrients do not promote the algal bloom.

The ammonium-nitrogen also affects the oxygen balance. The ammonium radical produces ammonia (NH_4) under the condition of high pH (hydrogen ion concentration). Ammonia imparts a toxic effect on the aqua-life, especially the fish.

Biological Treatment and Sludge Treatment

The objective of the treatment of municipal wastewater for biological pollutants is to isolate those deleterious substances in the wastewater that can cause hazard to the human beings or the environment, or to treat them in such a way that they lose their hazardous potential. This treatment operates under a principle controlled by the nature, 'that the naturally formed pollutants are converted into inorganic and organic end products by micro-organism' (Kunst *et al*, 2002). In the wastewater treatment plants, these conversion processes occur in the *activated sludge* tanks, where the wastewater and the activated sludge are mixed and aerated. Treatment and disposal of the sludge is expensive and it accounts for nearly half of the cost of the sewage treatment. In many treatment plants, the sludge solids produced by oxidation in secondary treatment are returned to the inlet and settled out with the primary sludge.

The oxygen required for the biological degradation in the wastewater is injected through aeration devices. *Aeration* provides oxygen from the atmosphere to the raw wastewater. Aeration is the practice of exposing small droplets of water to air to promote the absorption of oxygen from the atmosphere. This is achieved by splashing the raw water over trays in chambers to split the water stream into droplets. Conversely, the atmospheric air bubbles are blown into the raw water to achieve the same effect. Aeration of the raw water, coupled with release of other gases from it, effect beneficial changes, which would aid purification of the water. Aeration of water is an inexpensive and very effective means of controlling tastes, odours and corrosion. Therefore, aerators are always used when the raw water contains hydrogen sulphide (which gives a rotten egg smell to the water), carbon dioxide (which imparts corrosive tendency to the water), algal growth, and iron and manganese in solution. Aeration is achieved by specially designed nozzles, which direct thin jets of water against metal plates to produce a fine spray. This spray exposes the droplets of the water to the atmosphere.

The wastewater is purified by micro-organisms, the wastewater's organic substances being taken up and, in different ratios, respired or converted into settable biomass. From the activated sludge tank, the wastewater-cum-sludge mixture flows into secondary clarifier, in which the activated sludge separates itself from the purified water. The sludge settling in the secondary clarifier is returned to the activated sludge tank as re-circulation sludge, while the purified water flows off. As the activated sludge increases during the biological process, the surplus sludge is extracted, dried and used subsequently in other plants where water treatment is needed.

The primary purpose of municipal wastewater treatment besides the removal of the organic suspended materials is the isolation of nutrient components such as nitrogen and phosphorus. A certain amount of nitrogen removal takes place in biological wastewater treatment plants due to the uptake of nitrogen into the waste sludge produced in the process. Nitrogen is the component of waste biomass produced as a result of the biological treatment of carbonaceous matter. Organic nitrogen is also a component of the non-biodegradable particulate matter present in most of the wastewaters. This material is generally *flocculated* (the process of *coagulation of fine particles into coarse or big masses*) and incorporated into the biological treatment plant, and subsequently removed with the aid of the waste sludge. However, residual nitrogen is always present in this treated wastewater. If this recycled water is to be employed for toilet flushing and washing of automobiles and railway wagons, the residual nitrogen might not cause any damage. However, in the recycled water for some of the industrial applications, biological nutrient-removal processes such as nitrification and denitrification processes, which are popular in European countries, can be employed.

Nitrification is the biological conversion (in this case, oxidation) of the nitrogen in ammonia to nitrate-nitrogen. The group of bacteria called as autotrophs brings about nitrification. Autotrophic micro-organisms oxidize inorganic constituents to obtain energy for growth and maintenance, while they obtain carbon for the production of a new biomass by reducing carbon dioxide. Organic matter is not required for the growth of autotrophic bacteria. Nitrification is a two-step reaction. The first step is the oxidation of the ammonia nitrogen to nitrite-nitrogen by the bacteria of the genus

Nitrosomonas sp. The second step is the oxidation of nitrite-nitrogen to nitrate-nitrogen by the bacteria *Nitrobacter sp.*

The overall reaction involved in the nitrification process, including the synthesis of the new biomass, as discussed by Kunst *et al* (2002) is as follows:

$$NH_4 + 1.83\ O_2 + 1.98\ HCO_3 \rightarrow 0.98\ NO_3 + 0.021\ C_5H_7NO_2 + 1.88\ H_2CO_3 + 1.04\ H_2O$$

Oxygen is required to oxidize the ammonia nitrogen; 4.6 milligram of oxygen is required for each milligram of the nitrite-nitrogen generated. Bicarbonate alkalinity is also consumed in the reaction both to neutralize the acid produced (ammonia nitrogen is a base and nitrate-nitrogen is an acid) and as required for the synthesis of the new biomass from carbon dioxide that is present in bicarbonate alkalinity. 7.2 milligram of alkalinity as calcium carbonate is required for each milligram of nitrite-nitrogen produced. Biomass-yield values are in general low for the autotrophic bacteria. The yield value for the nitrifiers (both Nitrosomonas and Nitrobacter) is 0.15 mg of bacteria as total suspended solids per milligram of nitrate-nitrogen generated.

The growth of nitrifying bacteria is affected by a number of factors. As such, its growth rate is generally lower than that of the heterotrophic bacteria that oxidize carbonaceous organic matter in the biological wastewater treatment systems. The specific growth rate of micro-organisms in biological wastewater treatment systems is expressed as the minimum mean solids retention time, which represents the solids retention time corresponding to the maximum specific growth rate for the micro-organisms. When the process operates at minimum solids retention time, the micro-organisms grow at their maximum rate and are simply washed out of the system. The solids retention time has to be longer than the minimum, if the organisms are to grow and survive in the system. Most of the wastewater treatment plants are operated with a solid retention time between ten and fifteen days.

Nitrifiers' growth increases in higher temperatures comparable to those prevailing in India, and therefore, the solid retention times of lower order are sufficient for biological wastewater treatment systems. Further, the nitrifying bacteria's growth also depends on other factors such as dissolved oxygen, pH and, the organic and inorganic compounds. As nitrification is an aerobic reaction (requiring the presence of oxygen), presence of dissolved oxygen is vital for the reaction. The activity of the nitrifying bacteria is reduced if the dissolved oxygen gets reduced to about two milligrams per litre, and totally inhibited in the absence of dissolved oxygen. The optimum pH for the growth of nitrifying bacteria is about 7.5. Under the reduced pH condition, the activity of nitrifying bacteria drops, and inhibited if the pH drops below 6.5. Since the organic and inorganic dissolved substances inhibit the growth of the nitrifying bacteria, the treatment of industrial wastewater should take adequate measures to tackle management of these substances at the time of treatment.

Denitrification is the microbial reduction of the oxidized nitrogen compounds such as nitrates and nitrites into elementary nitrogen, which gets isolated and removed as gas from the wastewater. Many of the heterotrophic bacteria in biological wastewater treatment systems are capable of using either dissolved oxygen or nitrate-nitrogen as a terminal electron acceptor. Dissolved oxygen is preferred if both terminal electron acceptors are present. This is because more energy can be obtained by the oxidation of carbonaceous organic matter using oxygen as the terminal electron acceptor rather than nitrate-nitrogen. Dissolved oxygen and nitrate-nitrogen provide the same biochemical function. When nitrate-nitrogen serves as the terminal electron acceptor, it is converted to nitrogen gas, which is released into the atmosphere. This reaction causes the removal of nitrogen from the wastewater system. Theoretically, about 3.6 milligram of alkalinity as calcium carbonate is produced per each milligram of nitrate-nitrogen that is reduced to nitrogen gas. Denitrification also results in a reduction in alkalinity consumption due to the removal of the acid nitrate. Some 3.6milligram of alkalinity as calcium carbonate is produced per each milligram of nitrate-nitrogen reduced to gas nitrogen.

In the municipal wastewater treatment plants, denitrification rate varies considerably as different fractions of the organic matter contained in the water are oxidized. In the beginning, a relatively high denitrification rate occurs as the readily biodegradable organic matter is oxidized. When this matter is exhausted, the denitrification rate is reduced to that produced when the more slowly biodegradable organic matter is oxidized. When all the biodegradable organic matters are oxidized, the denitrification rate is extremely low and is achieved only by endogenous respiration. This means that the denitrification rates will initially be high as the readily biodegradable organic matter is oxidized. The denitrification rate will decrease with time, as at first the readily biodegradable organic matter is depleted, and followed by the slowly biodegradable matter. If the organic matter present in the wastewater is to be used as the primary carbon source, it must be used effectively if a relatively high degree of denitrification is to be achieved. This is accomplished by process configurations that use the influent wastewater for denitrification first. A number of biological nutrient-removal procedures have been designed and are put in to use in full-scale wastewater treatment plants. Most common is the single activated sludge biological nitrification or denitrification process.

Though there are several variants of process configuration, the process design follows two different approaches namely:

1. Pre-denitrification.

2. Simultaneous or intermittent denitrification.

The pre-denitrification process takes place in two sections of the activated sludge tank. They are—*anoxic* or denitrification zone and *aerobic* or nitrification zone. Anoxic zone is a region of the biological reactor where dissolved oxygen is excluded and nitrate-nitrogen is provided to serve as the terminal electron acceptor. Denitrification occurs in anoxic zones. An aerobic zone is the one that is aerated to provide dissolved oxygen as the terminal electron acceptor. When dissolved oxygen is provided in the aerobic zone, both the oxidation of carbonaceous organic matter and nitrification can occur. Denitrification, the second step in a biological nitrogen-removal process, occurs in the anoxic zone. Nitrate-nitrogen is provided to the anoxic zone by the mixed-liquor recycle stream from the aerobic to the anoxic zone.

Advanced Treatment

This part of the treatment involves the phosphorus and nitrogen removal in the Activated Sludge Systems. Phosphorus occurs both as chemical and biological form in the Activated Sludge Systems of the wastewater treatment plants. It is necessary that both forms were removed.

Chemical phosphorus removal: Most wastewater treatment plants employ the activated sludge process for the removal of chemical phosphorus. The total phosphorus-removal rate obtained in a conventional activated sludge plant is less than twenty percent, and is even less in wastewater treatment plants where anaerobic digester supernatant is recycled to the head of the plant. Since it is not possible to achieve one-milligram phosphorus per-litre effluent limit with conventional biological wastewater treatment processes, additional or alternative treatment methods must be employed. Phosphate removal in activated sludge plants is accompanied by chemical precipitation. Depending on the point of chemical addition at which chemicals are added, the precipitation can be realized as the primary precipitation (dosing point is upstream of primary clarifiers) or simultaneous precipitation (dosing point: within the activated sludge plant) or tertiary precipitation (dosing point: in a separate part of the plant downstream of secondary clarifiers). Simultaneous precipitation is the most common alternative because both process control and process stability are high and the additional construction costs are minimal. The chemical precipitation increases the volume of the sludge produced and often results in sludge with poor setting and dewatering characteristics. Further, precipitation with metal salts can lower the pH.

Biological Phosphorus Removal: The biological phosphorus removal system offers the following benefits:

1. It reduces or eliminates the need for chemical addition.
2. The production of sludge is reduced.
3. Sludge settling and dewatering characteristics are improved.
4. It reduces the oxygen and alkalinity requirements.

The biological phosphorus removal works on the principle that a group of micro-organisms contribute to the uptake of soluble ortho-phosphate for the biological phosphorus removal. These organisms are called *phosphate-accumulating organisms*. Under cyclic anaerobic or aerobic (presence of oxygen) or anaerobic/anoxic (presence of nitrate) conditions, they are able to seclude or sequester in the anaerobic reactor readily degradable organic substrates, which are stored as poly-hydroxy-alkanoates, PHA in brief (Kunst et al, 2002). The formation of PHA coincides with the degradation of intracellular poly-phosphate and the intracellular consumption of carbohydrates such as glycogen. After conversion to ortho-phosphate, the degraded poly-phosphate is released to the bulk phase.

Subsequently, in an aerobic reactor, the stored PHA is used for catabolism and metabolism. The dynamics for the internal storage products poly-phosphate and glycogen are diametrical to the anaerobic phase. Recent research results show that at least some of the phosphate-accumulating organisms are able to use nitrate as an electron acceptor, which results in an anoxic uptake of phosphate and denitrification by the phosphate-accumulating organisms. Owing to the growth of the phosphate-accumulating organisms in the process, an overall uptake of phosphate occurs, and the phosphate-rich micro-organisms are eventually separated by sedimentation.

Nitrogen removal: The biological nitrogen and phosphorus removal is a rather complex process. Anaerobic, anoxic and aerobic environments are needed and the operating conditions must ensure that nitrifiers, denitrifiers and phosphate-accumulating organisms are present in sufficient number while undesirable species (for example, filamentous bacteria) are not present in the system. In various subordinate processes, the micro-organisms compete for the same resources such as organic substrate, which is needed both for denitrification and uptake of phosphate. Furthermore, the holding time of the process must be shared between the sequential processes of nitrification and denitrification.

Sludge

The process of the activated sludge formation too is a complex process due to the different compositions of the influent wastewater, due to rain and various types of pollutants in the municipal wastewater. In the European wastewater treatment plants, the biological phosphorus removal combines nitrogen removal by nitrification and denitrification, with biological phosphorus removal. All are based on the sequential exposure of micro-organisms to anaerobic and aerobic conditions in the biological reactor. Often, in many plants, the sludge solids produced by oxidation in secondary treatment are returned to the inlet and settle out with primary sludge. The extent and type of sludge treatment would depend on the final sludge disposal route.

The disposal of the sludge that comes out of the wastewater treatment plants is a sensitive issue that requires considerations such as public health, environment and eco-balance. Most often, in many countries in Europe, and in India too, this sludge is treated by anaerobic mesophilic digestion and is disposed of in the agricultural fields. Some countries resort to ocean dumping, but this practice is not desirable due to large-scale and continuous pollution of the oceans it would cause and the resultant threat to the safety of the marine life. Further, the Third North Sea Conference in 1990 stipulates that the sea disposal of sludge would cease after the year 1998 (Try and Price, 1995).

Anaerobic digestion facilities are provided in most of the large wastewater plants in Europe. Sludge is heated to around 35°C for about two weeks so that organic solids are biologically broken down. This sludge is inoffensive and can be employed in agriculture where there are no limits on the levels of nitrogen and heavy metals in the soil. Methane gas produced is burned to heat the sludge in the digester, and is often used in combined heat and electrical power generation units (Try and Price, 1995).

The sludge can be employed mainly in three applications: 1. As landfill to level the uneven land. 2. In forestry. 3. In agriculture. The sludge that comes from the municipal wastewater treatment plants is indeed beneficial for the growth of plants, but it contains heavy metals and microbes. The heavy metals are transferred to the ground from the sludge, and then from the ground to the plants that grow on these grounds. As these heavy metals are harmful or even at times toxic, it is not desirable that the humans and the animals consume such plants' products. Further, special treatment of the sludge for removal of heavy metals needs to be done for the disposal of the sludge in agricultural fields, and this treatment escalates the cost of the whole exercise. Therefore, the disposal of the sludge in the agricultural fields is controversial and not desirable. The microbes do not cause harm if the sludge is subjected to proper composting. The composting is not a major issue in the hot, temperate, arid and semi-arid interior (away from the coast) and central parts of India. However, in the areas that experience humid weather such as India's west coast areas, Sub-Himalayan belt and northeast India, composting of sludge would indeed be a problem. Use of the sludge in forestry, where there is no possibility of the transfer of heavy metals and microbes to the humans and animals, appears to be best solution for the disposal of sludge. This would have an added advantage of better growth of the forest plants like trees that find utility as timber.

Yet another form of sludge disposal is by *incineration* wherein the sludge is reduced to ashes. Incineration requires the sludge to be dewatered up to about 30%. Incineration of sludge would be required in plants that produce large-scale recyclable water, and therefore, incineration facility would become an integral part of the wastewater treatment plants.

MUNICIPAL WASTEWATER TREATMENT SCENARIO AND PROSPECTS IN INDIA

In India, the stage of wastewater management is still rudimentary, barring rare instances of preliminary wastewater treatment prevalent in just a few cities, industrial units and rivers. The statistics on the amount of usable recycled water generated and put to its possible use in the day-to-day activities such as toilet flushing, automobile and railway wagon cleaning and industries in India is not exactly known. But, a rough estimate on the basis of the number of wastewater treatment plants, which are present and functional in various parts of the country, would allow us to infer the undisputable fact that this amount is too meager even to take a note considering the size of the country and the potential of this technology in yielding the large quantity of recyclable water. The recycled water has great potential for application in many more activities such as development of urban greenbelts, social forestry, parks, drip irrigation, horticulture, paper and pulp industry, and even building construction.

Currently, only in three or four Indian cities where wastewater treatment plants are functional, recycled water is used for the purposes of urban forestry and maintenance of parks. In all other cities and towns, fresh water from rain source or groundwater source is put in use in the activities where water quality is not an important issue. This has created an extreme stress on the source for the potable and good quality water for direct human consumption, and this stress is, therefore, avoidable. A cursory study of approach to the wastewater management in India gives an impression that there is no uniform national policy applicable in the whole country, concerning the management, treatment and use of the recycled water. At present, these aspects are looked after by the municipal corporations or municipalities in tandem with the Pollution Control Boards, which function under complex sets of standards that are not uniform and do not conform to the acceptable levels. Further, the existing standards are unequal, and their enforcement is complicated since these standards are applied by the enforcing agencies with several extraneous considerations with different yardsticks to the corporate units, municipal bodies and industrial sectors. Most often, the rules concerning the wastewater and its management are more flouted rather than complied with, resulting in haphazard disposal of wastewater.

The industrial effluents and municipal wastewater are directed into the streams, rivers, lakes and other water bodies with impunity. This has been causing a host of maladies such as pollution of these water bodies, derangement in the quality of groundwater and, above all, permanent loss of the wastewater resource that has a potential for reuse in supporting activities if managed properly by recycling. A thorough discussion on the points relating to the pollution of the rivers and water bodies, the pollutants, and the remedial measures concerning the pollution is not one of the objectives of this book. However, where absolutely necessary, a few points concerning the pollution of the freshwater bodies by the wastewater, rendering both the waters useless, are highlighted. The pollution of water in various sources such as rivers, streams, lakes and groundwater, grossly reduces the availability of freshwater suitable for the main needs of human activities such as utensil washing, personal hygiene, cooking and drinking.

It is well known that the major rivers of India receive large quantities of wastewater, both municipal and industrial. The quantities of this wastewater are so enormous that a quantitative estimation of the wastewater that needs treatment for reuse and avert pollution of the rivers is not possible. Further, the biological oxygen demand (BOD) in the river waters after the wastewater joins those rivers gets alarmingly increased. The extent to which the holy Ganges (Ganga) River is polluted due to the flow of the wastewater in to it from several locations on its course has been discussed and debated on several occasions in the past few decades and also gained media publicity with notoriety, and therefore not repeated here. The quantity of murky wastewater, which flows into the Jhelum River (in Kashmir) has reached such alarming proportions that the very life of the river is in danger. Perhaps what sustains the Jhelum River is its youthful state, as it flows through steep slopes of Himalayan terrain, which causes its frequent rejuvenation, dumping of coarse suspended matters and partial natural cleansing as it flows downstream.

If the water of flowing rivers is contaminated by wastewater from biological sources, it has a unique capability of undergoing natural cleaning as it flows through the river channels. This natural cleaning takes place mainly by three processes.

1. Natural dilution of wastewater-contaminated river water occurs, if water from tributaries flow into the streams and rivers. Natural dilution takes place provided the water from the tributaries are uncontaminated and purer than that of the main river or stream.
2. The water flowing in the rivers absorbs atmospheric oxygen as a result of which the dissolved oxygen (DO) levels of the contaminated river water increases. Higher DO levels oxidize the biological pollutants to generate sludge in the naturally flowing river system. The sludge thus generated would be dumped at places where the energy of the flowing water drops.
3. The natural undercurrents generated in the flowing wastewater-contaminated river water function as natural aerators and inject oxygen into the flowing water. This process oxidizes the contaminated water and boosts the DO levels.

On account of the natural processes described above, the BOD levels are maintained in the flowing rivers in way that such river waters can support life. These natural processes, however, do not take care of the industrial pollutants. In order to drive the point as to how the treatment of the municipal wastewater improves the quality of the effluent water in such a way that a handsome part of the treated water can be used for supportive human activity, and the rest may be diverted by channels to join the freshwater bodies such as rivers and lakes averting pollution of the water bodies, the cases of two major rivers, namely Yamuna River and Damodar River, and an urban lake namely Hussain Sagar Lake in Hyderabad Metropolis are discussed here. Figure 15.2. shows the locations of the Yamuna and the Damodar Rivers and some of the towns situated along the course of those rivers. The importance of many of these towns in terms of industries is well known, and is discussed in brief as necessary.

THE YAMUNA RIVER

The Yamuna River originates at the Yamunotri glacier in Northern Himalaya. While flowing in the Himalayan terrain for about 200 km in the Uttaranchal State, most parts of its channel are controlled by faults, the geologically weaker planes. In its higher reaches, this river receives snowmelt water from the glacial snowpacks. The Yamuna River and one of its important tributary, namely Hindon River, receive enormous quantities of wastewater. The Yamuna River gets no fresh water inflows the moment it reaches the plain terrain at Tajewala. Instead, several sewerage drains and industrial wastewater canals are linked to the Yamuna River, as a result of which only murky, toxic and sewage wastewater from both urban and rural places (Panipat, Sonepat, Gurgaon and Faridabad) situated all along its course drain into it. The Najafgarh drain, which joins the Yamuna River at the outskirts of Delhi, brings unimaginably low quality and high quantity of wastewater into the Yamuna River. The Hindon River, on its part, brings enormous amounts of industrial wastes, toxic substances from factories, sewage water of Ghaziabad town and several other smaller towns on its course, and eventually joins the Yamuna River near Delhi (see Figure 15.2).

Within the Delhi Metropolis, between Wazirabad and Okhla, five major and several minor drains discharge wastewater (both municipal and industrial) into the Yamuna River. There is a wastewater treatment plant near Wazirabad. Excepting the treated water influents into the river from this treatment plant, the other major and minor drains discharge raw untreated wastewater into the river. The Delhi Metropolis generates about 3,300 million litres of municipal wastewater daily during other-than-monsoon periods and about 4000 million litres daily during monsoon period. It is estimated that about ten percent of this quantity flows into the Yamuna River. The quality of water in the Yamuna River near and in Delhi City is a source of embarrassment to all those in authority, who are concerned with environment, water resources and protecting the rivers. The Yamuna Action Plan, which was started with a view to improve the quality of the Yamuna River, has not been able to yield much desired result, but could only marginally improve the effluents that are led into the river. Delhi *Jal* (Water) Board has only a limited capacity for the treatment of wastewater (Hindustan Times, 2003) and the efficacy of this treatment facility is highly suspect, if only the quality of water that flows in the Yamuna River in major stretches of Delhi City is any indication.

The water quality modeling work of Kazmi and Muttamara (1998) based on the data obtained from the Pollution Control Board for the year 1995 has shown that during the months of low inflows of water into the river (from November to June), the dissolved oxygen (DO) in the Yamuna River water

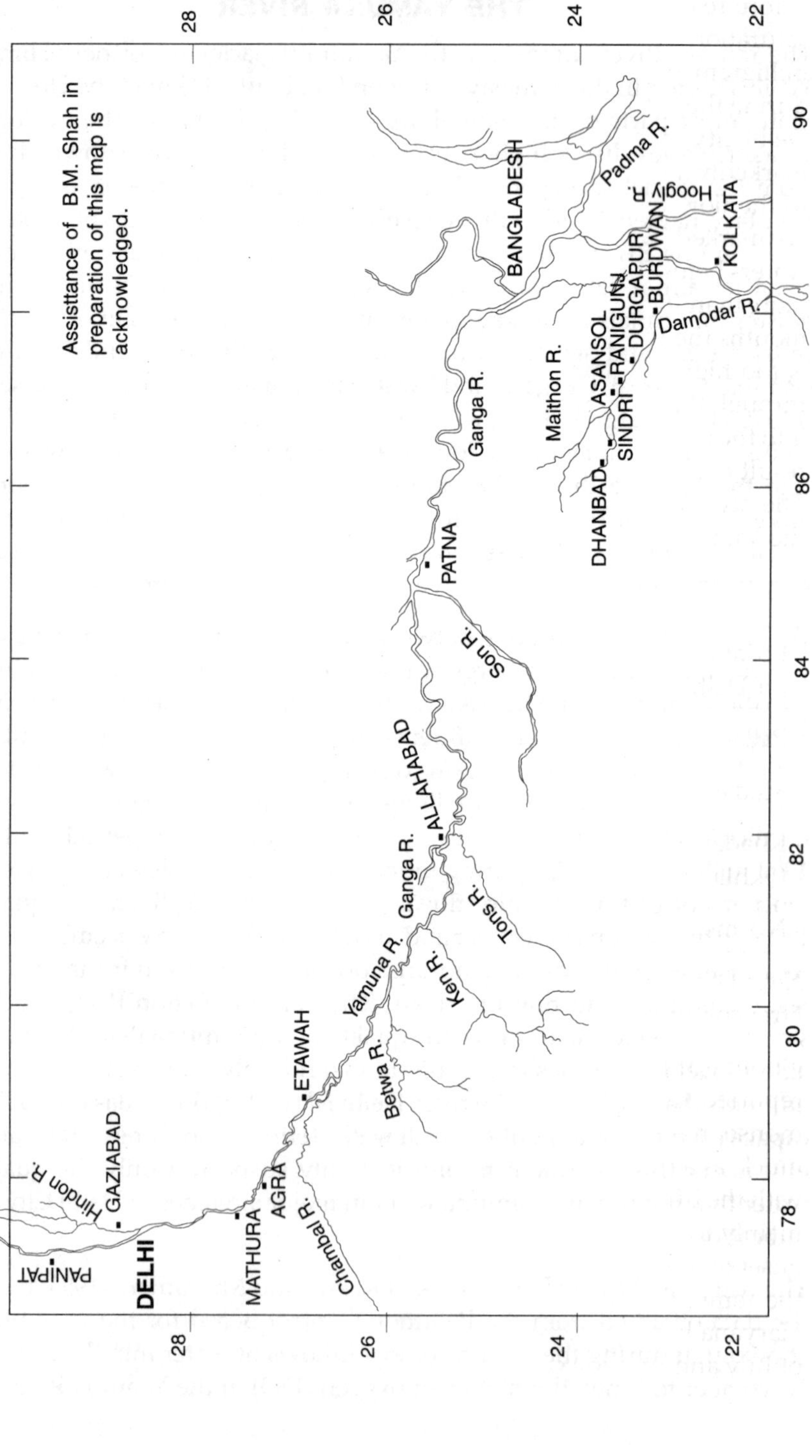

Figure 15.2. Map of the Yamuna and Damodar Rivers.

before reaching the Delhi Metropolis limit (for example, at Baghpat) is at saturation or well above the saturation level (approximately 700 to 1400 milligram per litre). However, the dissolved oxygen level is almost zero during the same period (from November to June) when the water enters the Delhi city. The levels of dissolved oxygen in the Yamuna water at Delhi markedly increase during the monsoon months (from June to October), due to the flow of fresh rainwater charged with dissolved oxygen. Also, there are marked variations in the Biological Oxygen Demand (BOD) of Yamuna waters before and after the water reaches Delhi. Whereas it is quite high at more than 1000 milligram per litre during November to June, in the monsoon months the BOD gets reduced to 400 to 600 milligram per litre, which itself is too high. Evidently, the effluents that contribute to the Yamuna River through the drains and sewers need to be treated at all places before it is led into the river. This observation has recently been confirmed by an interesting result of studies reported in the daily newspaper Hindustan Times (2003). The accompanying Table, quoted from Hindustan Times (2003) compares the Yamuna River water quality near Delhi, before and after monsoon.

	Before Monsoon				*After Monsoon*			
Location (in Delhi)	pH	DS	BOD	Coliform	pH	DS	BOD	Coliform
Palla	7.7	216	3	34,000	7.6	200	3	24,000
Madanpur	7.26	400	20	80,000	7.8	425	48	69,000
Khadar Okhla	7.19	569	52	85,000	7.16	510	48	89,000
Norms	6-9	---	3	5,000				

Note: DS: Dissolved solids (milligrams/litre); BOD expressed as milligrams/litre; coliform expressed as numbers/100 millilitre.

Quoting the officials of the Central Pollution Control Board (CPCB), it is reported that the wastewater that is added to the Yamuna River during monsoon time through the drains and sewers is heavily laden with silt and muck, as a result of which the quality of the Yamuna water has deteriorated, with the BOD levels remaining same or going up further. This was attributed mainly to choked drains and sewers, which were not cleaned prior to the onset of monsoon. What is more disturbing is that the high levels of BOD in the Yamuna waters in Delhi have not changed even after the Government of Haryana has released an extra 200 cusecs of water in the monsoon months of July and August 2003.

Indeed, the situation pertaining to the Yamuna River water quality is extremely complex, though it is highly certain that the wastewater treatment plants at several points will grossly improve the quality. The effect due to the contribution of low quality effluent water with high BOD by Hindon River into Yamuna River near Delhi is not exactly known, but it is expected to be significant, considering the pollution levels of the Hindon River itself due to the large quality of industrial and municipal wastewater this river receives. Therefore, the waters of the Hindon Rivers at various points, most importantly near its confluence with Yamuna River, need to be mechanically treated to improve the quality of influent water into the Yamuna River.

The water quality modelling exercises of Kazmi and Muttamara (1998) has specifically dealt with the determination of biological oxygen demand, dissolved oxygen and coliforms, both in the waters of Yamuna River in several places and also in the drains that feed the Yamuna River under free flow conditions. The Mike 11 model developed by the Danish Hydraulic Institute using hydrodynamic set was used for temporal calibrations. These studies have revealed that the quality of water in the Yamuna River in upper segments like Baghpat town can be improved by treatment from Class B to Class A, so that the water can be used after disinfection. But, the quality of the Yamuna River water within the Delhi city could not be improved after treatment even to Class D. This would imply that the Yamuna River water requires rigorous measures of treatment, and that the wastewater that joins the Yamuna River at multiple points through drains and sewerage lines within the Delhi Metropolis from Wazirabad to Okhla needs to be treated drastically with the aid of the several water treatment plants. Just one treatment plant near Wazirabad is ineffective considering the several sewers that join Yamua River. Each and every drain and sewage that joins the Yamuna River has to be identified and the water treatment plant has to be set up for each of them. Whereas a part of the recycled water may be distributed by separate pipelines to the Delhi residents for toilet flushing, garden maintenance and automobile washing, the other part of the treated water can be led into the Yamuna River. This exercise can thus serve a multipurpose of easing the water demand of the Delhi residents, the stress on the need of water for the supply by the concerned water supply authorities (Delhi *Jal* Board), groundwater recharge and inflow of moderately good quality of water into the Yamuna water.

It is equally important to consider the employment potential arising out of the setting up of the treatment plants and subsequently through the maintenance and supply of treated water to the masses. Though the project of setting up wastewater treatment plants at several points within the Delhi Metropolis with the said multipurpose objective would cost enormously, the one-time investment in the project and the maintenance cost of the treatment plants would be totally justified if the benefits that would accrue are seriously considered. Though it is evident the facts discussed here may be alive in the minds of the concerned planners and also there is an awareness of these aspects among the scientists and technologists of the country, it is not clear as to why there should be such long delay in implementation of progressive schemes such as wastewater treatment plants for the restoration of pride to the country's prestigious rivers and water bodies. We can stem the rot and degradation of the country's geographic heritages such as perennial rivers by timely implementation of such progressive exercises.

THE DAMODAR RIVER

The Damodar River originates from the hills of Palamau in the Chhotanagpur plateau (in Jharkhand State, eastern India) at about 600 metres above mean sea level and flows in a southeasterly direction for about 600 kilometres and joins the Hooghly River southwest of Kolkata in West Bengal State (see Figure 15.2). The Damodar River basin annually receives an average rainfall of about 120 cm, mostly during the monsoon months. The total yield of water in the Damodar River basin is estimated to be about 5000 million cubic metres at 75 percent dependability. Prior to the construction of dams and impounding water in the dams, the Damodar River was causing floods in its drainage basin, inundating several hectares of agricultural lands as well as the human settlements in the low lying areas on its course and, for that reason, this river was referred to as the *Sorrow of Bengal.*

In order to effectively manage the water resources of the Damodar River, an organisation known as the Damodar Valley Corporation was constituted by the Central Government through an Act in the Parliament in the year 1948. Subsequently, dams were built on the Damodar River and its important tributaries namely Barakar River and Konar River. Five major multipurpose dams, viz. Tilaiya and Maithon dams (on Barakar River), Konar dam (on Konar River), and Panchet and Tenughat dams (on Damodar River) were built in the Damodar River basin for flood control, supply of water to the several towns and industrial units, irrigation and hydroelectric power generation.

Availability of enormous quantities of seasonal water for industries and irrigation coupled with large coal reserves in the Damodar River basin and its environs has facilitated rapid industrialization in and around the Damodar River basin. Migration of a large population from all over the country to support the technical and labour force for the coal mines, coking industry, integrated iron and steel plants, hydel power stations, fertilizer and chemical industry units, paper mills and several other miscellaneous industries has, over the last five decades, led to a virtual population explosion in the areas in and around the Damodar River basin. The population is estimated to be about 14 million in the year 2000. This has resulted in a steep increase in the demand for good quality water for human consumption, besides an unimaginable demand for water in the said industrial units. Likewise, the demand for water for the agricultural use has also increased manifold on account of the spurt in the agricultural activities in the fertile lands of the Damodar River valley.

The crux of the matter pertinent to the present discussion, however, is the quality of water in the Damodar River that has deteriorated over the past few decades due to the excessive human influence and activities within the Damodar River basin. The proliferation of industries (including the coal washery and coking coal industries), coal mining activities, and the explosion of population have led to pollution of the Damodar River water by enormous amount of municipal and industrial wastewater that joins the river. Whereas on one hand the demand for fresh water supply has grown steadily over the years, concurrently, the availability of fresh water in the Damodar River basin area has dwindled to inadequate proportions, and the quality of the water in the whole basin has become dangerously substandard. The status of the Damodar River and its water quality scenario have been lucidly highlighted and discussed in an International Specialized Conference in the year 1998 by a few high rank technical officers of the Central Water Commission (Mittal *et al*, 1998). They have made a pragmatic projection of the demand for water in the basin to be about 11,310 million cubic-metres by the year 2041, and have categorically stated that a sustainable management of the available water resource in the Damodar River basin is essential for a healthy development of the river and its basin. They have also reported that the largest contribution of wastewater and Biological Oxygen Demand (BOD) to the Damodar River comes from the West Bengal State and large quantity of toxic load is contributed by the Jharkhand State. Unfortunately, these findings of Mittal *et al* (1998) highlighted in the year 1998 have not changed for better in the subsequent years, and the maladies that pose threat to the Damodar River continue even to date, if not become still worse in the recent times.

Large quantities of water is pumped out of the coal mines, and this water joins the Damodar River along with the pollutants such as coal dust, ash, sand, silt and natural metals associated with the coal. Further, the mine water contains bacteria and micro-organisms. The count of maximum total coliforms is at times as much as 2400. The higher ranges of the total suspended solids, total dissolved solids and sulphates in the mine water are determined to be to the tune of 1180 milligrams per litre, 860 milligrams per litre and 1400 milligrams per litre respectively, which seriously affect the utility and quality of the mine water. During the monsoon season, the rainwater that leaches the mine dumps gets concentrated in metals such as iron, manganese and copper, and contaminates the groundwater besides the water of the Damodar River. High concentrations of these metals can cause health hazards to both humans and livestock.

The study of Mittal *et al* (1998) has also indicated that the water discharged from the coal washeries and coking coal plants as effluents contains thiocyanate, phenols and ammonia, in addition to enormous quantities of metallic and non-metallic toxic load. Besides, the Total Suspended Solids in these effluent waters range from 800 to 3500 milligram per litre, alkalinity varies from 10 to 90 milligram per litre, and chlorides from 10 to 40 milligram per litre. The six thermal power plants, two integrated steel plants and several fertilizer and chemical plants (most important being at Sindri), which are operative in the Damodar River basin also contribute enormous amounts of effluents containing several *heavy metals* (among mercury, manganese, chromium, cadmium, arsenic, antimony, lead, nickel, barium and lead), fly-ash, phenols, industrial lubricant oils and greases, and a host of toxic chemicals, since most of these industrial units do not have proper plants for the treatment of the effluent water. Therefore, the Chemical Oxygen Demand (COD) in the Damodar River waters get enhanced to dangerous proportions at several locations, especially in lean periods of inflow. To make the matters worse, significant quantities of fertilizer and pesticide residues from the agricultural fields also pollute the river basin. These residues are concentrated in nitrogen, potassium, phosphorus and pesticide toxins.

Several cities and towns, the most important being Burnpur, Asansol, Bokaro, Sindri and Durgapur, are situated within the Damodar River basin and none of these towns has planned sewerage system and municipal wastewater treatment plants. From these cities and towns, the municipal wastewater having abysmally low or no dissolved oxygen (DO) directly finds its way into the river system thereby enhancing the Biological Oxygen Demand (BOD) of the river water to alarming proportions. The Biological Oxygen Demand of the river water at Durgapur city is estimated as a disturbing level of 15875 kilogram per day (Mittal and associates, 1998).

The Damodar River is yet another example of over-exploitation of the pristine wealth of river waters, without any ethical concern whatsoever to protect the life of the rivers that sustain the human life. It is believed that the Damodar River is the most polluted river of peninsular India, more polluted than the Ganga and the Yamuna Rivers. Evidently, from being a source of sorrow to the mankind in the past, the Damodar River of the present time faces sorrow by the maltreatment meted out to it by the mankind! Currently, its water is toxic, unfit for consumption by humans and livestock and even for irrigation because, the highly toxic heavy metals and other chemicals it contains would find their way to the human system through the media of plants and animals.

The Damodar River, though much less in length than the Ganga, Yamuna and Brahmaputra Rivers, has achieved the dubious distinction of being the most polluted river in India. The one and only way to save the Damodar River from total degeneration and decadence is to install a series of wastewater treatment plants for treating the industrial effluents generating from all the industrial units, as well as the municipal wastewaters generated in the cities and towns that are located in the Damodar River basin. The objective of installation of the wastewater treatment plants to recycle the industrial and municipal wastewaters is essentially two-fold: 1. To augment the water resources and ease the stress on water need in all the cities, towns and industrial units in the Damodar River basin, by providing the recycled water to the supportive human activities such as cleaning of automobiles and railway wagons, and to the industries in which water quality is not critical, such as paper and fertilizer industrial units. 2. To improve the influent water quality that drains into the Damodar River and its tributaries, so that the pollution of the Damodar River basin drastically gets reduced. Isolation of industrial wastes from the influents that join the Damodar River would also reduce the pollution levels in the river.

THE HUSSAIN SAGAR LAKE, HYDERABAD

The Hussain Sagar Lake is one of the biggest artificial urban lakes in Asia. It is situated right in the middle of the Hyderabad Metropolis, the historic Capital city of the State of Andhra Pradesh. This lake was built by impounding waters of a tributary of the Musi River, which, in turn, is a tributary of Krishna River. Like the city of Hyderabad that hosts it, the Hussain Sagar Lake too boasts a history of its own. This lake was built in the year 1562 by Ibrahim Quli Qutb Shah, the then ruler of Hyderabad and a monarch belonging to the 16th Century Qutb Shahi dynasty, to express his gratitude to Hazrat Hussain Shah Wali, after he had recovered from certain serious illness. The name of the lake is after Hazrat Hussain.

The lake was originally planned as a source of drinking water for the subjects of Hyderabad and irrigation of the nearby paddy fields. In the past, this lake was spread over an area of 550 hectares, but on account of various reasons, including encroachments, it had shrunk considerably, and today, it occupies an area of only about 340 hectares. A three kilometre long road built on the bund of the lake links the twin cities of Hyderabad and Secunderabad. During the 1980s, the State government had installed thirty-three statues of luminaries, freedom fighters and poets of Andhra Pradesh as a step towards beautification of this road and the city. Still later, a 16 metre tall and 350 tonne monolithic statue of Lord Buddha was installed in the middle of the lake. In recent times, several parks and green belts have been raised in the vicinity of the Lake, as also a road girdling the lake's perimeter, known as 'the Necklace Road', and linking the either end of the road on the bund have been built. Yachting and other water sports are conducted annually in this lake in recent years by the sailing clubs of national affiliations. The Hussain Sagar Lake has not only been a pride of the Hyderabad city for centuries, adding beauty to the city and attracting a large number of domestic and foreign tourists, but also has been responsible in keeping the day temperatures during the peak summer at tolerable levels.

Despite its utility, glory and legendary past, the Hussain Sagar Lake had suffered excessive human interference resulting in considerable deterioration of its water quality in recent past. The lake was receiving enormous quantities of municipal wastewater as a result of large-scale growth of population in the Hyderabad city in the last forty years. Further, large amounts of industrial wastewaters containing a host of dissolved substances and heavy metals also ended up in the Hussain Sagar Lake through a few first and second order streams that drained into the main stream which fed the lake. The quality of the influents into the Hussain Sagar has deteriorated so much that between 1980 and 1990 the lake water turned highly toxic and deleterious to the humans and livestock. Huge quantities of undissolved organic substances that were led into the lake through wastewater had promoted intense *eutrophication*, and proliferation of large colonies of hyacinths and weeds on the water surface of the lake, rendering it an ugly look for a large part the year. The dissolved oxygen (DO) levels had come down dangerously low and the Biological Oxygen Demand of the lake water had shot up alarmingly high. In a nutshell, the Hussain Sagar Lake was almost on the verge of extinction in the early 1990s.

It was then the Government of Andhra Pradesh realized the necessity to save the lake, and thus, the First Hyderabad Water Supply and Sanitation Project was undertaken by the Hyderabad Metropolitan Water Supply and Sewerage Board (HMWSSB), a service organisation under the Government of Andhra Pradesh. The HMWSSB functions with a declared Mission of '*Providing clean, wholesome potable water, effective disposal of domestic sewage and keeping a pollution-free environment*'. Under this First Hyderabad Water Supply and Sanitation Project, the HMWSSB has installed a wastewater treatment plant of twenty million litres per day capacity and commissioned it in the year 1997. This treatment plant processes the municipal wastewater, and currently, the processed water is led into the Hussain Sagar Lake. This exercise has not only infused a fresh lease of life to the lake, but also a significant part of the treated wastewater is used to raise several new greenbelts near the lake, as also maintain the existing urban greenbelts and municipal parks. The fact that the Hyderabad Metropolis could successfully face a fierce water crisis and could still sustain the green look of the city in the wake of the 2003 summer demonstrates the efficacy and impact that *just one* wastewater treatment plant alone can make, even on a land-locked urban centre that witnesses a semi-arid climatic condition. Let us now take a broad look at the wastewater recycling plant that feeds the treated municipal wastewater after primary treatment into the Hussain Sagar Lake.

As a basic measure to stop the inflow of municipal and industrial wastewater into the Hussain Sagar Lake, 35 million litres per day of municipal wastewater from the northern parts of the city, 20 million litres per day of municipal wastewater from the western and northwestern parts, and 15 million litres per day of industrial effluents also from the northern industrial areas of the city were barred from entry into the lake. This is achieved by providing Diversion Units for diverting the wastewater to sewerage lines that were laid specifically for this purpose. Further, municipal wastewaters from the western and southwestern parts of the city were diverted to another sewerage pipe, known as the 'Duplicate A Main Sewer'. These diversion units are provided with screens through which the wastewaters flow and subsequently enter the diversion canal system. This arrangement ensures the removal of undissolved coarse suspended materials and plastics and coarse heavy particulate matters from the sewage water before it enters the diversion pipelines. A typical diversion unit situated near the Hussain Sagar wastewater treatment plant is shown in Figure 15.3. The diversion units have a further objective of preventing the entry of wastewater directly into the terrestrial source, namely the Hussain Sagar Lake. However, during monsoon period, excess flow of wastewater drains into the lake after primary treatment and getting rid of plastics and coarse suspended materials.

The municipal wastewater from the Duplicate A Main Sewer is taken to the Sewage Treatment Plant of 20 million litres per day capacity and treated by the Biological Oxidation Method. Though, in terms of hydraulics, this treatment plant is designed for a peak flow of 30 million litres per day, the process design is made for an average inflow of 20 million litres per day. The wastewater treatment involves five stages:

1. Screening.
2. Degritting and Grit Washing.

3. Removal of the organic matter by adopting Biological Oxidation Method to decrease the Biological Oxygen Demand (BOD).
4. Solid-Liquid Separation.
5. Sludge Drying.

These processes are discussed in brief with the aid of illustrations in the following pages. Reinforced cement concrete (RCC) is used in civil works of the plant. An RCC screen inlet channel is provided for removal of floating materials such as plastic bags and bottles, leaves and other coarse undissolved materials. A Bar Screening Raking mechanism is provided in this channel, which leads the raw (untreated) wastewater to a Receiving Chamber by pumping (Figure 15.4). From there, the wastewater reaches a reciprocating type mechanism furnished for conveyance and washing of the grit (Figure 15.5). There is also a bypass screen channel, which operates when the main screen channels get choked or during their maintenance. An RCC grit chamber is provided in the plant for removal of the grit from the influent (Figure 15.6). In this chamber, degritting (removal of all grits) and other inorganic solids with specific gravity of 2.5 and more, from the influent takes place. Degritting is associated with grit washing so that the gritty matter is freed from the organic matter before disposal. The grit is scrapped from the bottom of the tank by the rotating type scrapper mechanism. The screened and de-gritted wastewater then reaches the Flow Chamber (Figure 15.7) where the inflow of the wastewater in to the plant is measured. The measurement of flow is essential to regulate the flow of wastewater into the aeration tank, as otherwise, the excessive flow will result, which would lead to hampering of the biological processes. From the flow chamber, the wastewater then flows into an Aeration Tank (Figure 15.8). Biological oxidation using Extended Aeration Method is employed in the aeration tank of the Hussain Sagar wastewater treatment plant for removal of the organic matter. The organic matter of the wastewater under treatment is oxidized in the metabolic activity of the micro-organism in the aeration tank. The oxygen required for the metabolic activity is provided by the Mechanical Surface Aerators. There are sixteen fixed vertical axis-type slow-speed mechanical surface aerators in the aeration tank, which keep the wastewater under treatment in the tank in a homogeneous condition and also supply atmospheric oxygen at the rate of 536 kilograms per hour in the course of treatment.

The plant has a Secondary Settling Tank as shown in Figure 15.9. Clear water after the removal of the sludge flows into the channel of this settling tank, whose floor slopes towards the centre of the tank. Below the floor of the tank is the sludge withdrawal pipe. The separation of solids from the liquid is effected by the separation of the sludge from the treated sewage. The sludge consists mostly of bacterial culture. The sludge-scrapping mechanism is fixed bridge type and centrally driven. The sludge comes from the sludge chamber of the settling tank to an RCC sump from where it is pumped back to the aeration tank for recycling. The excess sludge is pumped to a Decanter Centrifuge. A pump house consisting of dry well accommodates three return sludge pumps and pipes are connected to those sludge pumps (Figure 15.10.) and a RCC sump to receive sludge from secondary settling tank are also added to the plant. Further, a 1.5 metre long, 1.0 metre wide and 2.0 metre deep RCC filtrate sump is provided for collecting the clean filtrate from the platform, where two vertical wet pipe type centrifugal pumps are also installed. Excess sludge is dewatered by means of a Decanter machine, and further dried in open (Figure 15.11).

The dimensions of various chambers and channels in the wastewater treatment plant set up at Hyderabad in connection replenishing the water in the Hussain Sagar Lake correspond to those of the international standards earlier discussed. The dimensions of various parts of this wastewater treatment plant are given as follows in terms of *length x width x depth* (depth of the wastewater):

1. Inlet chamber: 4.5m to 3.5m x 2.5m x 1.5m
2. Main screen channel: 5.0m x 0.72m x 0.8m
3. Bypass screen channel: 5.0m x 0.72m x 0.8m
4. Degritting tank: 5.5m x 5.5m x 0.7 m
5. Flow measuring channel: 10.0m x 1.0m x 0.77m.
6. Aeration Tank: 77m x 77m x 3m. Water detention time: 21 hours. Number of Aerators: Sixteen. Oxygen transfer capacity: 536 kilograms per hour.
7. Secondary settling tank: Circular with 42m diameter and 2.75m deep.
8. Decanter centrifuge: Capacity 15 cubic metre per hour.

The quality of treated water that comes out as output from this primary Sewage Treatment Plant is clear, colourless and free from solid wastes, undissolved materials and organic substances (see Figure 15.12). The treated water is led into the Hussain Sagar Lake through a channel as shown in this figure.

In order to compare the quality of the output (the treated water) from the influent into the treatment plant, two sets of analytical data are furnished.

Figure 15.3. Typical Diversion Unit with screening facility near the Hussain Sagar plant.

Figure 15.4. Receiving Chamber of Hussain Sagar plant in which wastewater is pumped.

Figure 15.5. Reciprocating type mechanism in the Hussain Sagar plant for wastewater conveyance and grit washing.

Figure 15.6. Grit Chamber of the Hussain Sagar plant.

Figure 15.7. Flow chamber of the Hussain Sagar plant.

Figure 15.8. Aeration Tank of the Hussain Sagar plant.

Figure 15.9. Secondary Settling Tank of Hussain Sagar plant, where clear water flows into the channel after sludge settlement and removal.

Figure 15.10. Return sludge pumps with pipes of the Hussain Sagar plant.

Figure 15.11. Dewatered sludge from Hussain Sagar plant undergoing open drying.

Figure 15.12. Output drainpipe feeds treated water into a canal, which in turn, channels it into the Hussain Sagar Lake (not seen).

The data in Table 15. I are the analytical results in the influent and treated water samples collected in the peak summer season, the 28th May 2003. Table 15. II brings out the analytical results of the influent samples collected after mild dilution of the wastewater by the monsoon rains on 24th July 2003, and also of the water sample after treatment.

Table 15-I: Analytical details of the Influent and the Treated Water generated from the Sewage Treatment Plant of HMWSSB near the Hussain Sagar Lake, Hyderabad.

Sample Details: Summer Season samples. Date of the Samples: 28th May 2003. Daytime temperature: 41.3° C. Rainfall: Nil.

Tests	*Influent sample from Receving Chamber*	*Results: Treated Effluent from Outlet Chamber*	*Reduction (in %)*
1. pH 6.97	7.48	—	
2. Suspendend solids, mg/l	474	22	95
3. Phosphates as PO_4, mg/l	24	7	71
4. Phosphates as P, mg/l	7.83	2.28	71
5. Nitrates as NO_3,	4	35	—
6. Total Kjeldahl N, mg/l	56	7	88
7. Total Nitrogen, mg/l	57	15	74
8. COD, mg/l	860	58	93
9. BOD for 3 days at 27°C., mg/l	390	9	97.7
10. Dissolved oxygen, mg/l	2.05	3.2	—

Table 15-II: Analytical details of the Influent and the Treated Water generated from the Sewage Treatment Plant of HMWSSB near the Hussain Sagar Lake, Hyderabad.

Sample Details: Monsoon Season samples. Date of the Samples: 24th July 2003. Daytime temperature: 28.6° C. Rainfall (in the season): 53mm.

Tests	*Influent sample from Receving Chamber*	*Results: Treated Effluent from Outlet Chamber*	*Reduction (in %)*
1. pH	7.16	7.44	—
2. Suspendend solids, mg/l	186	20	89
3. Phosphates as PO_4, mg/l	11	3	73
4. Phosphates as P, mg/l	3.59	0.98	73
5. Nitrates as NO_3,	9	30	—
6. Total Kjeldahl N, mg/l	20	3	85
7. Total Nitrogen, mg/l	22	10	55
8. COD, mg/l	300	32	89
9. BOD for 3 days at 27°C., mg/l	140	5	96.4
10. Dissolved oxygen, mg/l	2.30	4.3	—

The commissioning of the wastewater treatment plant in Hyderabad for the purpose of accreting water to the Hussain Sagar Lake after primary treatment marks the beginning of a new era of environmental protection actions in Hyderabad achieved by the HMWSSB. Besides the environmental protection of the Hussain Sagar Lake and its neighbourhood, the wastewater treatment has unarguably augmented the water resources of the metropolis, though only to a limited extent, if one subscribes to the idea that the '*water conserved is water resource augmented*'. Because, the HMWSSB makes use of the recycled treated water, especially in the water-stressed summer seasons, for maintaining parks and greenbelts, which otherwise would have consumed significant quantities of pure rainwater or groundwater that could be preserved or diverted for the use of the city dwellers. The positive impact made by the one single wastewater treatment plant on the water quality of the Hussain Sagar Lake, the environment in and around the lake and on the quantity of the water that could be conserved, has indeed been markedly encouraging. One can now witness gross reduction in the amounts of hyacinths and weeds in the lake, on account of a considerable fall in quantity of organic matter nutrient that enters the lake, as a result of its removal followed by treatment. Furthermore, during peak summers, the treated water would have significantly percolated and accreted to the groundwater resource of the city.

Nevertheless, much more needs to be done in the sector of wastewater recycling as a means of augmentation of the influent water into the Hussain Sagar Lake. The drop in the water levels in the lake during the last one-decade, and the marked reduction in the area of the lake, as also the volume of water in the lake are matters of serious concern, which require sincere attention. Considerable quantities of murky municipal wastewater continue to flow into the Hussain Sagar Lake through three rivulets on the northern side of the lake. The wastewaters in these rivulets have to be treated before let into the lake. Therefore, yet another wastewater treatment plant of a capacity comparable to the existing plant appears necessary for the Hussain Lake, because the lake requires more inflows of good quality water if it has to retain *at least* the present extent in terms of its area. Besides this, the Hyderabad Metropolis too requires more and more quantities of water whose quality is comparable to the above-mentioned (Tables 15-I and 15-II) treated water for supporting the miscellaneous activities of human life, such as toilet flushing, servicing in automobile and railway workshops, and maintenance and further development of the urban greenbelts.

TREATED WASTEWATER AS AN ALTERNATIVE SOURCE FOR WATER RESOURCE AUGMENTATION

The potential of the treated wastewater as a source for augmentation of water resources as a logistic support for miscellaneous human activities is, indeed, tremendous if we take a perspective view at the technology and efficacy of wastewater treatment, the quality of the output after treatment, and the large segments of human activity in which the treated water can be effectively put to use. The quality of water, in general, is fundamentally a matter of aesthetics, and a large part of the Indian society in particular is fussy and fastidious about the quality of the water they use. The masses of India are most unlikely to accept to use the treated water even after the tertiary stage of biological treatment (to eliminate micro-organisms) for the mainstream and day-to-day activities such as personal hygiene, swimming pools and other recreational purposes. Therefore, the scope of wastewater treatment in the context of Indian scenario can be restricted to only primary level of making it free from undissolved coarse materials, grits and solids, and coarse organic matters. The treated water thus generated may be put to use only for the miscellaneous activities such as developing greenery around the country, arresting the soil erosion with the aid of soil conservation techniques by growing grasslands and shrubs, wasteland reclamation and other minor human activities where water quality is unimportant, such as toilet flushing, automobile and railway wagon servicing, as also select industrial sectors such as paper and pulp industry, petrochemical industries and, above all, for the protection of our rivers, lakes and other water bodies. Thus, the wastewater treatment serves a dual purpose of water resource augmentation and protection of our rivers, lakes, canals and a host of small water bodies. If the case history of HMWSSB in improving the quality of the influents into the Hussain Sagar Lake can be cited as an example towards the said dual purpose, India would need umpteen wastewater treatment plants of similar capacity and capability distributed across the whole country, with the objective of providing an alternative source for water to the nation's masses and improving the water quality in our rivers, many of which are considered as sacred.

It is reliably learnt that the efforts to improve the quality of the water let into the Ganga River, which is regarded as sacred by millions of people in our country, have suffered setback due to the failure in proper implementation of the wastewater treatment and the installation of the treatment plants. Though the 'Clean Ganga' movement was started with much gusto and fanfare, it later unfortunately turned out to be a whimper due to the failure on the part of the organisers to successfully translate the thought into action, the primary reason being our inability to install and maintain the wastewater treatment plants. As a result, the water quality in Ganga River near Haridwar, Kanpur, Patna and Varanasi continues to be dangerously substandard with very high levels of Biological Oxygen Demand. The story is no different in the case of the Pampa River, a dip in which is prescribed mandatory for the pilgrims of the holy shrine of Shri Ayyappa in Sabarimala, Kerala. Though it can be acknowledged that the sanitation officials keep a strict vigil during the peak of the pilgrimage season to this shrine by getting the coarse undissolved pollutants such as leaves and plastics removed on a continuous basis, the grits and organic matter in the waters of Pampa River have gone up so steeply that the Biological Oxygen Demand would appear intolerably high. The problem is compounded by the low or lack of inflows of water in this river in dry season. It is not known whether the sanitation authorities have at least made any efforts to analyze the waters of the Pampa River for Biological Oxygen Demand. It is not a surprise then that many of the most devout among the pilgrims to Haridwar, Varanasi and Sabarimala turn hesitant to take a holy dip in our rivers. The tourism industry is already badly hit in recent times, especially in the peak summer season, due to the acute water scarcity faced by most of our tourist centres. If the current trend continues, even the domestic tourism, and particularly the religious tourism, would suffer setback in the years to come.

The urban lakes in the cities such as Bhopal and Sagar in Madhya Pradesh, which once served as drinking water source to those cities, have now become veritable destinations of municipal wastewater. The urban lakes in the Bhopal city, for example, receive municipal wastewaters from almost fifty inlets from different parts of the city. Razdan (2001) has focused the spotlight on the current sorry state of affairs in the case of the beautiful lakes of Udaipur, which was once known as the Lake-City of Rajasthan. Indeed, just like the ancient and sacred rivers of our country, the condition of the lakes that once quenched the thirst of billions of people for centuries has turned appalling and pathetic. It is not known whether the authorities show some concern to take cognizance of the shrinkage in the areas of these lakes, as well as high levels of water pollution in them. It is learnt that some rudimentary form of sewage treatment system had existed in the case of the urban lakes of Bhopal city, but that has become defunct. The acute scarcity of water in most part of the year faced by the residents of these cities sends signals for some serious measures towards augmentation of water for supply to the people. If sincere efforts are not made to improve the quality of influents into these lakes in the lines of the Hussain Sagar Lake, not only these heritage lakes would become extinct in near future, but also the residents of these cities would have to live in perpetual water scarcity.

It is pertinent to note at this juncture that if the wastewater or sewage treatment plants in general, and in the case of 'Clean Ganga River' or Bhopal's urban lakes in particular, have stopped functioning, it is, perhaps, due to improper design and building of the plants, coupled with lack of proper maintenance procedures. The case of Hussain Sagar Lake's sewage treatment plant can be cited as a model and fine example for the success and durability of the technology of wastewater treatment. The technology of wastewater treatment plant cannot be faulted for the imbecile handling of plant design, construction and maintenance. The technologists, entrepreneurs, engineers, contractors and grass-root level workers associated with technological missions such as installation and maintenance of wastewater treatment plants and their ancillary or supporting technology units need to demonstrate utmost integrity while transferring the technology to the utility of the common man, and a high order character ethics while translating their plans into action.

The recycled water, unarguably, is a convenient and safe source to augment a supplementary type of water resource. With a one-time investment towards setting up of the wastewater treatment plant, and subsequently the maintenance cost of the plant, considerable quantity of usable water can be generated from the water that is considered as useless. The wastewater treatment plants are technological means to face the wrath of the Nature, when the Nature lets down the mankind in terms of failure or weak monsoons resulting in scanty rainfalls. The mankind regularly faces much hardships caused by the Nature and successfully tides over most of them using the technologies that are innovated from time to time. These technologies are essentially boons to improve the quality in the lives of the masses. If the technologies and the expertise of our technocrats are not properly employed, such acts lead us to a total failure and we cannot blame the technology for our failures or other extraneous reasons. It would be prudent on the part of those in the Governments to install more technological machinery such as wastewater treatment plants and provide the quality of life that is expected by the citizens, instead of blaming the Nature over which neither the Governments nor its subjects has any control. The prudence further lies in those in the Governments to realise the enormous employment opportunities that could be provided to the youth of the country, be it the civil and mechanical engineers, unskilled and semi-skilled workers such as plumbers, masons and other artisans, as well as the supporting staff who would be employed to collect the revenue that would be due to the concerned authorities from the people. The people on their part would reap the benefit of the technology in the form of treated water for their various day-to-day activities other than direct oral consumption.

While municipal wastewaters are of almost same type everywhere, the wastewaters that emanates from the industries contain a host of substances, which are mostly dissolved wastes. Let us now take a brief review of the treatment processes to isolate some of the industrial wastes from the wastewater to convert it into reusable water by recycling.

II. WATER RESOURCE AUGMENTATION BY TREATMENT OF INDUSTRIAL WASTEWATER

As has already been discussed in the context of the pollution of the Yamuna and Damodar Rivers due to the influx of industrial wastes, no river is free from the malady of getting affected by the inflows of waters contaminated by the wastes of industrial units. It is emphasized that the suggested scope of recycled water is restricted in this book to only for subordinate use such as supply to the toilets, maintenance of greenbelts and parks, and servicing of automobiles and railway wagons, and not the main human lifeline activities. As a principle of effective water resources management, the good quality water from the rain and groundwater source, which are currently used for the said purposes can be conserved for direct human consumption, and the recycled water can be substituted for these purposes in the place of good quality water resources. Since the quality of the water for the above-mentioned marginal uses is not a critical factor, it is sufficient to purify the wastewater that is contaminated with industrial wastes to just about tolerable levels without offensive odour, colour, turbidity, suspended particles and harmful dissolved matters.

The quality of water for direct use of the humans, such as oral consumption, dishwashing and personal hygiene is a matter of aesthetics, and people anywhere in the world would not like to use the recycled water for the said purposes, however consciously it is treated to make it cent percent pure, even after disinfections and treatment for the removal of biological impurities. Treatment of the wastewater containing industrial wastes does not warrant highly sophisticated technologies and heavy investments towards them, if the treated water is meant exclusively for use in functions such as toilet flushing, gardening, and washing of automobiles and railway wagons. The water for these purposes need to be free from coarse floating materials, suspended sandy and gritty matters, toxic materials, corrosive substances, discolouring agents and harmful chemicals that would react with or remove automobile paints. The only important management practice in the use of the treated water obtained from the wastewater contaminated with industrial effluents is to pay special care to lay separate pipelines for its supply without its getting mixed with potable quality water, and ensure that the recycled water is not mistakenly used or deliberately administered/sold with a profit motive by unethical persons for the mainstream human activities.

Wastewater treatment for industrial pollutants

The industrial wastewater contains an array of pollutants, and those pollutants depend on the products that the industries deal with, or the products and/or the by-products that are generated in the industries. An exhaustive list of all the possible pollutants in the industrial wastewater would be too long. Further, elaborate discussion on treatment of industrial wastewater is not within the scope of this book. It is recommended in this book that the water obtained by treatment of industrial wastewater be used only those segments of human activities where quality of water is not important. Therefore, only the commonest among the industrial pollutants are dealt with here.

Phenols

The most common pollutants that require treatment in the industrial wastewater are the phenols. Phenols are a class of aromatic organic compounds, which have weak acidic properties and can form salts. They are released from a wide variety of industrial units such as those dealing with coal, resins, plastics, dyes, textiles, timber, soaps and detergents, iron and steel and petroleum refineries. Phenol is a corrosive and poisonous substance with a carbolic smell, and is used as a disinfectant and also in the manufacture of plastics and dyes. The industrial units that deal with chemicals and dyes release a lot of chlorophenols too. Though the chlorophenol is useful in the manufacture of dyes, it imparts an offensive bad odour and colour to the water. The phenols can be detected by spectro-photometric method using 4-aminoantipyrine. Both phenols and chlorophenols are undesired substances in the good quality water, but their low concentrations (in the order of about 0.05 to 0.1 parts per million) in the water meant for flushing and the servicing of automobiles and railway wagons are not likely to affect the things with which the contaminated water comes in touch with.

Higher concentrations of chlorophenols have to be removed from the wastewater by photo-degradation, microbial degradation and activated carbon absorption and biological treatment, which are expensive. Phenol serves as a bacterial food without toxic effects at levels of about 500 milligrams per litre. At levels of phenol, which are below certain threshold levels, the bacteria use the phenol as food, but above the threshold limits the bacteria find the phenols too toxic. Due to these qualities of the phenols and phenol-tolerant bacteria, biological treatment of the phenol-polluted industrial wastewater using certain bacteria, followed by flocculation of the bacteria, which nourished on the phenols, are used as effective methods in isolating significant quantities of phenols for water recycling.

An interesting study by Namasivayam and Thamaraiselvi (2001) on the utility of the hydroxides of Fe(III)/Cr(III)—trivalent iron and chromium—as well as red mud, which are obtained as by-products in the fertilizer and bauxite industry respectively, has revealed that these substances are capable of removing chlorophenol from aqueous solutions. Use of such industrial waste by-products is considerably economical in industrial wastewater treatment and can be adopted in treatment of large quantity of wastewater to produce reusable water for subsidiary human activities.

The Damodar River, as had been already described, receives considerable quantities of phenols due to inflows that the river receives from the industrial units, coal washeries, iron and steel industries, chemical industrial units and so on. Therefore, the tributaries which bring industrial wastewaters and dump them in the Damodar River will be ideal surface water bodies, which can be experimented upon for the removal of phenols by using the above-mentioned industrial wastes such as Fe(III)/Cr(III) hydroxide and the red mud.

Pesticides

The pesticides are organic compounds used to arrest, destroy, repel and control the insects, weeds, pathogenic organisms, rodents and other forms of life that cause harm to plants, agricultural produce and also humans. They are used mainly for agricultural purposes as insecticides, algicides, fungicides and weedicides. Major types of pesticides (Sawyer and McCarty, 1978) are:

1. Chlorinated pesticides.
2. Organic phosphorus pesticides.
3. Carbamate pesticides.

The examples for chlorinated pesticides are DDT, Benzene hexachloride (otherwise known as BHC), endrin, dieldrin, dioxin, aldrin, chlordane, toxaphene, heptachlor, di- and tri-chlorophenoxyacetic acid. Among the BHC, there are several stereo-isomers, but the gamma-isomer BHC (gamma-BHC) called as lindane is the most effective insecticide, which is used widely in India.

All chlorinated pesticides are capable of causing harm to humans and the environment, and therefore, many of them are either banned or, there are enforcements in India and world over in their restricted usage. Parathion, malathion, systox, chlorthion, disyston, dicapthon and metasystox are some of the important organic phosphorus pesticides. Most of them, except malathion, are highly toxic to the humans and animals alike. Among carbamate pesticides, isopropyl N-phenylcarbamate (IPC; a herbicide), captan (a fungicide), and ferbam and sevin (insecticides) are important. They have low toxicity to the humans and animals. WHO (1993) is of the opinion that though the degradation products of pesticides may be a problem in the drinking water, there are inadequate data on their identity, presence and biological activity of the pesticides.

The pesticides gain access to the groundwater through the percolating medium of water that acts as their carrier, and to the surface waters through runoff. Both percolating water and surface runoff carry pesticides that are directly applied in agricultural farms. The chemical industrial units producing pesticides as well as residual pesticides from agricultural applications are the major contributors of pesticides in the water systems. Some of the pesticides, especially chlorinated pesticides, are resistant to biological degradation and persist in soils and water for long periods of time. Most, if not all, of the pesticides are toxic to the life as a whole. They get hydrolyzed at pH values above neutral and lose their toxicity, but under dry conditions, they remain unaffected in the water system for long durations.

Residual pesticides in the water meant for use by humans is, indeed, a veritable challenge of the modern times in water resources and water quality management. Though the higher range of pesticides in the drinking water is a subject of dispute, as different quality control agencies fix different amount as higher limit, one thing that is certain is that the pesticides should not be present in the drinking water and the products derived from water that are meant for oral consumption. Biological treatment and flocculation techniques are not effective in the case of the pesticide residues in water, because pesticides are non-biodegradable. The expensive treatment such as reaction with certain specific *activated carbon* is the only way that can reduce the pesticides in water (Smethurst, 1988). The masses of the Indian population will be grateful to the research institutions, if researches are stepped up and appropriate substances that are less expensive than the activated carbon are discovered to isolate pesticides from water at comparatively lower costs.

The international norms stipulate 0.0001 milligrams per litre as the limit for various pesticide contents in the potable grade water. Extreme caution has to be exercised while handling the pesticide-contaminated industrial wastewater meant for recycling after treatment. It should be ensured that the recycled water does not mix with the water resource meant as protected water supply for potable purposes. The standard of water for the potable purpose needs to be maintained at high standards so as to satisfy the human tolerance limits of pesticides, whereas for the other subsidiary purposes of life such as the provision of water for agriculture, industries, automobile servicing and so on, there can be some compromises in the augmentation and management of water resources. The water from pristine source such as rain has to be effectively harnessed directly for direct human consumption and the treated wastewater can be diverted for the subsidiary activities where water quality is not an issue. Clearly, use of good quality rainwater and groundwater for the subsidiary activities like urban greenery maintenance and servicing of automobiles and railway wagons amounts to misuse or mismanagement of the good quality water. Management of water resources meant for different purposes of human life needs different approach, and we can adopt different norms for the quality of water meant for use in different purposes of human activity, which practice does not seem to be in vogue as of now.

Alcohols

Industrial alcohols often occur as water pollutants in the vicinity of the distilleries and wineries, which deal with the manufacture of liquors and wines. Besides, the alcohol-based drug-manufacturing units and the industries that deal with the molasses, solvents and sugar also emanate alcoholic pollutants. The residues remaining after the distillation of the desired product, namely the ethyl alcohol, constitute some of the most potent industrial wastes, which have to be removed for treatment and recycling. There are four common types of alcohols of commercial importance; they are methyl, ethyl, isopropyl and butyl alcohols. Alcohols are soluble in water. Most of the alcoholic wastes from the industrial wastewater can be removed by biological treatment, since the micro-organisms oxidise the alcohols under aerobic conditions, producing carbon dioxide and water as end products.

Other hydrocarbon compounds

A wide variety of hydrocarbon compounds find its way into the industrial wastewater from the industrial units that deal with or produce or create by-products such as solvents, fats and oils, polymers, medicines, aromatic and alcohol-based substances, foods and beverages, preservatives, antiseptic compounds, petrochemicals-based products (paraffin, waxes, nylon, tyres) and synthetic fibres. The list will be too long if we attempt to take an exhaustive overview of the industries, their activities, their product lines and the raw materials they use for their products. Among hydrocarbons, besides the above-mentioned alcohols, aldehydes, ketones, organic acids, esters, ethers, alkyl halides, amines and amides, nitriles, benzene-series hydrocarbons (benzene, toluene, xylene and ethylbenzene), carbohydrates (such as sugars—mono-, di- and poly-saccharides), proteins, fats and oils, dyes and amino-acids. For further information and an overview of the specific pollutants that make into the various industrial wastewaters that emanate from different industrial units, the reader may refer to Sawyer and McCarty (1978) and other enduring works on the environmental chemistry. Many of the hydrocarbon pollutants generated by the industries, such as for example formaldehyde, were once considered to be too toxic for treatment by biological methods. Through dilution of such wastes to reduce the concentration of the formaldehyde below 1500 milligram per litre, it was found that the micro-organisms could use the formaldehyde as nourishment and oxidise it to carbon dioxide and water. This experience has led to the concept of *toxicity thresholds* in the industrial wastewater and waste treatment practice (Sawyer and McCarty, 1978). This means that below certain concentrations all materials are non-toxic. The use of completely mixed activated sludge in the wastewater and industrial waste treatment in the polluting units employs this concept.

Iron and manganese

These two metals, if present in water as pollutants, cause discolouration and turbidity of water, stain the cloths while laundering, impart strange tastes even in low concentrations, affect the plumbing fixtures and facilitate the growth of some species of bacteria. Their presence in water causes more nuisance rather than health hazard. Iron exists in soils and minerals contained in the rocks mainly as insoluble ferric oxide and iron sulphide. When iron occurs in the form of carbonate (siderite) it gets dissolved. Under reducing or anaerobic conditions, the ferric iron is reduced to ferrous iron, which is dissolved easily. Same is the case with manganese, which exists in the soils as insoluble manganese dioxide; under reducing conditions, manganese dioxide is reduced from an oxidation state of IV to II, thereby facilitating solution of manganese in water (Sawyer and McCarty, 1978). For public water supplies, the upper limit of iron and manganese are stipulated as 0.3 and 0.05 milligram per litre respectively. However, for subsidiary human activities, little excess of these limits will not be harmful, and therefore, in the treatment for the removal of iron and manganese, barring for the potable purposes, fastidious care need not be taken for their total removal.

Treatment for the removal of iron and manganese from wastewater involves charging it with oxygen by aeration, and increasing the pH (Smethurst, 1988) to over 7.0. This changes the iron from the soluble ferrous state to the insoluble ferric state by oxidation. The iron is precipitated as rust-coloured silt, which can be settled and isolated in the settling basins. Manganese can also be removed in the same way, but the pH is kept at a higher level at a little more than 7.5. Sand filter can be used as filters for the insoluble precipitates; the sand filter becomes a catalytic filter after the sand bed becomes seeded with iron and manganese. Interestingly, if the procedure for the isolation of iron and manganese is reviewed in the backdrop of the municipal wastewater treatment described earlier, it becomes clear that when the industrial wastewater is subjected to treatment in the aeration chamber as in the case of municipal wastewater, at a pH of about 6.5, most of the iron and manganese get removed and settle along with the sludge in the settling chambers.

Heavy metals and trace inorganics

Some heavy metals and inorganic substances are present in the wastewater as traces. Metallic elements such as barium (Ba), cadmium (Cd), chromium (Cr), lead (Pb) and mercury (Hg) are considered as the heavy metals in the wastewater. Further non-metals such as arsenic (As), selenium (Se), fluoride and cyanides are included as inorganic substances in the wastewater and these substances also occur in traces. The source for the trace heavy metals and inorganic substances with the exception of fluoride and selenium is industrial wastes from manufacturing units dealing with these metallic and inorganic substances, manufacturers of batteries, paints, plastics, linoleum, papers, alloys, laboratory chemicals and reagents, as well as amalgam producers (using mercury) and metal finishing units. The Environmental Protection Agency (EPA) has stipulated the following upper limits for the primary *drinking* water standards:

Metals	*Milligram/litre*	*Non-metals*	*Milligram/litre*
Ba	1.0	As	0.05
Cd	0.01	Se	0.01
Cr	0.05	Cyanide	0.2
Pb	0.05	Fluoride	1.4—2.4
Hg	0.002		

Source: Sawyer and McCarty (1978).

All these heavy metals and trace inorganic substances are potentially harmful to the humans and the livestock. They may cause muscular, cardiovascular, brain and kidney disorders (Ba, Pb and Cd), skin and liver damage (Cr) and a host of other health problems in humans. Therefore, removal of these heavy metals and trace inorganic elements is of great importance if it is found that the polluting industrial units are situated close to the drinking water source. However, for the subsidiary human activities such as servicing of automobiles and railway wagons, little higher concentrations of these pollutants will not cause any hazards. Use of proper water for appropriate activities being the right and a matter concerning health of the masses of population, there should be fool-proof arrangements in the society and the water supply departments to ensure that the recyclable treated water with slightly higher concentrations of these pollutants do not mix with the drinking water source.

Chapter 16

In this chapter on Seawater Desalination technology, we review the:

- Continents, Oceans, Seawater Properties and Seawater Composition.
- Thermal Desalting processes such as Multi-Stage Flash, Multiple Effect Distillation and Vapour Compression.
- Membranes and Membrane Separation processes like Reverse Osmosis and Electro-Dialysis.
- Seawater Desalination technology transfer for water resource augmentation.

SEAWATER DESALINATION

The municipal and industrial wastewater treatment is capable of cumulatively augmenting the water resource only to the extent of restricted quantities of recycled water. In addition, this recycled source of water can provide solely for the supportive activities such as miscellaneous human needs and a few specific industries where the quality of water is not an issue. The arid and semi-arid climatic parts of our country experience widespread and frequent failures of (or weak) monsoon, as has been observed in the first three continuous years of the twenty-first century, in the western, north-western and central parts of India. The resultant scanty rainfalls during the weak monsoons cause acute short supply of fresh water for drinking and domestic needs. Even in the coastal towns and cities of India, the quantity of the fresh water available in dry season drops so much that the water supply authorities of these cities and towns had to resort to restricted water supply. In recent years, the residents of all the metropolises, cities and towns of India face restricted water supply for extended parts of the year, almost from February to June. The residents of Mumbai and its neighbourhood, a city in the west coast that receives an annual rainfall of over five times above the national average rainfall, had experienced an official ten percent cut in water supply in the summer of 2003. The Mumbai scenario offers just a glimpse of the gravity of situation pertaining to the dwindling freshwater resource in our country. The residents of the country's capital, Delhi have to manage major parts of the days in summer without water due to heavily restricted water supply. The availability of potable water in the summer months in India's west coastal towns such as Kollam, Alapuzha, Kochi and

Ernakulam in Kerala, and in Mangalore in Karnataka is no better. The situation in the east coastal cities of Chennai, Nellore, Ongole and Visakhapatnam is even worse in this respect. In July 2003, it was observed that about forty days *after* the onset of the monsoon, several domestic wells in different parts of the Chennai city remained dry. If this is the situation in the coastal belts of the country where the monsoon rains are comparatively more than the interior rain-shadow zones of the terrain, the water resources in the latter tends to be precariously grim. Barring the terrains located close to the channel-fills, valley fills and paleo-channels where the groundwater occurs at shallow depths of about 35 metres, in the other locations consisting of sheet-rocks, soil-covered pediplain and pediment zones and areas where the natural drainage density (of river system, such as first and second order streams) is low in the watersheds, the depths of groundwater table have gone down to below 200 metres in the land-locked cities and towns in interior parts of peninsular India. This situation is strikingly similar in the case of several central African countries also.

In recent years, certain amount of earnings in a month in each urban family is set aside from the monthly budget for buying water for drinking and cooking purposes. It is paradoxical that major parts India, a peninsular country, face acute scarcity of fresh water though it has very long coastlines on either side, and a large terrain of the country is virtually surrounded on three sides by sea or ocean. This demonstrates our inability to tap one of the commonest resources that we are blessed with, namely the seawater after appropriate processing, to make our own lives a little more comfortable than now. This also demonstrates our dismal record of water resource management by tapping the dominant component of Earth, namely the seawater that comes totally free of cost. This is in spite of the fact that we possess the necessary wherewithal, such as the surplus funds flowing from financial sectors like banking and insurance, knowledge-workers in the form of mechanical and civil engineers, entrepreneurs who can mobilise the funds and the workforce, and the cheap labour force available in plenty in the country.

Whilst the population of the country continues to grow, the shortage of fresh water is bound to increase further, and therefore, large parts of India will have to face water-stressed condition if adequate measures are not taken at once to augment the water resources. As the country's water resources get scarcer and scarcer by every passing year, and concurrently the population continues to grow exponentially, we need to look at and resort to the *non-conventional* ways to augment our water resources and provide this fundamental necessity to all the people. Our water resource management policy has necessarily to address this aspect, and along with recycling of treated wastewater, we have to add more and more sources of usable water by generating it by non-conventional ways, in order to make our water resource management as holistic and complete.

Under the present circumstances as well as looking at the future trends in respect of the scenario of water resources in India, it is certain that one non-conventional way is to harvest the vast, inexhaustible reservoir of water in the oceans and seas. Since the salt concentrations in the seawater exceed the limits that humans or livestock can tolerate, the water of the seas must be purified before it can be used either as potable water or as source for the crops and the livestock. The methods of purification of the seawater are based on the technology pertaining to membranes or to mechanical engineering. India is neither short of technocrats qualified and vastly experienced in these two aspects of technology nor is under the crunch of funds that are required for the purpose. What India requires is a firm resolve to put the available expertise and experience into practice of seawater desalting technology. The following quote from Abdul Kalam (1998) is fully appropriate at this juncture: "*We are not alone in thinking of a vision for 2020. Many others…are striving hard to make a better future for themselves. One common feature is the greater emphasis on technological strengths in multiple areas such as manufacturing, advanced materials, electronics and information technologies. In addition, these countries share the concern for clean and environmentally friendly technologies*".

It is high time that India too had applied technologies and installed a chain of seawater desalination plants along both the east and the west coasts of the country in select locations proximate to the cities and towns that are chronically affected by freshwater scarcity. That would pave way for displaying the country's ability to keep pace with the technological advancements concurrently with the benefit of building adequate water resources. The desalted seawater can be used as a supplementary source to the national water resources and employed as a suitable substitute in an array of functions from oral consumption to personal hygiene, from food and diary industries to drugs manufacture, and for filling in the swimming pools to establishing ocean theme parks. As we can realise, that would mitigate the stress on the availability of pure and pristine rainwater source we get from the annual monsoons, which can be conserved and diverted for the purposes in which water quality is critical. Such steps would substantially mitigate the water-stressed situation in the central African and South American countries too, which face drought-like conditions frequently due to a variety of reasons earlier discussed including the effects of El Nino.

Before we take an overview of the seawater desalination technology, a recollection of information on the architecture of oceans, seas and seafloor *vis-à-vis* the continents that constitute the landmasses, as also the chemistry of the seawater, will indeed be beneficial in appreciating the values and merits of seawater desalination as a means to augment our water resources.

CONTINENTS, OCEANS AND SEAFLOOR

Of the 510 million square-kilometres area of the Earth's surface, just about 149 million square-kilometre, which works out to about 29.22 percent is land, and the rest 361 million square-kilometres of area, making roughly 70.78 percent, consist of seas and oceans. Along with lakes, rivers and other water bodies, these seas and oceans constitute the Earth's Hydrosphere and occupy about three-quarters of the Earth's surface. The surface of the Earth's crust is located at different levels in different places. Broadly, these surfaces can be recognised as two levels, namely the Continental Platform and the Oceanic or Deep-Sea Platform (Arthur Holmes, 1945). The part of the land that connects these two platforms is a gentle slope called as the *Continental Slope,* which actually is situated within the sea or ocean. The margin of the continental platform at its contact with the continental slope is also a slope, but it is much gentler than the continental slope, and is also submerged under the water of the ocean or the sea; this part is called as the *Continental Shelf.* It extends beyond the shoreline to an average depth of about 200 metres. The average depth of the sea is estimated as 3,800 metres. The actual ocean basins start at the edge of the continental shelf. The sea/ocean basins mostly overflow and, at times, the seawater inundates scores of million square-kilometres of the continental platform. Figure 16.1. illustrates the

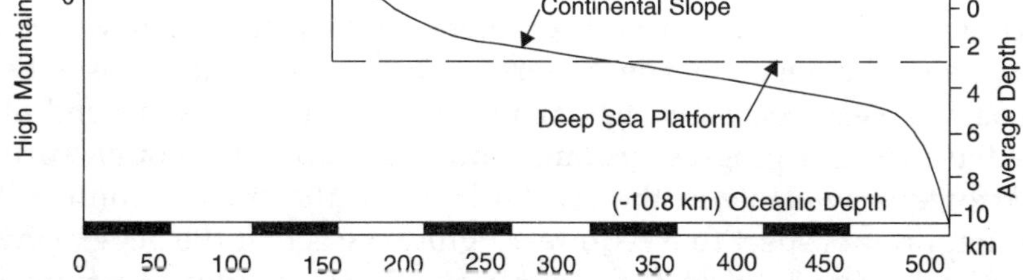

Figure 16.1. Conceptual diagram to show the highest peak on the Earth's surface to the deepest part of the ocean, as well as other components like continental and deep sea platforms, continental shelf and continental slope. (Redrawn from Holmes, 1945).

relation between the continental and oceanic platforms and also brings out the conceptual structure of the Earth's crust in continents and the oceans.

Some of the seas such as Baltic Sea and Hudson Bay lie on the continental shelf regions. It is estimated that if significant quantities of ice-sheets now covering the continents of Antarctica and North America were to melt away, the sea level would rise alarmingly, and the seawater would submerge considerable parts of the continents. The architecture of the continent—ocean interface as described may not apply to the land-locked seas such as the Mediterranean, Caspian, Aral and Dead Seas. Though the basins of these seas might consist of slopes of varying degrees, well-defined zones like continental shelf and continental slopes might be absent or not so distinctly developed in these seas. That however has not hindered the countries like Israel and a few others from Middle-East, which have resorted to desalting of the water from the Mediterranean Sea to develop their water resources. The positive impact on the economic growth and the standard of human life made by the desalting technique and the desalination plants is unimaginable.

Peninsular India has very long coastlines both on the eastern and western sides. Continental shelf is extensively developed all along the east coast and major parts of western coast, and it extends for significant but variable distances within the sea. This aspect, coupled with the sub-tropical offshore weathers has offered congenial conditions for the proliferation of a large variety of marine life. This marine wealth has supported intense fishing and seafood industries as an unorganised sector in the coastal belts of India. The favourable geographic setting, rich marine fauna and extremely friendly climate in the coastal areas have bestowed enormous employment opportunities to the less privileged uneducated sections of the Indian society in the fisheries and its ancillary industries such as manufacture of fishing nets, motor boats and so on. Fish harvest and the export of seafood have significantly contributed to the increase in the Gross Domestic Product of the country. Such positive aspects of intense human activity, however, are not without the negative side. Severe water scarcity in the coastal belts, especially the east coast area, has come as a bane in recent years and as an impediment to our progress through industrial growth, though surplus work force is available in the coastal belts. In the current context, the most important aspect that requires cognisance is that the geography of the coastal belts, both on the eastern and western sides of peninsular India, makes it extremely suitable for the installation of any number of mechanical plants that are capable of converting the saline seawater into usable grades of water with varying salt-contents, yet suitable for various human activities.

SEAWATER CHEMISTRY

The seawater is a solution of salts of roughly constant composition, dissolved in variable quantities of water. A total of ninety-two elements as compounds have been detected in the seawater, and they show a vast range of concentrations. These elements occur as electrically charged atoms or groups of atoms, known as the *ions*. Water is a *polar* substance, and therefore, it is an excellent solvent. In water (H_2O), oxygen and hydrogen share electrons in covalent bond; oxygen atom partially acquires negative charge and hydrogen atoms partial positive charge. The partial charges of water molecules facilitate the ionic compounds such as salts dissolve in water easily. Six ions or electrically charged atoms or groups of atoms constitute more than ninety-nine percent of all the dissolved salts. Seawater has more density than the fresh water because of these dissolved ions. These ions are made up of atoms that are denser than hydrogen and oxygen, the two components that make water. The six major constituents are as follows: Sodium (Na): 30.61 %; Magnesium (Mg): 3.69 %; Potassium (K): 1.1 %; Chlorine (Cl) as in chloride: 55.04 %; Sulphate (SO_4): 7.68 %; and Calcium (Ca): 1.16 %.

Though the seawater is a good solvent, there is a limit to how much of a salt it can hold. In the case of halite and sylvite (chlorides of sodium and potassium; NaCl and KCl respectively), this limit is high, but for the salts of iron and silver this limit is low. When the water dissolves as much of a salt or compound as it can hold, it is called 'saturated', with respect to that particular salt or compound. If more of that salt or compound is added to the water, it will not dissolve, but will just sink to the bottom. If some of the water is removed by freezing or evaporation, the dissolved ions will form solid crystals and sink to the bottom. This process is called as precipitation.

Salinity

The quantity of salts that are present in the seawater is expressed as *Total Dissolved Salts* (TDS). The oceanographers express the salt content in the seawater in terms of *Salinity*, which is the amount in grams, of total dissolved salts present in one kilogram or one litre of water. The salinity of seawater is a measure of the amount of dissolved salts. Fresh water has a salinity of near zero, while the oceans range in salinity from about 3.3 % to 3.7 %. That is, of every 100 grams of ocean water, about 3.5 grams of them are actually made up of dissolved ions, while 96.5 grams of them are water. In other words, the normal seawater has a salinity of 35 grams per kilogram or litre of water. The seawater from Wormly in southern England is regarded as the international standard for the seawater composition.

Principle of Constant Proportions: It is interesting to note that the composition of the seawater is governed by the *'Principle of Constant Proportions'*, which states that the compositions of the dissolved matter in the seawater do not vary much from place to place. In other words, the seawater and its dissolved ions show same ratios. For example, if there is seawater with salinity of 32 parts per thousand, it will have the same ratio of chlorine to sodium (Cl/Na = 55.0/30.6 = 1.8), as the seawater with a salinity of 38. The water with a salinity of 38 will have more of chlorine and sodium and the other ions, but they will still have the same relative proportions. This principle holds good spatially all over the globe, as the ions exist in the same ratios. This principle equally holds well in terms of temporal relation too since the ancient seawater had the same proportions of different dissolved salts in it as that of the modern seawater. This is because of the fact that the salts come from weathering of continents and deep-sea vents, and the inputs are extremely slow in the order of about 100,000 years, compared to the mixing rate of the ocean, which is of the order of 1000 years. It is, thus, possible to measure just one component of the dissolved material and then estimate the total amount of dissolved material, which gives the salinity.

The seawater contains small amounts of dissolved gases such as nitrogen, oxygen, carbon dioxide, hydrogen and rare inert gases. The surface seawater is usually saturated with atmospheric gases such as oxygen and nitrogen. The amount of gas that can dissolve in seawater is determined by the water's temperature and salinity. Increasing the temperature or salinity reduces the amount of gas that can be dissolved.

A number of streams and rivers carry freshwater and wastewater from continental landmass and empty these waters into the sea at the edge of continental shelf. As a result of this slight dilution of seawater by the river waters, the salinity of the seawater for some distance in the continental shelf tends to be slightly less than that of the seawater farther the coastline. This slight variation may be more pronounced during the rainy seasons when comparatively more continental water is emptied into the sea.

In addition to the above-mentioned six elements, some more elements are also present in the seawater in minor concentrations. These concentrations are expressed in terms of *parts per million* (ppm). If the water has a concentration of 10,000 ppm of dissolved salts, then one percent of the weight of the water comes from the salt solution. The normal seawater salinity of 35 grams per kilogram or litre of water can be otherwise expressed as 35,000 *ppm* of TDS. The fresh water for human consumption needs to have less than 1,000 *ppm* of TDS. The other elements that are present in the seawater in traces but in significant quantities, besides the six elements quoted above, are listed in the accompanying Table.

Elements	*Amount (ppm)*	*Elements*	*Amount (ppm)*
Hydrogen, H (as in H_2O)	110,000	Lithium, Li	0.170
Oxygen, O (as in H_2O)	883,000	Carbon, C	28.0
Sodium, Na (as in NaCl)	10,800	Nitrogen, N (as ion)	15.5
Chlorine,Cl (as in NaCl)	19,400	Fluorine, F	13.0
Magnesium, Mg	1,290	Silicon, Si	2.9
Sulphur, S	904	Phosphorus, P	0.088
Potassium, K	392	Argon, Ar	0.450
Calcium, Ca	411	Rubidium, Rb	0.120
Bromine, Br	67.3	Strontium, Sr	8.1
Boron, B	4.45	Iodine, I	0.064

ppm = milligram/litre or 0.001 gram/litre. Source: Karl K. Turekian, Oceans, 1968, Prentice-Hall.

Besides these elements, about fifty other elements are present in the seawater in traces or less than 0.01 ppm or, at places, in *parts per billion*. The quantities of various trace elements control the biochemical reactions in the ocean waters. The biochemical reactions are responsible for the proliferation of different types of chlorophylls and phytoplanktons, which, in turn, control the concentrations of marine life, especially that of the fish. The concentrations of these trace elements frequently fluctuate due to the variations in local environmental conditions; further, human interference such as oil spill from the marine vessels and the act of dumping of different types of wastes (including the nuclear wastes) in the oceans also alter the concentrations of the trace elements. The higher concentrations of the harmful among these trace elements frequently become the focal point of debate on the quality and population of marine life, ecology and toxicity of the ocean waters, and ocean environment.

The processes by which the salts find their way into the seawater are extremely complex and hinge on several factors such as chemical reactions, climates on the landmass, temperatures of the seawater, convection system that controls the circulation of the seawater from depths to the surface, types of rocks on the landmass and the ocean floor, and importantly, the human interference and influence on the natural processes that take place both on the land and the oceans. However, in simple terms, the salts that are dissolved in the seawater can be broadly attributed to three main sources. They are: 1.Volcanic eruptions. 2. Chemical reactions between the seawater and the hot newly formed volcanic rocks of spreading zones (mid-oceanic ridges). 3. Chemical weathering of the rocks on the continents.

1. *Volcanic eruptions* generate large volumes of gases that eventually reach the oceans. The most important of these gases form sulphates and chlorides by reactions with various metals. Submarine eruptions at lava spreading ridges inject gases directly into the oceans. Gases from sub-aerial volcanoes are dissolved in rainwater and subsequently find their way through rivers into the seas and ocean.

2. *Chemical reactions* between the hot seawater and the oceanic crust of basaltic composition lead to removal of magnesium and sulphate from the seawater. The chemical reactions are also responsible for the transfer of other elements like lithium and rubidium from the rocks to the seawater. The seawater undergoes continuous circulation and reacts with the oceanic crustal rocks, which results in exchange of several elements from the rocks to the seawater and *vice versa*.

3. *Chemical weathering* of the rock formations on the land gives rise to many salts. Physical weathering and decay of the rocks on land accentuate the chemical weathering. These weathering phenomena are facilitated by the climatic conditions prevailing on the landmass. As the rocks are weathered to form soils, they release soluble constituents such as silicates and elements like sodium, potassium and magnesium. Through the natural drainage system of the landmass such as rivers and their tributaries, these soluble constituents and elements join the seas. The river waters also carry into the sea the bicarbonates (HCO_3), the by-products generated by weathering of calcium carbonate-rich rocks such as limestone, marble and calc-silicate rocks. Once these constituents enter the seawater, various salts are formed by an array of chemical reactions. These salts remain dissolved in the seawater, while the seawater alone without the dissolved salts continues to take part by evaporation in the Hydrological Cycle (see Figure 9.1).

Most reactions occur between waters that are entrapped in the earth's crust and the minerals that are present in the rocks. For example, plagioclase feldspar is a common rock-forming mineral, which is a silicate of sodium, calcium and aluminium; on chemical reaction with carbon dioxide in the presence of water, plagioclase feldspar is weathered into clay minerals (hydroxy alumino-silicates) with the release of sodium and calcium ions, along with bicarbonate radical (HCO_3^-) and ortho-silicic acid (H_4SiO_4). For example, the albite, a sodic plagioclase feldspar is chemically altered as per the following chemical reaction:

$$2NaAlSi_3O_8 + 2CO_2 + 1H_2O \rightarrow Al_2Si_2O_5(OH)_4 + 2HCO_3^- + 4H_4SiO_4$$

(albite) (hydroxy alumino-dilicate)

Sodium and chloride are removed from the seawater when some of the seawater evaporates to form the evaporite deposit of the common salt (sodium chloride, NaCl). Deposits of potash (potassium salts such as potassium carbonate, potassium hydroxide, potassium chloride and so on) also form in similar way by isolation of potassium from the seawater.

Limestone, dolomite, marbles and other rocks containing carbonates of calcium and magnesium get completely or partly dissolved by constant exposure to mildly acidic water, by the following chemical reactions:

$$CaCO_3 + H_2O + CO_2 \rightarrow Ca^{2+} + 2HCO_3^-$$

$$MgCO_3 + H_2O + CO_2 \rightarrow Mg^{2+} + 2HCO_3^-$$

This also explains as to why the terrain or land comprising limestone, dolomite and marble become cavernous, and full of potholes and deep cavities on surface.

When the salts and the evaporite deposits on the earth surface are unlifted by tectonic means, they are again subjected to solution, reactions and chemical weathering phenomena, and the weathered and dissolved salts are eventually recycled into the sea again.

Density

The *mass* of unit *volume* is known as density. The density is expressed in *Syst'eme Internationale d'Unit'es* (SI units), an internationally agreed coherent system of units, as kilograms per cubic-metre. In *c.g.s* units, it is given in terms of grams per cubic-centimetre (*cc*); in these units, density is numerically equal to the *specific gravity*. The density of pure water is one gram per cubic centimetre, and therefore, the specific gravity of pure water is numerically 1. The seawater is denser than the fresh water because of the ions dissolved in the seawater are made up of atoms that are denser than hydrogen and oxygen, the components of water. The seawater density depends on the temperature, salinity and pressure. Colder and saltier seawaters are denser. High pressure increases the density. It is estimated that the densities of seawater in SI units are in the range of 1020 to 1050 kilograms per cubic-metre, with most of this range being due to the hydrostatic pressure. The range of densities at the sea surface varies between 1020 and 1030 kilogram per cubic-metre.

Temperature

The fresh water freezes at 0°C and boils at 100°C. But, the dissolved ions in seawater make it hard to freeze, since the ions interfere with the water molecules at the time of their transformation into ice crystals. Therefore, the freezing point of the seawater is considerably lower than that of the fresh water, and it depends upon the ionic constituents and their proportion in the seawater. Similarly, the dissolved ions raise the boiling point of the seawater to significantly greater than 100°C. Interestingly, while freezing as well as evaporation of the seawater, the ions stay in the liquid water, because both ice and water vapour are neither solvents nor can they hold the ions. Hence, when the water freezes at the poles of the Earth (that is Arctic and Antarctic ice sheets), the ice is pure, but when these ice sheets melt and the melt-waters join the oceans, they regain the salinity.

Conductivity

The conductivity of the seawater depends mainly on the temperature, to a less extent on salinity and just marginally on pressure. If the temperature of the seawater is known, the conductivity can be used to determine salinity. The salinity as computed through conductivity is closely related to the actual dissolved constituents, and more independent of salt composition. Therefore, the temperature has to be measured at the same time as conductivity to remove the temperature effect and obtain salinity. The accuracy of salinity determined from conductivity is 0.001 to 0.004. Again, the accuracy depends on the accuracy of the seawater standard used to calibrate the conductivity-based measurement.

In laboratory measurement of conductivity of the seawater sample, an instrument called as the 'auto-salinometer' is used. It gives the ratio of conductivity of the seawater sample to a standard solution. The standard seawater solutions are either seawater from a particular place, or a standard potassium-chloride (KCl) solution made in the laboratory. The latter provides greater accuracy and has recently become the standard. Since the conductivity strongly depends on the temperature, the measurements have to be carried out in temperature-controlled conditions. Saline water having a conductivity of less than 2250 microSiemens/centimetre is suitable for irrigation purposes.

SEAWATER DESALINATION

Desalination is a technology by which alternative source of pure water is prepared by mechanical means, and this pure water is capable of augmenting the potable-grade water source of genuinely aesthetic nature. Desalination provides a means of upgrading poor quality saline waters, and it refers to a suite of technologies designed to produce fresh water from seawater or brackish groundwater. Seawater Desalination is the process by which the salt-contents or the 'total dissolved salts' in the seawater are removed in appropriate quantities in order to effect purification of the seawater in such a way that the purified water can be effectively put to use in a variety of applications. The quantities of the dissolved salts for removal from the seawater depend specifically on the level of purity and the purpose for which the water is needed. Since the desalination process basically involves the elimination of the salt from the seawater, this process is frequently referred to as the '*desalting*' of the saline water. The type of process to be employed in the seawater desalination industry is also determined by the degree of purity of the water required. Water desalination technologies can be broadly categorized on the basis of the energy or power system used to run them. Accordingly, there are mainly two types of energy system, namely (1) Thermal and (2) Electric Systems, which are globally employed to operate the desalination plants. Furthermore, there are a few minor desalination technologies that employ solar energy or a combination of the thermal and electric energies. Though such technologies have distinct advantages, the disadvantages weigh more in them, and therefore, they have not become popular. On the basis of these two energy systems, the desalination plants also can be grouped into two types, namely:

(*i*) *Thermal Desalting,*

(*ii*) *Membrane Separation.*

The technologies utilizing thermal energy are dominantly three, and they are known as:

1. *Multi-Stage Flash* (MSF).
2. *Multiple-Effect Distillation* (MED).
3. *Vapour Compression* (VC).

On the other hand, the electric energy-based desalination technologies function on membrane system, and there are mainly two types of these:

1. *Reverse Osmosis* (RO).
2. *Electro-Dialysis* (ED).

The major factors that determine as to which type of treatment is best suited to a particular application include the levels of salinity, temporary hardness, the presence of colloidal suspended matter, dissolved metal ions, oxidizing agents, hydrogen-sulphide and the temperature of the input water.

Seawater desalination technologies are in vogue in the Arab countries of West Asia for the last half-a-century. The desalination using membrane technology began in the 1940s and continues with several hundred desalination plants in operation throughout in this region. These plants range in capacity from a few cubic-metres to over 10000 cubic-metres per day. These countries put together account for about sixty percent of the desalinated water produced in the world. A small country in this region, Syria alone produces over two million cubic-metres of desalinated water. The largest plant is in Baghdad, Iraq, which desalinates 24000 cubic-metres of water per day. If one goes by the number of different desalination plants installed in these countries and the cumulative capacity of these plants, it can be judged that the Multi-Stage Flash types are the most popular, followed by Reverse Osmosis types. The solar or combination energy types are too rare in West Asia. The choice of the category and type of the desalination to be installed depends upon various factors the most important of which are the economic status, type of energy that can be made available to the plant continually, and the quantity of the output required.

I. THERMAL DESALTING

The Thermal Desalting technology functions under the scientific principle of *'Distillation'* and for this reason, the thermal desalting processes are frequently referred to also as *'Distillation-Desalination'*. This is the oldest and most commonly used method of desalination of the water. The simplest way to desalt the seawater is to boil it in a way to effect evaporation into steam. The steam gets converted to pure water as *'distillate'* when condensed. All the thermal desalting technologies apply this principle in their plants. The difference between the plants, however, lies in the process by which the distillate is obtained from the steam or vapours. The world's first distillation desalination plant was installed at Curacao, The Netherlands Antilles in 1928; this plant had a capacity to produce 60 cubic-metres of potable-grade water per day. The ship-board distillation plants were designed and developed during the World Wars, to produce potable-grade water from the seawater for the consumption of the warring naval personnel on board the vessels. Commercial development of the land-based seawater distillation plants has taken place in the late 1950s based on the technology developed for the industrial evaporators such as the sugar concentrators in the sugar mills.

The apparatus or the part of the plants that contains the input seawater in the thermal desalination or in the distillation plants is referred to as a *'still'*. In a simple still, seawater can be boiled continually releasing steam, which on condensation, forms and collected as pure water. Many stills can be connected together making the process more efficient. In order to achieve this, each still needs to be operated at different pressures. At the sea level, the seawater boils in the range of 100—120°C (or 212—249°F) depending on the types and proportions of the ions present in it. However, in vacuum, the seawater can boil at much lower temperatures. Therefore, a series of vacuum chamber stills to boil the seawater can be connected together, making the process to yield significantly large quantities of steam and subsequently pure water distillates. This explains, in short, the principle of the thermal desalting or distillation-desalination technology. The type of plants to achieve this effect is variable depending upon the technique employed in the distillation process. Broadly, the three forms of distillation in the thermal desalination plants, namely Multi-Stage Flash, Multiple Effect Distillation and Vapour Compression show subtle variations in the way they produce the steam. Let us now take an overview of these technologies.

Multi-Stage Flash Distillation

The plants of Multi-Stage Flash (universally known in abbreviated form as 'MSF') distillation technology is comparatively younger than the Multiple Effect Distillation technology, but the former has become more widespread on account of its comparatively better efficiency and also it is more economical than the Multiple Effect Distillation technology. The plants employing the Multi-Stage Flash technology are the most dominant in the West Asian countries like Saudi Arabia, United Arab Emirates, Kuwait, Qatar and Oman; a bulk of desalted water produced in these countries comes from the Multi-Stage Flash technology. The Multi-Stage Flash operates under the principle that if the raw seawater is heated to its boiling point and held under a reduced pressure until it is released into a vacuum chamber, the water 'flashes' (almost like exploding) into water vapours in the form of dense steam. The major difference between boiling and flashing is that the boiling is carried out under a constant input of heat as in the case of cooking on a stove; on the other hand, flashing involves heating the water to its boiling point, shortly holding it under pressure and subsequently introducing a vacuum. The process of distilling the seawater with a vessel operating at a lower pressure has been in vogue since the 1950s when units, which used a series of stages set at sequentially lower atmospheric pressures were designed and developed. In these units, the raw feed water could pass from one stage to another, and boiled repeatedly without additional heat.

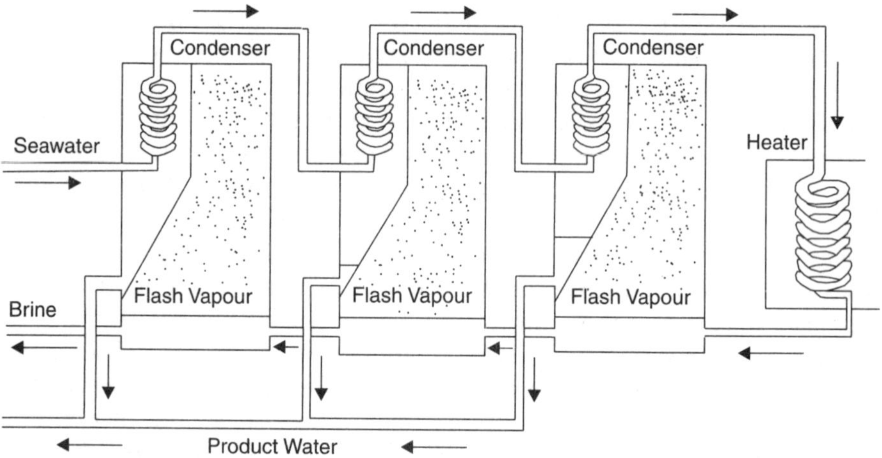

Figure 16.2. Multistage Flash Distillation.
(Redrawn from: http://www.e-monitoring.com/bws/deslination Technology/multiStage Flash Distillation.htm)

The flow diagram of a conceptual Multi-Stage Flash Distillation plant is shown as Figure 16.2. In the Multi-Stage Flash process, the seawater is heated in an apparatus called as the *brine heater*. The heating is achieved by condensing the super-hot steam into a bank of tubes that passes through the apparatus, which in turn heats the seawater. The heated seawater subsequently flows into another vessel or apparatus called *'evaporator'* or *'Stage'* in which the ambient pressure is adjusted in such a way that the entering water would immediately boil. Connecting multiple *stages* at successively lower pressures is the fundamental principle behind the Multi-Stage Flash. The sudden introduction of the heated water into the chamber causes it to boil so vigorously and rapidly that it almost explodes or flashes into steam. Usually, at a time, only a small percentage of the raw water is converted into steam or water vapour, and this depends on the pressure maintained in the evaporators. This is because, the flashing would continue only till the water becomes cooler to the boiling point. Depending upon the quantity of the output desalted water required, the Multi-Stage Flash distillation plants can any number of stages, as many as forty.

A high-pressure steam-driven ejector or condenser vacuum system provided in the plant creates vacuum in the stages. The distillation process operates from a low vacuum in the first stage to a high vacuum in the last stage, with stage-to-stage pressure differential being the key to the repeated flashing. The spent high pressure steam and the non-condensable components are released from the ejectors into the condensers, where the steam is condensed and the non-condensable substances are vented to atmosphere.

In the next step, the 'flashed' water vapours or steam are drawn to the cooler tube bundle surfaces where they are condensed and collected as pure, salt-free water as *distillate*. In this process, the steam or water vapours are transformed into fresh water by the process of condensation, in tubes of *heat exchangers*, which pass through each of the stages or evaporators. The tubes are cooled by the input feed water that flows into the brine heater. The loss of heat by the tubes is directly gained by the input feed-water, because in an exchange of heat, the hot tubes transfer their heat to the input feed-water, which gets warmed up in such a way that the thermal energy required in the brine heater to raise the temperature of the raw seawater input is considerably reduced.

The Multi-Stage Flash distillation plants function efficiently at temperatures in high ranges of 110—120°C. At this temperature range, however, the plant tends to suffer undesired effects such as deposition of the scales in the brine heaters, stages and heat-transfer surfaces, as also corrosion of the metal surfaces of the plant by the input seawater. The problem of scaling can be overcome by appropriate chemical treatment systems, of which the acid-treated brine re-circulation is capable of permitting the plant operation at higher evaporation temperatures, but at the same time reducing the amount of feed water chemical treatment. Further, the thermal efficiency of the distillation plant is affected by the vast difference in the temperature ranges between the brine heater on one side, and the condenser in the cold end of the distillation plant on the other. This setback can, however, be overcome by maintaining the operative temperature of the plant at constant levels of about 120°C.

In the plant, the filtered raw seawater is pumped through the condenser tubes of the heat rejection stages and is discharged. The cool feed water that flows through the condenser tubes dissipates the heat in the last stages. A portion of raw seawater is withdrawn as make-up water and the reminder is discharged to the sea. The make-up water is de-aerated for removal of the non-condensable gases, and is treated with sulphuric acid for scale-control purposes. The de-aerated feed water is introduced into the brine section of the last stage, and mixes with the re-circulating brine flow. The mixture of the re-circulated brine and de-aerated feed water is pumped through the condenser tubes of the heat recovery stages, and the brine heater by the brine re-circulating pump. In order to maintain a material balance in the distillation system, and control the maximum brine concentration, a portion of the concentrated brine from the last stage is discharged to the ocean as blow-down. The blow-down flow rate is controlled by the brine level in the last stage. As the brine flows through the condenser tubes of the heat recovery stages, it is progressively heated in each stage by flashing brine vapour condensing on the outer tube surfaces. The brine then passes through the tubes of the brine heater for final heating up to the maximum temperature.

The distillate formed is collected in a distillate trough, which is common to all the stages and the distillate flows from the first stage to the last stage. From the last stage, the distillate is pumped to the distillate tanks.

A known amount of sulphuric acid is continuously injected into the make-up flow to neutralize the scale-forming components in the seawater, thus preventing the formation of scales in the stage and the brine heater condensing tubes. A known quantity of anti-foam agent can be injected into the feed water stream to reduce the foaming characteristic in the evaporator.

Multiple Effect Distillation

The Multiple Effect Distillation (MED) process is the earliest form of desalting of the saline water, and it has been in vogue in the industrial distillation as well as seawater desalination for over a century. The most popular uses for this process is the evaporation of sugar cane juice in the production of sugar, and the manufacture of salt using evaporative process. The plants in the Multiple Effect Distillation derive the name from the principle that a series of boiling chambers is used to produce the distilled water. Though the Multiple Effect Distillation is an older technology than the Multi-Stage Flash technology, it has not been extensively employed for the production of pure water. On account of better and higher efficiency, and the cost factor, the Multi-Stage Flash plants have substituted the Multiple Effect Distillation plants in many places globally.

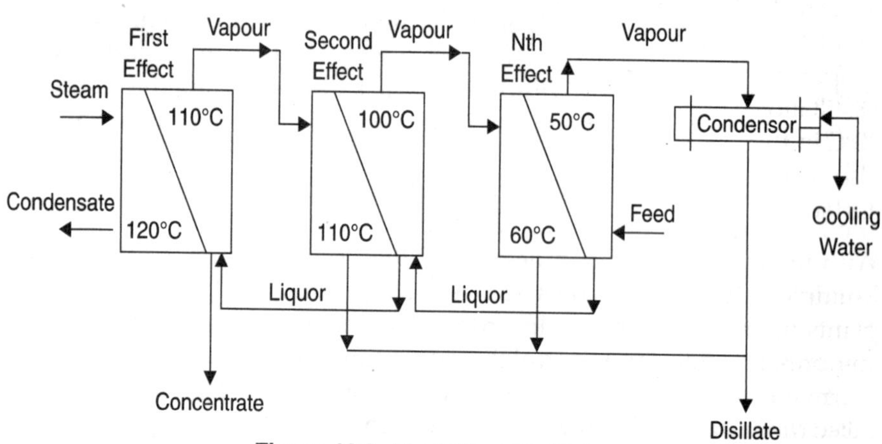

Figure 16.3. Multi-Effect Distillation.
(Redrawn from: http://cape.uwaterloo.ca/che100projects/sea/med.html)

A schematic diagram illustrating the concept of Multiple Effect Distillation plant is shown as Figure 16.3. The Multiple Effect Distillation takes place in a series of apparatuses or what are known in this technology as the *'effects'*, and uses the principle of reducing the ambient pressure on different effects. This aspect facilitates the seawater input to be subjected to multiple-level boiling without providing supplementary heat after the first effect. The energy used for evaporation is the heat of condensation of the steam. In these plants, the steam or water vapours are condensed on one side of a tube, while the saline raw water is evaporated on the other side.

In the Multiple Effect Distillation plants, the seawater is pre-heated in tubes and then let in the first effect, where the temperature is raised to boiling point. Then the seawater is either sprayed or distributed on the surface of evaporator tubes as thin films to promote rapid boiling and evaporation. The steam emerging out from a boiler heats the tubes; in turn, the steam is condensed on the opposite sides of the tubes. The condensate from the boiler steam is recycled to the boiler for reuse.

Only a small portion of the seawater applied to the tubes in the first effect is evaporated. The remaining raw feed-water is fed to the second effect where it is again applied to a tube bundle. The steam or water vapours generated in the first effect heat these tubes. This vapour is condensed to fresh water while cooling, to evaporate a portion of the remaining seawater feed into the next effect. This process and cycle continues for several effects in large plants, which have up to sixteen effects.

The remaining seawater in each effect has to be pumped to the next effect so as to apply it to the next bundle of tubes. Further condensation takes place in each effect on tubes that bring the feed-water from its source through the plant to the first effect. This exercise heats the feed-water before it is evaporated in the first effect. Some of the recent plants of Multiple Effect Distillation are designed to function at a maximum temperature of about 70°C in the first effect. This reduces the potential for scaling in the plant, as also increases the need for more area for heat transfer in the form of tube surfaces.

With the introduction of compression technology, the performance of the Multiple Effect Distillation plants has increased to a large extent and the plants have shown improvement both in performance and economy. The compressors can be either mechanical compressors driven by electricity, or thermal compressors utilising motive stream. In the former, the units are called Mechanical Vapour Compressors and in the latter, the plants are referred to as the Thermo-Compression Distillers. In the last fifteen years or so, the interest in the Multiple Effect Distillation plants has undergone revival since the concept of operating the Multiple Effect Distillation plants at lower temperatures (and therefore at lower cost) is evolved.

In recent times, new type of plants operating at low-temperature by the horizontal-tube Multiple Effect Distillation process has been successfully developed and deployed in the Caribbean Islands. These plants are very strong, easy to operate as also economical, as they can be made of aluminium or other low-cost metals. It appears that similar plants may be suitable in Indian conditions too.

Innovations in the Multiple Effect Distillation plants to suit the manufacturers' needs, and depending upon the types of energy locally available are possible, in view of the latest technologies such as Thermo-Compression Distillers and Mechanical Vapour Compressors. The knowledge-workers in India such as mechanical, civil and electrical engineers and entrepreneurs need to make innovations in various parts of the Multiple Effect Distillation plants, as also employ these technologies, so that large quantities of seawater can be desalted in shorter times at much less cost, and the water produced thus be accreted to our water resources.

Vapour Compression Distillation

The technology of Vapour Compression Distillation, also known as Compression Distillation, is used for small to medium scale seawater desalination plants, which can produce potable-grade water in the range of 6000-12000 litres capacity. This process uses its own steam after it has been compressed as its primary heat source. The heat required for evaporating the water comes from the compression of vapour instead of the direct exchange of heat from the steam produced in a boiler. The Vapour Compression Distillation plants function under the principle of reducing the boiling point temperature by reducing the pressure. In these plants, one of the two methods is used to condense the vapour so as to produce enough heat to evaporate the incoming seawater—either by a *mechanical compressor* type or a *steam jet*.

The mechanical compressor runs on electricity. Each unit of the mechanical compressor distillation is a compact plant and occupies very limited space. The plant does not require an external heating source as in the other plants; this facilitates the sole use of electrical power in this plant to produce water by distillation. However, their maintenance cost is said to be little higher than the other distillation systems. The mechanical compressor distillation system operates at low temperature ranges of about 60—70°C. This is a great advantage for this plant because, at those low temperatures much below the boiling point of the seawater, the nuisance of scaling and corrosion

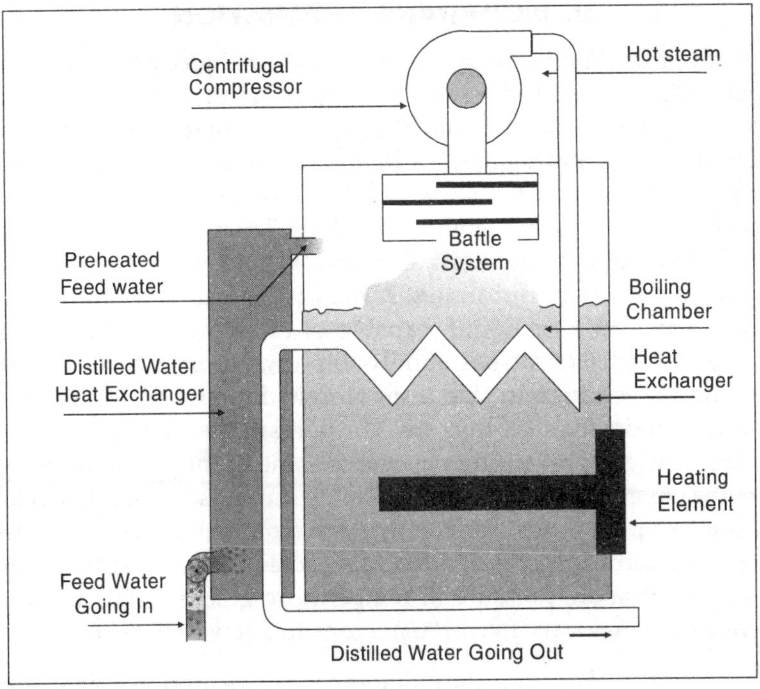

Figure 16.4. Mechanical Vapour Compression. (Redrawn from: http://norlandintl.com/multiple_effect_vapor_compression_distillation.asp)

are averted. The Vapour Compression Distillation units are built in a variety of configuration to promote the exchange of heat to evaporate the seawater.

A compressor produces vacuum in the chamber and then compresses the water vapour taken from the chamber and condenses it in the interior of tube bundles in the same chamber. The seawater is sprayed on the exterior of the hot tube bundles where it boils and partially evaporates, generating additional pure water. The principle by which the Mechanical Vapour Compression plant functions is diagrammatically shown in Figure 16.4.

A modified type of Vapour Compression Distillation plant is the Steam Jet-type Vapour Compression plant, or otherwise called as Thermo-compressor. In this plant, an attachment in the plant called as 'venturi orifice' provided at the steam jet creates and extracts the water vapour from the main chamber of the plant and creates a lower ambient pressure in it. The steam jet compresses the water vapour that is extracted from the main chamber. The extracted water vapour as well as the steam are condensed on the tube bundle walls, which incidentally provides the requisite heat also to evaporate the seawater that is let in from the other side of the tube walls in the plant's chamber.

II. MEMBRANE SEPARATION

The advent of the 'ion exchange membranes' has indeed been an event in the technologies related to all the types of membrane separation technologies. Membranes are thin films of porous material, which basically function as barriers that separate or divide two phases of liquid and restrict the transport of various chemicals, ions and possible micro-organisms in a selective manner from one side to the other. Though the membranes are mostly made up of polymer films, they can be made from ceramics, carbon fibres and porous metal substrates. A membrane can be homogeneous to heterogeneous, solid or liquid, capable of allowing anionic or cationic substances through it or can be neutral or bipolar. The membranes are always porous, the pores range in size from atomic dimensions (of less than 10 angstroms) to more than 100 microns. The pores of the membranes can serve as physical barriers, preventing passage of some of the materials present in fluid media, such as salt, colloidal substances and bacteria, while allowing free passage of pure fluids. Passing through a membrane can take place by convection or diffusion of individual molecules, or induced by an electric field or concentration, pressure or temperature gradient. The thickness of the membranes may vary from 100 microns to a few millimetres.

Semi-permeable and ion-specific membranes can be used to isolate, separate and purify a variety of fluids, including water, and such process is known as the 'Membrane Separation'. In this process, a semi-permeable membrane separates an influent stream that contains undesired chemicals and ions, into two effluent streams of fluid, known as the '*permeate*' and the '*concentrate*'. The fluid that passes through the membrane is permeate. The stream of fluid that contains the constituents that have been rejected by the membrane is the concentrate.

In contrast to the earlier discussed thermal desalting technologies in which distillation process plays the pivotal role, in the membrane separation technologies, the function of ion separation from the raw seawater is the key process involved in the desalination technology. In the past, distillation was the only way to isolate and eliminate the salt from the seawater. In current times, membrane technologies have become not only alternative to the distillation technologies, but they have become less expensive too. Membranes can now remove the salt from the seawater with a fraction of the energy consumption used in the distillation technologies. Furthermore, the membrane technology plants operate without much technical complexities allowing flexibility in the plant design, and the results are produced without interfering with, or causing hazard to, the environment. The end products are of much better in quality than that of those of the distillation desalting processes. The only disadvantage of the separation by membrane technology is the cost factor, as the product generated by the membrane separation is more expensive than the distillation processes; but on the long run, the industrial researches in membrane technology are capable of producing cost-effective and less expensive membranes. Improvisation in the membrane technology depends on the introduction of better quality membranes. Under such circumstances, the quantities of permeate would grossly go high, and the cost of energy required to produce pure water would significantly decrease. That would mean lowering of the operating cost of the plant, and therefore, the consumers would get the pure water at lower cost.

Types of Membrane

The membrane separation technology is useful in a wide variety of applications like, for example, effluent treatment, manufacture of drug and pharmaceutical products, potable water production and desalination of water. The specific aim of the membrane technology is to isolate particulate or dissolved solids from fluid media, remove specific substances from the fluids, reduce hardness and obtain purest form of water free from bacteria, micro-organisms and dissolved salts. There are various types of membranes meant specifically for each purpose. Basically, the end-use determines the membrane types that are to be used in various applications. Let us now take a review of the different types of membranes.

(*i*) **Micro-porous Membranes:** The pores in these membrane types vary in size from one nano-metre (1 nm) to 20 microns. These membranes are like fibre-filter, as they isolate the undesired materials by sieving operation that is determined by the pore diameter or particle size. Materials such as ceramics, graphite, metal oxides and polymers are used in making the micro-porous membranes.

(*ii*) **Homogeneous Membranes**: These are dense film through which a mixture of molecules is transported by pressure, concentration or electrical potential gradient. With the aid of these membranes, chemical substances of similar size and diffusivity can be separated if their concentrations differ significantly.

(*iii*) **Asymmetric Membranes**: These types of membranes are 0.1 to 1.0 micron thin skin layer on a highly porous 100 to 200 microns-thick substructure. The skin layer acts as a selective membrane. The nature of membrane material or pore-size of the membrane determines the separation characteristics. Likewise, thickness of the skin determines the rate of mass transport of the material. The porous sub-layer functions as a support for the thin fragile skin and it has little effect on the separation characteristics.

(*iv*) **Electrically-Charged Membranes:** They are ion-exchange membranes consisting of highly swollen gels carrying specific positive or negative charges. The electrically charged membranes are used mainly in the electro-dialysis.

(*v*) **Liquid Membranes**: A liquid membrane utilizes a carrier to selectively transport components such as metals ions relatively high rate across the membrane interface.

PROPERTIES AND MODULES OF THE MEMBRANES

The membranes in the industrial technologies can be cast as flat sheets, tubes and extremely fine hollow fibres. Membrane manufacture modules to accommodate such shapes, structures and any dimensions are currently available. The recent times have witnessed the innovation of spiral membranes, which are capable of functioning at high pressures of over 50 bars. The technical and economic factors for the selection, design and operation of the membrane modules hinge on the parameters such as supporting materials, pressure chambers in the separation plants, power consumption in pumping operations and the ease with which the membranes can be replaced in the plants.

The following membrane modules are largely employed in the industrial applications: 1. Plate-and-Frame Membrane Module; 2. Spiral-Wound Membrane Module; 3. Tubular Membrane Module; 4. Capillary Membrane Module; and 5. Hollow Fibre Membrane.

II. MEMBRANE SEPARATION PROCESSES IN SEAWATER DESALINATION

Membrane Separation technology finds utility in a wide variety of industrial applications. A discussion on all these applications would, however, be out of scope in the present context. The application of the membrane separation technology, which is capable of touching the day-to-day human life at every moment and of great significance to the modern civil life at comfortable levels, is in the Seawater Desalination processes. In short, in the seawater desalination processes using membrane technology, the membranes that would allow the passage of water excluding the salts and other contaminants are put to use. Pure water of high quality and potable grade can be produced from seawater as permeate, and the salt and other components of the input seawater are rejected as brine waste. There are several membrane separation processes for various specific industrial applications, but the membrane technologies for seawater desalination are mainly of two types. They are:

1. *Reverse Osmosis.*
2. *Electro-Dialysis.*

1. Reverse Osmosis

Reverse Osmosis is essentially a pressure-driven process for separating salt molecules and other solutes from water or any other aqueous fluid streams. This process is capable of removing particles that are as minute as 0.1 nanometre, such as bacteria and viruses. Reverse Osmosis process is used in a wide range of applications including removal of salts from salty, brackish and sea water (desalination), wastewater treatment and obtaining potable grade water and ultra-pure water for special industries like pharmaceuticals, dairy, beverages and food-processing. Unlike water filtration processes, which can remove only the suspended matter larger than one-micron size, the reverse osmosis can eliminate the dissolved solids, germs, bacteria, micro-organisms and viruses in the raw water that requires up-gradation to potable standards.

Normal Osmosis: In normal osmosis, *a solvent moves through a semi-permeable membrane into a solution of higher solute concentration and tends to equalize the concentrations of the solute on the either side of the membrane.* This is explained in Figures 16.5.a. & 16.5.b. In Figure 16.5.a., the mouth of a tube containing some amount of salt solution is closed with the aid of a semi-permeable membrane and the mouth is submerged in a beaker containing ordinary water. Initially, the level of the salt solution and the water are equal. Over a period of time, the water in the tube rises due to the passage of ordinary

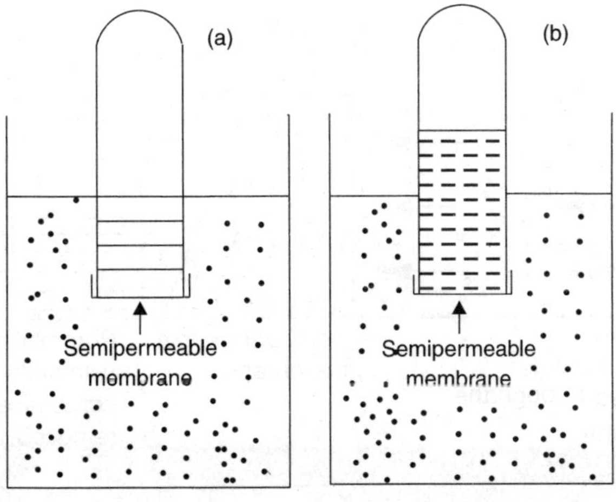

Figure 16.5.*a* & b. Normal Osmosis.

water into it (Figure 16.5.b.). It can be seen that the membrane allows passage of only water molecules, but not the salt molecules. The passage of the water molecules is attributed to the Osmotic Pressure, which occurs due to the *Brownian Movement*. (Erratic random movement made by the microscopic particles in a disperse phase is called as the Brownian Movement.) In the case of osmosis, the water molecules on both sides of the membrane are in constant and random motion. The pure water, which has less specific gravity than the salt water, flows over to the salt solution side because it tries constantly to dilute the salt solution. On the salty side of the semi-permeable membrane, some pores of the membrane get plugged with the salt atoms. But, on the pure-water side, the pores are open, and therefore, the pure water molecules pass from this side to the side of salt solution. This continues till an equilibrium state is reached. The height of the column of salty water rises as the water pressure rises, until it is equal to the osmotic pressure. At such point, the osmosis would stop.

Reverse Osmosis: In Reverse Osmosis for saltwater or seawater desalination, however, the objective is not to dilute the salty water, but to purify it by removing the salts. Applying a pressure to the side containing the saltwater or seawater does this. The applied pressure causes the water component alone of the seawater to flow to the other side of the membrane. Since the membrane prevents the salt molecules of the seawater from passing through it, the seawater gets purified and accumulated on the other side of the membrane. This sums up the process of reverse osmosis, which is achieved by applying very high pressure to seawater column against a membrane to counteract the osmotic flow. Schematic diagram of a reverse osmosis concept is shown as Figure 16.6. The notable features of the reverse osmosis process are:

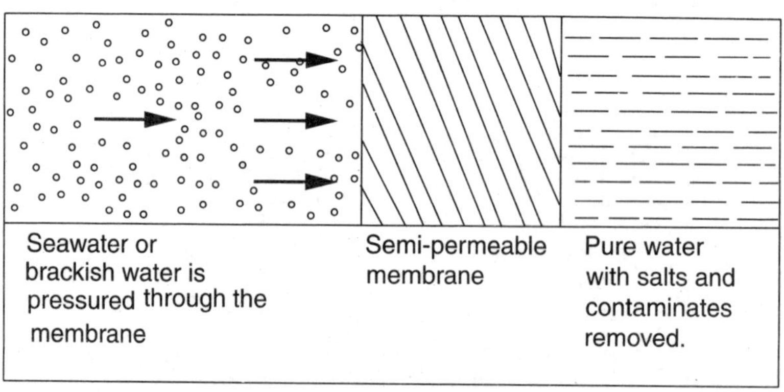

Figure 16.6. Conceptual diagram to explain Reverse Osmosis.

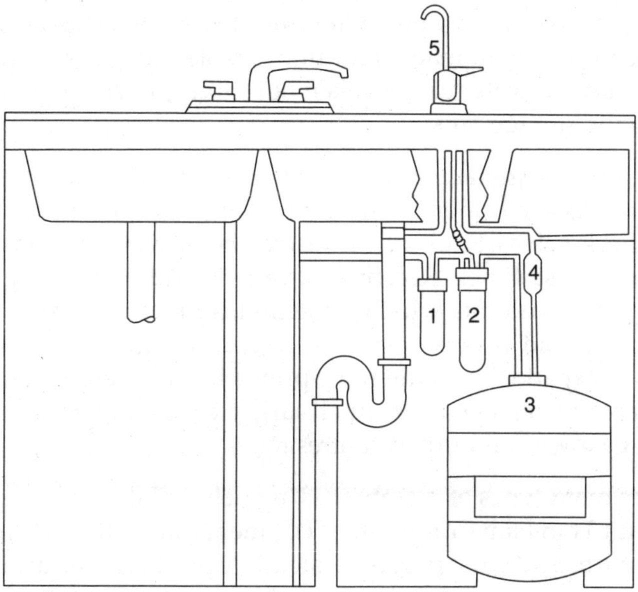

Figure 16.7. A domestic Reverse Osmosis system.

Index: 1. Particle Filter. 2. Reverse Osmosis membrane unit. 3. Pressurized treated water storage. 4. Carbon adsorption post-filter. 5. Treated water tap.

(Redrawn from: http://agbiopubs.sdstate.edu/articles/FS877RO.pdf)

(*i*) In this process no phase-change like from water to steam or water vapour is involved.

(*ii*) The process is relatively a low-energy application process.

A typical household reverse osmosis system is shown in Figure 16.7. It consists of pre-treatment particle filter, reverse osmosis membrane unit, a pressurised storage tank for the treated water, a post-filter and a delivery tap. These systems are capable of producing about 25 litres of potable-grade water per day. Booster pump can increase the output. Raw water first passes through the particle filter (or sedimentary filter as known in some of the units), which culls coarse solids that could clog the membrane. Next, the water flows into the spiral winding of the reverse osmosis membrane unit, which removes up to 98% of total dissolved salts. The membrane unit in these systems consist of cellulose-acetate or polyamide resins or thin-film composite (TFC) membranes. Most of the household reverse osmosis units operate on 25 to 50 *psi* pressure. The purified water moves on to the pressurised storage tank from where it can be drained out through the tap with adequate pressure. The carbon filter is provided in most of the domestic reverse osmosis system as a post-filter unit. This filter removes the residual organic chemical contaminants and undesirable odours from the purified water. The purified water can be taken out through a separate delivery tap attached to the storage tank.

No special controls are required for the household reverse osmosis purifier system units, as they operate by the use of pressure-sensitive controls, switches, check-valves. Shut-off valves may be provided to conserve water during sparing use times. Recent innovations add monitoring gauges or servicing lights to indicate whether the unit is functioning or not. Though this is a household reverse osmosis system, through innovations, this technology is expandable in the same principle and a larger design to cater to the potable-water needs of much larger community like a cluster of villages, large towns and urban centres too.

Of the three main types of membranes, namely cellulose-acetate, cellulose-triacetate and Thin Film Composite (TFC) membranes, the TFC membranes are the most efficient as the product water they produce even at less pressure is higher in quantity than the plants attached with the other two types; and the quality of water they produce is also better in terms of the total dissolved salts, nitrate and bacteria. However, the cost of TFC membranes is higher than the other two. The reverse osmosis membranes can be fouled and clogged by bacterial slimes, hard water scales, iron and silt. Pre-treatment solves most of the problems and the maintenance rate and cost can thus be reduced considerably. Disinfections can be done by using chlorine, formaldehyde or per-acetic acid. However, use of chlorine should be minimal, as chlorine spoils the membrane.

Eleven reverse osmosis desalination plants were installed on experimental basis to tackle drinking water problem in the coastal belt of Ramanathapuram district in Tamil Nadu (Ansari *et al*, 1998). The raw water is brackish groundwater, with total dissolved salts up to 10000 milligrams per litre. Filtration and pre-treatment using chemicals are performed before resorting to desalination by reverse osmosis. It was found that the total dissolved salt rejection had taken place to the extent of 91% and that the quality of the water satisfactory. Though the water quality has been good, the local population is satisfied with the water quality and the product water is socially accepted, it is found out from the study of Ansari *et al* (1998) that the operative costs of these reverse osmosis plants have been high due to extraneous reasons such as pump failure, operation and maintenance cost, non-availability of spares, lack of knowledge in villagers in membrane maintenance, and so on. It can be seen that these problems can be, to a large extent, overcome in India over a period of time, if several desalination plants using membrane technology and reverse osmosis method is adopted as a routine, when more and more people attain the skill and spares are made available as a constant flow as and when necessary.

Usually, the reverse osmosis membranes are made up of polymers, cellulose-acetate, cellulose-triacetate and matic-polyamide. These membranes are generally of two types namely Asymmetric or Skinned membranes and Thin Film Composite (TFC) membranes. The materials that support the membranes are commonly 'polysulphones', while the Thin Film Composite is made from various types of polyamines, poly-ureas and so on. The reverse osmosis membranes have the smallest pore structure ranging from 5-15 A° (Angstrom units) or 0.5 to 1.5 nanometres. These extremely small sizes of the reverse osmosis pores allow only the smallest organic molecules and unchanged solutes to pass through the semi-permeable membrane, along with the water. Between 95 and 99% of the inorganic salts and charged organics will be rejected by the membranes due to charge repulsion established at the membrane surface.

The main disadvantage of the reverse osmosis unit is that it uses a lot of water. It recovers only 5 to 15 percent of the water entering the system. The reminder is discharged as wastewater. Since the wastewater carries with it the rejected contaminants, the methods to recover this water are not practical for household systems. A reverse osmosis unit delivering 5 gallons of treated water per day may discharge 40 to 90 gallons of wastewater per day.

According to the website of the Technology Information, Forecasting and Assessment Council (TIFAC), Department of Science and Technology, Government of India, entitled 'Membrane Separation Processes—Technology and Business Opportunities', a complete Reverse Osmosis equipment comprising pre-treatment system, membrane module and other components 'is estimated to cost US $450 per cubic-metre of flow per day'. The operating cost for the reverse osmosis system is estimated as Rs.24 per cubic-metre of flow, excluding the cost of labour and chemicals used. In the case of the experimental studies conducted in Ramanathapuram district, Tamil Nadu, the cost factor has been worked out to be as follows: 'The average capital cost per cubic-metre and operation and maintenance cost per cubic-metre of product water from these eleven plants work out to Rs.27.40 and Rs.44.50 respectively when plants are utilized as per the design capacity' (Ansari *et al*, 1998). We should bear in mind that these costs are only based on experimental work laden with several constraints; but, when the technology of reverse osmosis in production of potable quality water is initiated as a regular and routine manner in an organised way, the cost should grossly go down.

Nanofiltration: This is a type of desalination using membrane separation process, which effects removal of organic impurities in the raw water at high levels and low levels of inorganic contaminants. In the plants that operate on nanofiltration technology, cellulose-acetate and aromatic polyamide membranes are used. These membranes have larger pore spaces, and therefore, they can separate impurities even at less applied pressure, but these membranes allow more salt in the product water. Less cost, more product water yield and slightly inferior quality of product water in terms of (higher) total dissolved contents, make the nanofiltration technique suitable to plants that can produce water for the purpose of swimming pools, personal hygiene and some specific industries where the water quality is not critical in terms of total dissolved salt contents.

Ultra-filtration and Micro-filtration: These two processes are also types of membrane separation technology with some variation, but function under the principle of reverse osmosis. Micro-filtration technology is, perhaps, the most widely employed membrane separation technology around the globe. These two are employed more in the industrial applications where concentration and fractionation of substances needs to be effected, such as fruit juices, dairy products such as milk and whey, beverages, paints, chemicals, protein fractionation and so on, rather than in seawater desalination processes.

Ultra-filtration is employed to separate a solution containing a mixture of both desirable and undesirable components, and when the desirable components can be present in the output product, say water or any other fluids. The process operates under 'differential pressure' to discriminate the components and disallow the undesired components in the feed fluid or solution from passing through the membrane. Polymer substances such as polysulfone, polypropylene, poly-tetrafluoro-ethylene, acrylics and nylons are used in the ultra-filtration.

In the micro-filtration process, the membranes are with pore-spaces of smaller than 0.3 microns, and therefore, they can be employed to sterilize the solutions to remove bacteria. The micro-filtration membranes, in fact, effect *sieving* to disallow the passage of the particulates of large size, while allowing the particulates whose diameter is smaller than the pore size, and the presence of such smaller particulates is allowed or desirable in the end product. The micro-filtration membranes are made from natural and synthetic polymers such as cellulose-acetate or nitrate, polyamides, polypropylene, poly-tetrafluoro-ethylene, as also inorganic materials such as metal oxides, glass and so on.

Though both micro- and ultra-filtration processes, along with nanofiltration are essentially technologies employing membrane separation, they are useful more in functions other than seawater or brackish water desalination. However, that would not mean that they should not be used for seawater desalination. Processes such as these should be encouraged to be put up in select industrial units, which now depend on the water supply authorities for their water needs, to install their small-scale plants to produce the desired quality of water suitable to their own needs.

2. Electro-Dialysis

One of the most sophisticated, modern and effective processes of seawater desalination is the Electro-Dialysis. Besides seawater desalination, the electro-dialysis process finds use in desalting of cheese whey, pharmaceutical and electroplating industries, and several other industrial applications too. This is a membrane-separation process in which ions are transported through semi-permeable membrane, under the influence of an electric potential. A direct current electrical source is connected to two electrodes immersed in the feed saline water. The charged ions in the saline water migrate towards an electrode of opposite charge. The membranes used in this process are cation-selective or anion-selective membranes, placed in the electric field. The cation-selective membrane permits only cations, and the anion-selective membranes allow only anions to pass through them. These two sets of membranes are alternately installed in the plant unit in one series. The cation-selective membranes are poly-electrolytes with negatively charged matter, which rejects negatively charged ions and allows positively charged ions to flow through them. Likewise, the anion-selective membranes allow only negatively charged ions to pass them, isolating the positively charged ions. Placing multiple cation- or anion-selective membranes in a row alternately allows positively or negatively charged ions to flow through them, thus removing all the ions from the seawater. This concept is diagrammatically illustrated in Figure 16.8. In some columns bound by membrane, concentration of ions takes place, and in other columns the ions will be removed. Therefore, a strong brine solution develops in the compartments that retain a high concentration of ions. The alternate compartments contain water depleted in ions, which is the processed water. The concentrated saltwater flow is circulated until it reaches a stage when the salts get precipitated. At this point, the flow of water is discharged.

This technique can be applied to remove those ions in the seawater, which can carry electrical charges; the particulates and ions that cannot carry electrical charge cannot be removed. The cation-selective membranes consist of sulfonated-polystyrene, and the anion-selective membranes are

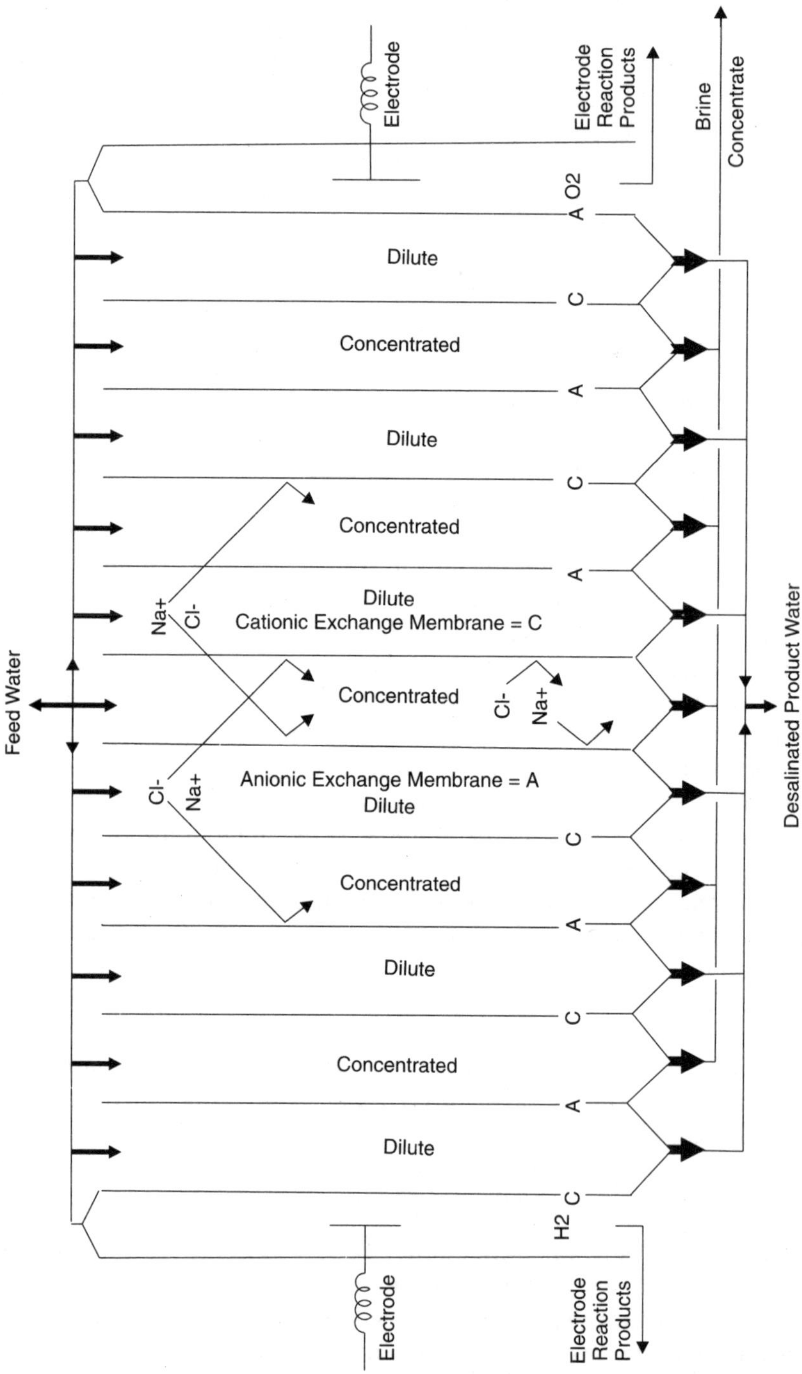

Figure 16.8. Principle of Electro-dialysis.
(Redrawn from: http://www.usace.army.mil/publications/armytm/tm5-813-8/c-7.pdf)

polystyrene with quaternary ammonia. The anion and cation membrane life is estimated at four years and seven years respectively.

One major problem in the electro-dialysis technique is the 'fouling' of the membranes, which results from the trapping of certain ions in the membranes' polymer network. Usually, prior to subjecting the seawater to electro-dialysis, pre-treatment to isolate the coarse suspended solids would be beneficial to avert such undesired results. Further, 'flushing' the membrane is also an efficient process to nullify the effect of fouling. This is effected by reversing the polarity of the direct current source, thereby reversing the movement of the ions, which then alters the configuration of the compartment. This process is referred to as *Electro-dialysis Reversal*. Such exercises have to be regularly performed periodically; lest, they clog the membrane pores and restrict the water separation process. Also present in the seawater at times are the substances capable of neutralizing the membranes, such as organic anions, colloids, iron-oxides and manganese-oxides. Since these substances render the ion-selective nature of the membranes less effective, their removal is necessary. Yet another problem the membrane units face is the scaling, due to the deposition of hard salt layer. The pre-treatment exercises, which can prevent these effects, comprise filtration of raw water through activated carbon (to remove organic matter), flocculation (to isolate colloids) and simple filtration techniques described in earlier chapter. Sodium hexa-meta phosphate is added in to the brine stream to prevent the precipitation of barium sulphate and the resultant scale formation within the membrane stack.

The website of Technology Information, Forecasting and Assessment Council (TIFAC) under the Department of Science and Technology, Government of India provides the information that the capital cost of an electro-dialysis system, including the pre-treatment unit, membrane modules, pumps and electrical unit, is estimated around US $275 per cubic-metre of flow. The operating cost of the electro-dialysis system is around Rs.21.50 per cubic-metre of flow, excluding the cost of labour and chemicals used. Though these costs are high in the present circumstances, on a regular use of the technology on a continuous basis at full efficiency, the costs are bound to reduce.

SEAWATER DESALINATION AS TECHNOLOGY TRANSFER FOR WATER RESOURCE AUGMENTATION IN INDIA

Seawater desalination is a proven technology worldwide and it has touched the day-to-day human life in many parts of the world. The technologies that are invented and innovated by the human brains need to be put to use for a better living of the human society at large. Various technologies brought forth into the world after painstaking researches have to be transferred into regular practice for the purposes human use to make them worthy to be called as technologies. Otherwise those technologies would remain only as fantasies written on paper. The responsibility for transforming the technological principles into practices lies equally with all the strata of our society, such as the governments and their administrative machinery, entrepreneurs, engineers, scientists, contractors, labour force, and above all the common man in the country with whom the responsibility of paying back the cost of the facility he enjoys actually rests.

A six-year spell of drought in and around California, USA has prompted the government and general population alike to unite and construct desalination plants to convert the saline water, including the seawater, into fresh water. California Coastal Commission has initiated measures for large-scale conversion of seawater to potable-grade water by several communities and industries. Several desalination plants having a wide range of capacity from 20 to 112,000 acre-foot per year are proposed or already installed at costs ranging from $1000 to $2200 per acre-foot. (One acre-foot equals approximately 326,000 gallons or more than 1,500,000 litres.) The technologies that are used are reverse osmosis and distillation-desalination.

The transformation of the life-style brought about by a large number of desalination plants in the Middle East countries is not repeated at this juncture. However, the enviable status of Israel in this respect cannot be glossed over too. Israel, like other Middle East countries in the neighbourhood, faces water shortage and the threat of desertification of vast area of its territory. The desert occupies already more than 50 percent of Israel's land area. Nevertheless, Israel's water policy takes into consideration the need to share the limited existing water resources with the neighbours. Israel "supplies water to the Jordanians and the Palestinians as well", though its own population is affected by water scarcity. The people of Israel have not increased the per-capita consumption of fresh water for urban demand, which stood at a constant 100 cubic-metres per year, despite huge population growth. Israel expects a crisis for water by the year 2010, when the population between the Jordan River and the Mediterranean Sea reaches 10 million, and therefore, the government have taken advance precautions to enhance the seawater desalination activity. Having realised that "the cost is beyond the capability of both Palestine and Jordan, who also suffer from severe water scarcity", and also realising the need to assist its neighbours, Israel has taken extra-ordinary measures to construct one 50 million cubic-metre desalination plant each at Gaza and Jordan Rift Valley, as well as another plant of 50 to 100 million cubic-metre capacity on the coast of the Mediterranean Sea. Quoting the Israeli Foreign Ministry, the Jewish Virtual Library states that Israel will '*over a period of 10-15 years, provide one billion cubic-metre of desalinated seawater, and provide a solution to the current water crises as well as meet the future needs of Israel, Jordan and the Palestinian Authority*'. Even with its limited water resource, Israel produces large quantities of citrus fruits and the milk yield of Israel's cattle is the highest in the world. Despite complex political and inter-racial situation prevailing in the region, the way Israel has augmented its water resources for the use of not only the country's population, but to provide for her inimical neighbours is very remarkable.

India, on the other hand, does not have the additional responsibility of providing water to any of its neighbours, as in the case of Israel. In contrast to Israel's coast formed by the land-locked seas like the Mediterranean Sea and the Dead Sea, India has a distinct advantage of having very long coastlines as well as continental shelf region that are of gentle slope. The coastal belts of India are capable of accommodating any number of desalination plants. Further, India has the potential to manufacture any kind of membranes that are used in the desalination plants.

The Technology Information, Forecasting and Assessment Council (TIFAC) states that the reverse osmosis "*technology has gained industry acceptance as a variable water treatment option for many different fluid separation applications. Low operating costs and the ability to remove organic contaminants and 95-99% of inorganic salts with minimal chemical requirements make reverse osmosis an attractive technology for many industrial applications*". With regard to the electro-dialysis technology, the Council has identified "*new areas of application…in the areas of chemicals and pharmaceuticals, food, industrial and municipal effluent treatment*". Further, the Council is of the opinion that steady growth in the electro-dialysis in the above-mentioned thrust areas "can be achieved through slight modifications in the conventional processes as well as extensive research and development work in those areas".

In our holistic approach to the water resource augmentation and management, we have to exploit the Nature's bounty that we are bestowed with in the form of enormous seawater. Instead of depending heavily on, and linking our economic progress to, the monsoon rains on which the humans have no control, we can write our own story of success as has been done by Israel, California Coastal Commission and other countries in the Middle East. We need to install a series of desalination plants that employ both distillation and membrane technologies and generate enormous quantities (comparable to or more than the quantities produced in California, Israel and other Middle East countries) of potable-grade water to augment our water resources. The water that is generated by desalination of seawater in India would relieve the stress on the only source of pure water—that is the rainwater, which can be budgeted, along with desalted water, for a variety of direct human consumption.

Besides, we should never lose sight of the extremely dry and drought-like conditions in the arid and semi-arid belts of central Indian peninsular region, northwest India and the interior parts of Gujarat, Orissa, Andhra Pradesh, Tamil Nadu and Karnataka. Making provision for certain adequate quantity of water from our water resource budget for the agri-horticultural process wherein fruit-bearing plants are grown by drip irrigation as well as for developing greenery in these arid and semi-arid belts can arrest such dry and drought-like conditions in those arid regions. It is in this context the availability of water for the said purpose during the water-starved long summer season holds the key for the economic and environmental development in this region, and the utility of desalination processes in augmentation of the water resources has to be taken note of.

Chapter 17

The objective of this chapter is to analyse:

- Energy requirement for water resource augmentation technologies—Wastewater Treatment and Seawater Desalination; justification to meet additional energy requirement by establishing new nuclear power plants.
- Fundamentals on Radioactivity and Nuclear fission.
- Overviews of Gas-Cooled Reactors, Heavy Water Cooled Reactors (CANDU), Light water cooled reactors like Pressurized Water Reactor and Boiling Water Reactors, and Fast Breed Reactors.

ENERGY MANAGEMENT IN WATER RESOURCE DEVELOPMENT AND MANAGEMENT

By now we are convinced that we require considerable quantities of additional source of water to augment our existing resources, if we simultaneously want to:

1. Keep our agricultural activity on track without interruption despite monsoon failure and drought-like condition.
2. Keep the wheels of industry running continually.
3. Attract foreign direct investment in the country.
4. Provide more protected water to our citizens for their personal hygiene.
5. Attract tourists to our country.
6. Develop greenery as well as clean surroundings, be it in the place we live or the places we visit as tourists.
7. Make an impact in the water-sports arena at international levels.
8. Quench the thirst of many who cannot afford to buy clean bacteria-free water to drink.

It is also clear that the only way to achieve these objectives is to resort to technological means and establishing innumerable wastewater treatment plants, distillation plants, and the units employing membrane technique and electro-dialysis. Augmented source of water from the wastewater plants is capable of cleansing our heritage rivers and lakes. Furthermore, these technological facilities should stand distributed somewhat evenly across the length and breadth of the country, in order to offer a regional balance; lest, that population of a region, which is deprived of these facilities would feel let down by the State and would turn rebellious. There is no need to emphasize that to establish these technological missions and also keep them continually functional, we need adequate electric power in large measures. Strong resource of electric energy and plentiful water resource are like two wheels that are kept in full balance to facilitate a turbo-jump of the economy and prosperity of a country. Therefore, in our efforts to augment and manage the water resources in the country, we need to concurrently generate enormous amounts of electric power too, to run our wastewater treatment plants and seawater desalination plants. Under the present circumstances, it would appear that the one and only way to increase our capability of power generation is to resort to tapping the nuclear energy in a big way. Let us analyze why.

The power generation using nuclear energy is a well-established technology in nearly 35 countries throughout the world. Though the world's first nuclear reactor was demonstrated at the University of Chicago in December 1942, the first fully engineered nuclear reactor had appeared in Western Europe in August 1947. It is more than 50 years old technology and is capable of working for twenty-four hours a day. During the 50 years of nuclear reactor history in Britain, not one member of the public has sustained harm through nuclear causes and only a few employees within the industry are said to have lost their lives through what *might have been* nuclear causes (Cole, 1988). Except two major mishaps, namely the Three Mile Island in USA in 1979 and the Chernobyl in the erstwhile USSR, no other fatal or traumatic incident relating to the nuclear leakage in power plants has taken place anywhere in the world in the recent or even distant past. In fact, compared to the nuclear power plants, which provide uninterrupted electric energy to billions of population across the world, the *presumably* innocuous industries (such as chemical, dyes, petrochemical and rubber industries, and thermal power plants), biological causes and excessive human activities (and the resultant impacts like greenhouse effect) have been inflicting more harm to the human beings *all over the world* by altering the biome characters and effecting climate changes, besides silently and slowly polluting the atmosphere and the water resources. The electricity generated by the modern nuclear stations is considerably less expensive than that generated by the modern coal- and oil-based power stations. Enormous savings can be achieved each year as a consequence of generating energy in the nuclear power plants. The nuclear energy saves a lot of fossil fuel (oil and coal) reserves. For every tonne of otherwise useless uranium used as fuel in one nuclear power station, some 20,000 tonnes of much more valuable coal and oil can be saved!

In view of several positive qualities associated with the nuclear power plants, as also consider a possibility of augmenting our energy resource by installation of a few more nuclear power plants, a brief overview of the nuclear power technology at this juncture is justified. Furthermore, it needs to be realised that India would undoubtedly require more power plants as pre-requisite to bolster the energy supply to the possible emerging water resource augmentation facilities. An in-depth description on the aspects of nuclear power plants is out of scope of this book, but reasonable highlights are made to offer an insight into the nuclear power technology to the readers, in the context of the energy requirement in water resource management. Equally rewarding will be to refresh the memory on some of the basic facts about the radioactivity, which are fundamental to the concepts on nuclear reactors, radioactive elements and isotopes, as well as the fission process that operates in those reactors.

It is acknowledged that all those departments, scientists and technologists concerned with nuclear energy in the country are striving hard towards generation of power by nuclear means. But, much more needs to be done in this sector in view of the already existing large gap in the demand *vis-à-vis* supply of power. All the more, when we establish numerous wastewater treatment plants and seawater desalination plants, the demand—supply gap is bound to widen, if we do not plan to augment the power resource to meet the additional demand. It is, therefore, hoped that we would accelerate our pace in this sector of activity, and resolve to achieve further commensurate to our capabilities and needs, because we need to not only provide electricity to the masses of our population, but also to numerous water augmentation plants to which the availability of energy holds the key.

BASIC FACTS ABOUT RADIOACTIVITY

The nuclear power plants function with the use of radioactive elements. A non-radioactive substance is inactive, as no change takes place in its internal (or atomic) structure; this condition is said to be '*stable*'. On the other hand, radioactive substances are highly *unstable*, as they undergo changes internally in their atomic structure. During the changes, they give off vast amounts of energy in the form of nuclear radiation. The radioactivity is measured in units of *curies* (Ci), or in SI units in *becquerels* (Bq). This energy is, indeed, dangerous to any one getting close to it, but when this energy is trapped or contained within tight spatial compartments that would not allow radiation to permeate through them, the radiation energy can be unleashed to tap its utility as a highly useful substance that can support the power generation plants. The radioactive substances become less active over a period of time and eventually become inactive and stable. The most important radioactive substance is the element uranium (chemical symbol, U), which is the heaviest element.

Instead of an out-of-place detailed discussion on the atomic structure and radioactivity, it would be suffice to refresh the memory in this context and recollect the following most important facts about them: (i) The smallest unit quantity of the elements is called an *atom*. (ii) The atoms of every element have their own atomic structure. (iii) An atom comprises nucleus surrounded by *electrons* (an elementary particle bearing negative electric charge). (iv) The nucleus itself is composed of positive-charged particles known as *protons* and neutral particles known as *neutrons*. (v) The number of protons in a nucleus defines the *atomic number* of the element. (vi) The total number of protons and neutrons (together constitute *nucleons*) in the nucleus of the atom is the *mass number*.

The absorption of an additional neutron by a nucleus is called as the *neutron capture*. The susceptibility of an element to the neutron capture is described by the term *neutron capture cross-section*. Atoms of the same element (having same atomic number), but differ in the mass number are called *isotopes* of that element. The different isotopes of an element contain different numbers of neutrons in their nuclei. The isotopes of an element are identical in chemical properties and in all physical properties, except those determined by the mass of the atom. Of the isotopes of uranium, about 99.3 percent of the uranium has the atomic number of 238, (therefore known as U^{238}), and the rest 0.7 percent has the atomic number 235 (therefore U^{235}).

The term *nuclide* is used while referring to the nuclear structure of an element without actually implying that it may or may not exist in many isotopic forms. In other words, the term nuclide refers to the nucleus of an atom of a specific isotope, characterized by its atomic number, mass number and its energy state. An isotope refers to a type of atom while a nuclide refers to its nucleus. When an unstable nuclide gets rid of its energy in the process of becoming stable and inactive, it emits nuclear radiation; this process is known as the *decay*. The time taken for half of the number of atoms in a nuclide to decay is known as its *half-life* period. The radioactive decay or the impact of atomic particles emitted by a radioactive nuclide at times leads to changing of one atomic element to another; this process is known as *transmutation*.

Nuclear Fission

The most important mode of nuclide decay, as far as the nuclear power is concerned, is the phenomenon called *nuclear fission* in which a nucleus splits or fissions into two nearly two equal parts known as *fission products*. Just only one of the ninety-two naturally occurring elements has the ability to display nuclear fission; it is the 235-isotope of uranium (U^{235}), which constitutes only about 0.7 percent of the naturally occurring uranium. Although U^{235} is spontaneously fissile (splits or fissions without external influences) it does so only infrequently, and it can be mostly ignored. On the other hand, U^{235} can be made to fission quite readily by forcing its nucleus violently unstable and it quickly breaks up into two nearly equal parts.

Under normal circumstances, the 92 protons and 143 neutrons (adding up to 235), which make up the nucleus of a U^{235} atom, are tightly bound together by a very powerful nuclear force known as the *binding energy*. When a bombarding neutron overcomes this binding energy, the U^{235} nucleus splits into two and the resulting fission products fly apart in opposite directions at tremendous speed, as though the nucleus has exploded. Eventually, these particles come to rest after multiple collisions with the atoms of the material through which they had travelled. In this period, they shed their kinetic energies in the form of frictionally generated heat. This precisely is the heat that is used by the nuclear power stations to boil the water, produce steam and drive turbines to generate the electricity. This is the fundamental principle of the nuclear power. Besides, the kinetic energies of the fission neutrons and from the nuclear radiation emitted by the radioactive fission products also produce more heat.

The total energy obtained from the fission of a single U^{235} atom is about 205 millions of electron-volt (MeV). About 87 percent of the total energy is released at the instant of fissioning and the rest from other radiations. Virtually all the energy possessed by the *neutrinos* (nuclear particles having no electrical charge and virtually no mass) is lost by their escape from the reactor assembly. However, it is the slow release of energy by the fission product decay, which necessitates a continued source of cooling for the core, long after the fission process of the reactor is terminated.

The U^{238} in the naturally occurring uranium dilutes the less abundant U^{235} by a factor of about 140:1. This means that 140 times more uranium is required to create a critical mass of U^{235} than that would be necessary, if the uranium was in pure U^{235} form. This is the reason why a nuclear reactor can never explode like an atom bomb. The atom bombs are made from pure U^{235} or pure plutonium-239 (Pu^{239}). On the other hand, the uranium as the nuclear fuel in reactors is heavily diluted by U^{238}.

Uranium is a fairly abundant element in the earth's crust. Huge deposits of uranium ores occur in Australia and the countries situated in the continents of Africa and America. Canada is the largest uranium ore mining country. In

India plenty of uranium ore is present in Bihar and Andhra Pradesh. The uranium ore concentrate is known as *yellow cake*, comprising uranium oxides, UO_2 or U_3O_8. The uranium metal production for the nuclear reactors is done from uranium oxides in several stages such as uranium tetra-fluoride and uranium hexa-fluoride (known in short as *hex*). The solid rods of pure uranium used in Magnox-type fuel elements are produced from uranium tetra-fluoride.

Enriched Uranium

As already mentioned, 99.3 percent of the natural uranium is the non-fissile isotope namely the U^{238}. The rest 0.7 percent is the fissile isotope U^{235}. It is this 0.7 percent of U^{235} content of natural uranium that all the nuclear power plants all over the world depend upon for their energy in the electricity generation. Basically, there are two types of reactors, one the Magnox type and the other is the CANDU type, which run on natural uranium as the fuel. The other type of reactors requires the U^{235} fuel to be enriched from the 0.7 percent level to about 2 to 3 percent. This is done artificially in facilities known as Enrichment Plants.

GAS-COOLED REACTORS

These are the some the of oldest type of reactors in the world. Among the gas-cooled reactors are Magnox reactors, which consist of the fuel elements clad in a special type of magnesium-aluminium alloy. These are in vogue since 1956 and are still operative. Magnox reactors are in use in Japan and Italy, besides Britain where it was installed first. Each fuel element comprises a solid rod of natural uranium as heat energy, which gives the energy equivalent to that derived from about 150 tonnes of coal. Many thousands of such elements are used in a reactor. Pumping of the pressurized carbon dioxide gas cools the elements, and the neutrons generated in the reactor are moderated by several tonnes of graphite blocks. The core of the reactor comprising fuel elements, coolant gas, moderator blocks and control rod assembly are kept in a gas-tight pressure vessel surrounded by a thick concrete container. This container provides radiation shield for the protection of the plant-operating personnel.

Improvised gas-cooled reactors, known as *Advanced Gas-cooled Reactors* (AGR), operate at higher coolant temperatures (>600°C) than the Magnox reactors, and therefore, the thermal efficiency of these reactors is better, matching with that of the coal-fired power stations. The differences between the Magnox and the AGR reactors are in the chemical and physical compositions of the uranium fuel and the material used for the cladding of the fuel assembly. In the AGRs too, like in Magnox reactors, graphite blocks are used for neutron moderation, and carbon dioxide as the coolant. The concrete pressure vessel is also similar, but thicker than, that used in the Magnox reactors. The fuel for the AGR is uranium dioxide (UO_2) in the form of pellets packed in thin stainless steel tubes called as the '*pins*'. Thirty-six pins are bundled together and fixed inside a graphite sleeve to constitute a fuel element.

HEAVY-WATER COOLED REACTORS, CANDU

The CANDU reactors are Canadian reactors and are thermal reactors with their core containing a critical mass of fissile material. These reactors too employ moderators for slowing down the neutrons produced during the fission, and neutron absorbers to control its power output. The heat generated in its core is removed by circulating coolant. The difference between the CANDU on one side and the Magnox and the AGR on the other side is that the CANDU uses heavy water (D_2O) as both coolant and moderator. Canada is the only country that has developed nuclear power stations using heavy water reactors. The heavy water, though highly expensive, is a good moderating material because it gives neutron economy by avoiding wastage. The CANDU reactors with its D_2O moderator, extracts roughly twice as much energy from each gram of U^{235} used than does a light-water moderated reactor. This is mainly because of the greater amount of plutonium (Pu^{239}) created from the U^{238}-content of the fuel (about 2.5 gram of Pu^{239} per one kilogram of U^{238}), much of which is incinerated in the core to add to the overall power produced for relative plutonium production rates. In other words, the conversion ratio of the CANDU reactor (or the ratio of fissile material of all types produced to that destroyed) is greater than that of the light-water reactor. The value for conversion ratio for the CANDU reactor is 0.79, while for the light water reactor this value is about 0.6.

A large cylindrical tank called as 'calandria', containing heavy water functions as the moderator of the CANDU reactor. Several hundreds of channels containing bundles of fuel elements are grouped closely to form the reactor core, and these channels pass through the calandria. The time spent by a neutron in the moderator before it escapes or is captured by an atom of fuel by the moderator or structural materials or control absorbers is about 50 minutes, a period which is about 500 times longer than that of a light-water reactor. This would mean that the response of the heavy-water reactor to a rapid transient is relatively slow and sluggish, which facilitates its control a lot easier and less critical. This is notable in the point of view of the safety of the plant. The fuel for the CANDU reactors consists of natural uranium dioxide (UO_2) in the form of small ceramic pellets packed in thin-wall zircaloy tubes (about 50 cm long and 1.5 cm diameter), which are bunched together to form a '*fuel bundle*'. A fuel bundle used in a CANDU reactor contains about 19 kg of uranium and is about 50 cm long 10 cm in diameter. It generates about one million kilowatt-hours of electricity (Dormuth et al 1995).

LIGHT-WATER COOLED REACTORS (LWR)

Nearly seventy percent of the world's nuclear reactors (concentrated in USA) are of the light-water cooled (LWR) types, in which ordinary water is used as both moderator and coolant. Any reactor using light water as a moderator would have to use for its fuel the U^{235}, which is artificially to a level of about four times above its natural abundance of 0.7 percent. Therefore, uranium enrichment plants are pre-requisite for the LWRs. The LWRs are mainly of two types, namely *Pressurized Water Reactor* (PWR) and *Boiling Water Reactor* (BWR).

Pressurized Water Reactor (PWR)

In the PWR, a cylindrical pressure vessel containing the fuel elements and control rods is completely filled with light (or *ordinary*) water, which functions as both moderator and primary coolant. This water is circulated in a closed circuit through the fuel (where it becomes very hot), and also through a heat exchanger in an external steam generator. In the heat exchanger, the hot water gives up much of its heat to a completely separate light water circuit, being the second coolant. The primary coolant water that circulates through the fuel is prevented from boiling by an external pressurizer, and it attains temperatures as high as 325°C. Prevention of the primary coolant from boiling is an essential requirement of the PWR. Boiling of this water would lead to a steam and water mixture, resulting in the reduction of efficiency of the plant, as also overheating of the reactor core. The secondary coolant water in the steam generator is not artificially pressurized and is allowed to boil and produce steam for driving the turbo-generator set. If the outlet temperature of the primary coolant is sustained at high levels, it facilitates production of high-temperature steam in the steam generator, and yields a good thermo-dynamic efficiency. The pressurizer is a part of the primary coolant circuit, and is partially filled with primary coolant water. The locally generated steam above the water level creates the pressure. The steam is generated in the pressurizer by an electrical immersion heater and its pressure carefully controlled so as to maintain a constant pressure in the pressure vessel. The control process partly involves a high-pressure injection nozzle located immediately above the surface of the boiling water. This is used to spray cold water in to the pressurizer whenever it is necessary to reduce the pressure quickly. The pressure created in this way is about 2200 pounds per square inch (psi) or about 150 atmospheres, which ensures that the primary coolant remains in the liquid phase when operated at temperatures in the region of 325°C. The pressurizer is about 16m in height and 2.2m in diameter, and it can hold about 30 cubic-metres of water, though its total capacity is about 50 cubic-metres. It is fitted with a pressure-operated relief valve, which is designed to blow off the pressures in excess of 2500 psi.

The PWR has a reactor pressure vessel made up of carbon-steel with an internal cladding of corrosion-resistant stainless steel. The vessel contains 334 cubic-metres of light water (pressurized to 2200 psi) and circulated by four pumps at a combined rate of about 19 tonnes per second. Each circulating pump is rated at 8000 horse-power and weighs about 94 tonnes. Under operating conditions, this vessel contains about 100 tonnes of UO_2 fuel, in the form of 193 fuel assemblies, each containing 264 fuel rods and 25 neutron absorber control rods arranged in a 17 x 17 matrix. The residence time of each fuel assembly in the core is about three years. Re-fuelling takes place approximately once in a year and lasts for about six weeks. The specific thermal power rating of the fuel in the PWR is 33 kilo-Watt per one kilogram of uranium, a factor which is about twice that of the Advanced Gas-cooled Reactors (AGR) described earlier. The whole set-up of PWR is housed inside a structure known as containment building whose purpose is to contain the whole of the primary coolant in the event of the accident due to the loss of the coolant, withstand the excessive pressures, and prevent unacceptable releases of radioactivity to the environment. Figure 17.1 is a cross-sectional diagram of a PWR's containment building. This building is fabricated from steel-lined reinforced concrete with about 1.3m thick wall and an approximate height of 53m, and an outer diameter of about 38m.

The PWR type of power station is preferred to that of AGR because of the following reasons (see Cole, 1988):

- (*i*) It costs 17 percent less per kilowatt to build.
- (*ii*) It will have a ten-year longer life.
- (*iii*) It will be marginally cheaper to run.
- (*iv*) It will have a lifetime advantage over the AGR by saving electricity enormously.
- (*v*) It will create at least 10,000 jobs over seven years.

The Boiling Water Reactor (BWR)

The BWR is similar to the PWR in that it uses light water for both moderator and coolant, and it uses enriched uranium fuel in the form of small ceramic pellets of UO_2 contained in Zircaloy tubes (fuel pins). The main difference

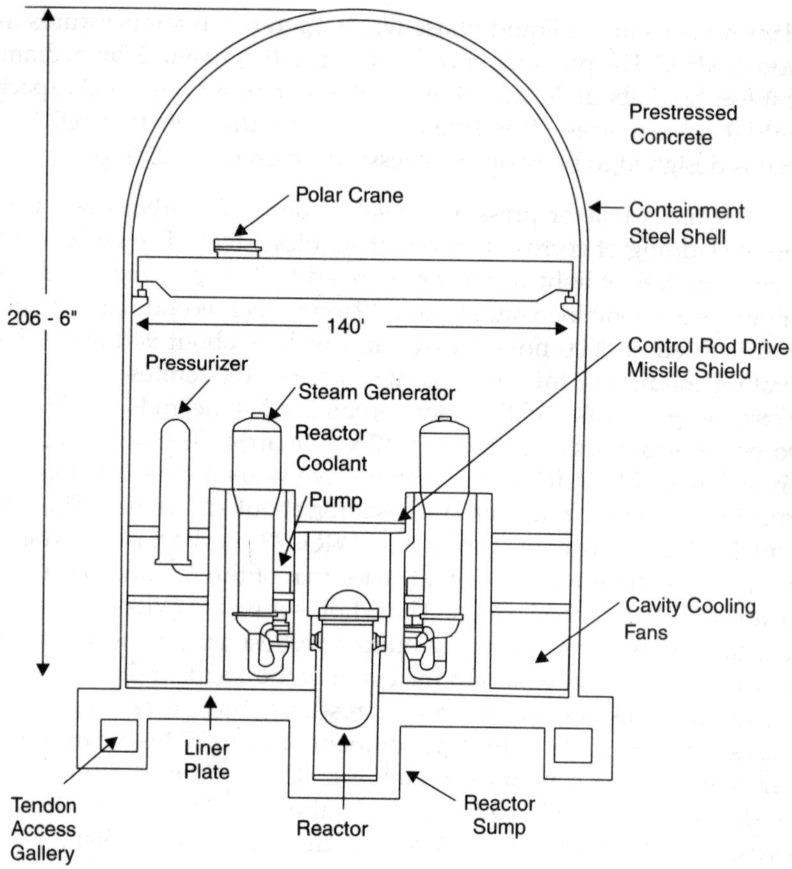

Figure 17.1. Schematic diagram for the cross section of a Pressurized Water Reactor (PWR) containment building (Redrawn from Cole, 1998).

between the two reactor types is that in the PWR the light water coolant is artificially pressurized to stop it boiling, whereas in the BWR it is not. As a result, the coolant is allowed to boil as it passes through the core and the steam is produced above the water level in the steel pressure vessel, which also contains the core. In the PWR the pressure vessel is completely filled with water, while in the BWR it is only partially filled. There is always enough water in the pressure vessel to ensure that the core is fully immersed at all times. The heat produced by the core causes the coolant to boil and the steam collects in the upper region of the pressure vessel above the water level. The steam is conveyed to the turbo-generator electricity generating plant, after which it is condensed and returned to the pressure vessel as warm water condensate. Some of the water contained in the pressure vessel is converted to steam and the remainder is re-circulated through the core by a number of jet pumps. These jet pumps force the water in a downward direction through the annular space formed between the wall of the pressure vessel and the core, and then up through the fuel elements, which make up the core where some of it is converted to steam.

The operation of the BWR is similar to that of a domestic pressure cooker in which the boiling point of the water content is raised above the normal temperature of 100°C by preventing the steam escaping and allowing the pressure to build up. The same phenomenon takes place in the BWR. The steam created in the top of the pressure vessel is kept in a closed circuit formed by the vessel itself and the turbo-generator set and the pressure builds up to a level of about 1040 psi. This self-generated pressure is sufficient to raise the boiling point of the coolant, and the temperature of the resulting steam, to about 288°C, which is about the same as the temperature of the steam produced in the PWR steam generator. Because of this similarity in the steam temperatures, the thermo-dynamic efficiencies of the PWR and BWR are about the same and close to about 33 percent (Cole, 1988).

The coolant in the BWR has to be contained in a thick-walled pressure vessel, since the coolant operates at a high pressure. The BWR's pressure vessel is made from 152 millimetre-thick low alloy steel plate and is of about 6.5m of diameter and 22m high. Since the wet steam can cause erosion of the turbine blades, a steam dryer assembly is provided in the BWR. Unlike in the PWR where the steam is produced from a secondary water loop that is made to boil, in the BWR the steam that is fed to the turbines is derived directly from the water that passes through the core of the reactor. Therefore the steam that is fed to the BWR turbines is radioactive, in contrast to the non-radioactive steam and water of the PWR turbines. This is a disadvantage associated with the BWRs.

The BWR fuel rods are about four metre long zircaloy tubes, each filled with UO_2 pellets of cylindrical shape 18 millimetres long and 10.6 millimetre in diameter. A zircaloy metal wrapper envelops the fuel rod clusters. A BWR contains about 800 fuel assemblies and more than 48,000 fuel rods containing about 170 tonnes of UO_2 (Cole, 1988). The BWR is housed within a containment building (as shown in Figure 17.2.) equipped with safety measures that combat the possible accidents due to the leakage of the coolants.

FAST REACTORS OR FAST BREED REACTORS

The thermal reactors described above are fuelled with natural uranium containing 0.7 percent U^{235} or enriched uranium containing up to about three percent of U^{235}. They require a moderator to slow down the neutrons produced during the fission. The use of enriched U^{235} in the reactors leads to the wastage

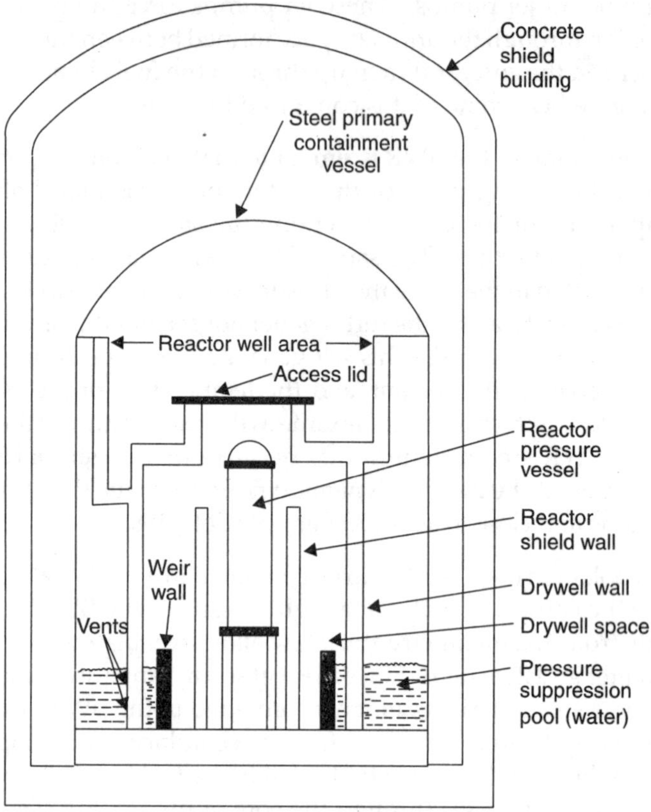

Figure 17.2. Cross-section of a Boiling Water Reactor (BWR) containment building (Redrawn from Cole, 1988).

of U^{238}. The *fast reactors* are those that do not use a moderator or slow down the neutrons produced in their core, and the fuel they use is fissioned directly by the fast neutrons emitted during the fissioning process.

As has already been stated, the U^{238} is a non-fissile substance and it makes up most part of the naturally occurring uranium ore. However, the U^{238} can be made to fission with the aid of highly energetic fast neutrons, whose energies are in excess of about 1.1 millions of electron-volt (MeV), which is the fission threshold of U^{238}. Though many neutrons emitted during the fast fissioning of U^{238} have energies in excess of this figure, most of them are involved in multiple non-fission producing collisions with other U^{238} atoms. Consequently, they lose a lot of their energy. Therefore, even if an external source of fast neutrons were used to cause fissioning of a few atoms of U^{238}, the average energy of the resulting fission neutrons would be below the critical 1.1 millions of electron-volt (MeV) threshold required to start and sustain the chain reaction. For this reason, it is not possible to maintain a fast neutron chain reaction in pure natural uranium. The reason for only few neutron collisions with atoms of U^{238} result in fission is because of the small fission cross-section of U^{238}. The U^{238} has the ability to produce the fissile isotope of plutonium, Pu^{239}, by neutron capture. The Pu^{239} has a higher fission neutron yield than the U^{235}. The Pu^{239} is considered superior to U^{235} as a fissile material in a fast neutron flux. The fast fission neutron yield of U^{238} is less than that of Pu^{239}, but higher than that of U^{235}.

If the nucleus of a fissile element absorbs a neutron, it either fissions and emits neutrons, or it is transformed into a higher order isotope by neutron capture. The fission process is important of the said two processes from the point of view of the nuclear reactor. The capacity to fission coupled with the magnitude of the neutron yield by fission jointly determines the utility of a particular material as a source of the reactor fuel. In a fast reactor, more of the available neutrons result in fission of the nuclear fuel than in a thermal reactor. This facilitates good neutron economy. U^{238} is a better nuclear fuel than U^{235} for both thermal and fast reactors, because its capture to fission cross-section ratio is much lower and its neutron yield is much higher. Though the U^{238} is a non-fissile element in a thermal reactor, it can be made to fission in a fast reactor. The relatively high capture cross-section of U^{238} makes it an ideal material from which Pu^{239} can be produced for use in the fast reactor.

A non-fissile element such as U^{238} can be transmuted into a fissile material like Pu^{239}, and therefore, the former is known as a *fertile material*. Likewise, Thorium-232 (Th^{232}) is also a fertile material, since it transmutes via Th^{233} and protoactinium-233 (Pa^{233}) to fissile material uranium-233 (U^{233}) by capturing a neutron. In a fast reactor, the fertile U^{238} is used to produce Pu^{239} by packing the depleted uranium around the core. The stray neutrons that escape from the core are captured by the surrounding material and used to transmute the U^{238} to Pu^{239}. This process is known as the *breeding*. The material used for breeding (the U^{238} in this case) is termed as the *breeding blanket*. The breeding material that is packed around the core is called as the *radial breeder*. A *fast breeder reactor* (FBR) is known by that name since the reactor uses un-moderated (or fast) neutrons and is able to breed the fissile Pu^{239} from practically a non-fissile substance such as U^{238}. The term *burn-up* is used to express the amount of thermal energy extracted from a specified quantity of fuel, in terms of thermal megawatt-days per tonne of fuel (MWd/t). Liquid sodium, a soft silver-white metal, is used as the coolant in the fast reactors all over the world, as well as the Prototype Fast Reactor (PFR), since it has a low neutron capture cross-section, a density approximately same as that of water, high thermal conductivity and specific heat. Further, the sodium has a melting point, which is slightly below the temperature of boiling water, and a boiling point that is much higher than the normal operating temperature of the reactor core. This means that pressurization to suppress the boiling of the coolant is not required, which is a great advantage. The disadvantage of sodium's property of explosion when it comes into contact with water can be overcome by adopting separate heat-exchange steps, as provided in the PFR (see Cole, 1988 for further details).

DISPOSAL OF NUCLEAR FUEL WASTE

Disposal of the fuel waste coming out of the nuclear power plants is a hotly debated topic and it has ramifications in terms of international politics and environmental issues. Safe disposal of nuclear waste is necessary to ensure that the possible residual low intensity radiation risks emanating from the discarded spent nuclear fuels do not cause any harm to the posterity, as also to avert the possible burden on the future generation of the user country arising out of the nuclear fuel waste produced by the current generation, considering social and economic factors of the future. Broadly three types of disposal methods are considered in the case of nuclear fuel waste by the researches. They are:

1. Removal of the nuclear waste from the Earth and disposal into the space.
2. Transmutation of the nuclear fuel element in which the residual elements in the spent fuel is changed by nuclear methods, in order to reduce the radio-toxicity of the waste.
3. Geological disposal, which is isolating the waste in geological medium, such as rock beneath the deep sea-bed, permanent ice sheet or sediments or rock on the land, in such a way that their maintenance and administration would not be required in long term (Dormuth *et al* 1995).

Researches on the disposal of nuclear fuel bundles of about 50 cm long and 10 cm diameter used fuels of uranium in CANDU reactors were conducted. Plutonic rocks such as granites are considered as repository at depths of 500 metres. Titanium shells of 6.35 mm thick wall and of 72 used fuel bundles capacity are emplaced in borehole drilled into the floor of a disposal room and is separated from the rock by clay-based buffer material. The rooms, tunnels and shafts are sealed with clay-based backfill and other repository seals (Dormuth *et al* 1995). While such safe disposal methods for the spent nuclear fuels are present and further innovative methods are under study in this matter, the public antipathy for the nuclear power plants should fast disappear and lead to acceptance of the nuclear reactors as a way of life.

OPTION OF THE NUCLEAR REACTOR TYPE, PUBLIC CONCERN AND SAFEGUARDS

Despite several attractive and positive features associated with the nuclear power plants, it is a fact to be borne in mind that they are built with extremely complicated devices. They require long planning and their site locations too need long ground surveys to ensure their safety from seismicity and other natural hazards. They take much longer time to build than the coal- or oil-fired power plants of the same size or production capacity. They need to be put on extremely rigorous experiments and tests for a certain period of time, especially to test for possible leakage of radioactive rays. They take long time to qualify for commissioning due to the various intensive tests conducted on them. They have to be monitored for possible radioactive leakage every moment by qualified technical personnel, even when the plant functions normally. They are about 50 percent more expensive to build than the thermal power plants. Nevertheless, when completed, a nuclear power station requires only a small fraction of fuel quantity that a coal- or oil-dependent power plant needs, in terms of the cost of the fuel. For example, a coal-fired power station of 2000 megawatt-electrical (Mwe) capacity consumes about 20000 tonnes of coal per day, which works out to 5 bags every second, or a trainload every hour (Cole, 1988).

The nuclear power cannot be made available to the public in a short time. Its planning, site location, construction and commissioning are extremely long-drawn processes. Their installation might have to face resistance from the *ill-informed* biased environmentalists and general public, lop-sided propaganda in the media on only the negative aspects of the nuclear plants, and the periodic changes in the international nuclear policies to safeguard the use of nuclear facilities. In spite of stiff opposition from the protagonists of a nuclear-free world, till 1986, there were 396 nuclear power reactors with a combined generating capacity of over 272,000 Mwe, operating all over the world. More and more reactors are under construction, and more and more countries, which are currently not nuclear, are contemplating to go for nuclear power plants. With the gap between demand and supply of the electrical energy steadily widening in the countries world over, more and more non-nuclear countries are bound to jump into the nuclear bandwagon in the near future.

In spite of the nuclear power making inroads into the peoples' lives in several affluent countries of the world and improved the quality of their life, the people in several other parts of the developing world face the scarcity of electric power, besides the shortage of water. And, India is no exception to this. There has, indeed, been a quantum growth of power generation in India since Independence; we have been generating power by hydroelectric and thermal means, and to a small measure from the nuclear power plants all these years. However, the development and all the progress made in the power sector are neutralized by the need of power by the huge population that India has, and the concurrent development in the activities, which use profuse electrical energy. This situation is exactly similar to the demand-supply status of the water resource in India.

The reason for the seasonal power scarcity that prevails currently in India is due to a variety of reasons, the most important of them being a marked improvement in the electricity-dependent lifestyle of an average middle class Indian citizen, coupled with a lack of proper management of the power that is generated. With a virtual spurt of the consumer market in India in recent years, more and more population belonging to the major middle class segment of the Indian society can afford to buy electrical appliances that function with the aid of heater elements, such as room-heaters (in winter), geysers and cloth-irons. Air-conditioners, air-coolers, microwave ovens, rice-cookers, domestic water pumps and elevators have become affordable necessities in today's modern life, be in metropolises or other cities and smaller towns. Besides, the power supply authorities currently are under great stress round the year to provide adequate electricity to the metropolitan transport system such as electro-motive train units, commercial establishments that require huge quantities of power to illuminate their premises, industrial belts, hospitals, agricultural pump sets and a host of other activities that require considerable electrical energy. There is always a gap in the supply against demand of electricity in the country round the year, but the situation gets worse in summer times. Frequent tripping at the supply stations and sub-stations due to inadequate power inputs has become a bane to the power supply authorities. As a cumulative effect of all these, the people at large face both scheduled and unscheduled power-cuts frequently, especially in the peak summer times.

Under the existing power-deficient situation, if India were to establish more wastewater treatment plants in different parts of the country and a chain of seawater desalination plants on the either coastal belts, we would need to generate more electric energy on a continuous basis to run those plants for several years to come. Planned installation of several wastewater treatment and desalination plants over a period of time is bound to widen the electricity demand—supply gap, because large amounts of electricity would be continuously required to run those plants. It is important to note that the power-dependent critical segments of the wastewater treatment plants such as the sludge pumps and the aerators should *not* stop functioning under any circumstances. Otherwise, there would a host of adverse effects in the wastewater treatment process, since anaerobic conditions and the resultant putrefaction of the wastewater and sludge would be created in the above-mentioned critical plants of the plant.

Dams of adequate height need to be constructed in the case of hydroelectric (hydel) power projects, so that enormous quantity of water can be impounded in the reservoir. The disadvantages of the hydel power projects and their supporting water reservoirs—such as loss the of fertile land due to submergence, water disputes between riparian States, problems related to the displacement and rehabilitation of large population from the areas of submergence, and the *possibility* of damage to the impounding dams due to reservoir-induced (or –related) seismicity (RIS)—have already been discussed in an earlier chapter. Besides, the weak monsoons and the resultant poor rainfall may lead to inadequate inflows due to scarce surface runoffs into the reservoirs. The reservoir levels in many major hydel projects in recent years (for example, between 2000 and 2003) have gone alarmingly low, necessitating restriction on the electricity generation in those hydel projects. An option for more hydel projects for power generation is fraught with umpteen uncertainties and, therefore, it would not be an ideal solution for the augmentation of power resources.

The enormity of the coal requirement for the purpose of fuel in the thermal power plants has been discussed a little earlier. The major disadvantage of the thermal plants is that when the fossil fuels (coal and gas) are used to boil the water to convert it into the steam that runs the turbines, those fuels are permanently burnt and are lost. If adequate steps to conserve the coal reserves were not taken, the reserves of coal wealth in the country would fast deplete, since the thermal power plants are the major consumers of coal reserves. For this reason, additional thermal power stations too would *not* be a viable solution to bolster the energy supplies to the wastewater treatment plants and the seawater desalination plants.

Ideally for a country of India's size, and in a country that abounds in vast continuous stretches of wastelands punctuated by natural forests, installation of a more number of nuclear power plants would be the only pragmatic solution to generate additional quantum of electricity to augment the energy source that would keep the wastewater treatment plants and desalination plants continually operational. If the proposed wastewater treatment plants and desalination plants are to be distributed evenly all over the country, then the electricity generated by these nuclear power stations can feed the wastewater and desalination plants through the national grid of electricity. It may not be possible to draw distinct lines of separation in the matter of power distribution to the various segments of human, agricultural and industrial activities (as we discussed earlier in the case of the water resource management). Therefore, the power that is generated in excess of requirement for the wastewater treatment plants and desalination plants can be diverted for distribution to the other activities.

Many people world over dread the power projects based on nuclear reactors. The scenario in India is no different. Often, these peoples' anxiety and fear can be ascribed more to their ignorance rather than rational thinking, and the diffidence on the capabilities of the experts in the country. The most frequently asked questions about the nuclear power plants are—(i) whether the effluent water emanating from the nuclear power plant is radioactive; and (ii) whether radioactive leakages are common hazards associated with nuclear power plants. If the claddings provided to the fuel and the fuel assembly of the nuclear reactors are in tact, the water that comes in contact with them *does not* become radioactive (see also Cole, 1988), and therefore the effluent water from the atomic power plant would *not* be radioactive. As has been discussed earlier, there are several in-built safety measures in the nuclear reactors; if these safety measures are implemented and the safety procedures are consciously followed, there is absolutely *no* possibility of radioactive leakage from the nuclear reactors. Safety measures are prevalent in adequate measures in every plant, and the nuclear reactor and power plants are no exception. The fact is that the nuclear industrial units have the maximum number of safety devices compared to all the industrial plants.

To a large extent, the self-styled environmentalists and the imbecile media have been largely responsible for spreading canards about the *possible* ill effects of radioactive leakage and disasters in the nuclear industry. They have shirked in their responsibility of educating the people about the positive aspects of the nuclear power projects, which create totally pollution-free environment, enrich the water resources by recycling of wastewater, provide us with plenty of uninterrupted power, and also help conserve our riches of fossil fuels, agricultural land and natural forests. We have to take up the challenge of reaping the benefits bestowed on us by the Nature in the form of bountiful atomic minerals and establish more and more nuclear reactors and atomic power plants. That would not only help us augment our power and water resources, but also demonstrate that we are independently capable of handling some of the complex but utilitarian technologies such as the nuclear technology for power generation.

The world now acknowledges and respects India as a nuclear power to be reckoned with. We have shown to the international agencies that monitor the activities of the other countries on the nuclear front that we have the capability in the field of research and development in the nuclear science. We have achieved these by the two Pokhran nuclear experiments. We have achieved these by installing a small number of nuclear power plants that are fully operational now. We have achieved these by demonstrating that we can handle complex devices such as atomic reactor aided power plants without glitches and mishaps. Our prudence lies in cashing in on this nuclear status, and establishing our unquestionable credentials in nuclear technology. Setting up of many more nuclear power plants is the best way to do this.

The choice on the type of nuclear reactors (such as Magnox, gas-cooled, CANDU, water-cooled and fast-breed reactor types) for power generation would depend primarily on the type and the availability of nuclear fuel on a continuous basis. Further, much depends on our capacity to convince the international atomic regulatory agencies on our requirement of energy resources for the water resource augmentation. Availability of the nuclear fuel has to be ensured on a priority basis throughout the life of the nuclear power plants. This requires a thorough exercise of making an inventory of our existing reserves of uranium and thorium, and also gear up our efforts to locate more mineralised belts of atomic minerals.

Many countries have developed strategies for their economic growth taking a distant vision for the year 2020 (Abdul Kalam, 1998). Those countries propose to achieve their envisioned growth mainly through the areas of agriculture and industries. Further, infusion of advanced technologies in all spheres of their activity is the foremost item in their agenda for the vision of their country for 2020. Not to be left far behind, India too has similar agenda for 2020 and is continually striving in that direction. However, much needs to be done in the area of water resources augmentation, to which availability of power is crucial. It would *not* an avaricious proposition if we envisage that in the year 2020 at least a minimum of 70 percent of the power supply available on the national grid to various segments of activity in the country, including the power supply to a large number of wastewater treatment and seawater desalination plants, should come from the *nuclear source*. If we work towards that mission, in the year 2020, India would possess adequate amounts of water and electricity, which would be able to support the needs of agriculture and industries of all kinds, and also the requisite quantities of protected water source to meet the daily demand of potable grade water by every citizen of the country.

Chapter 18

In this concluding chapter, we shall:

- Review the responsibility of various strata of the society in water-related matters.
- Consider important aspects of river interlinking.
- Discuss broad aspects of Water Policy.
- Make a Mission Statement concerning water resource.

WATER RESOURCE AUGMENTATION AND MANAGEMENT: WHOSE RESPONSIBILITY?

Civil engineers carry out planning and installation of the water work facilities and the distribution network of water supply setups, under the aegis of the municipal or the corporation authorities. The civil engineers are involved in the planning, designing and construction of water filter and water treatment plants too. Execution of the road network in both urban and rural environs is also done by the civil engineers. Besides, the urban development authorities employ a host of civil engineers to look after a variety of functions relating to town planning, road and communication networks within the metropolises, cities and towns, provision of amenities, as also monitor whether the constructions are carried out as per the approved plans. Despite the presence of such a large army of civil engineers in various organizations dealing with the urban, town and rural development matters, it is strange that year after year, so much of storm water and surface runoff water flood the human habitats during the monsoon period. Further, by the end of the monsoon and even in the course of the dry spells during monsoon period, the cities, towns and vast stretches of sub-urban lands wear a parched look, since the storm water and the surface runoff water are mostly barred from percolating to the groundwater system due to the absence of groundwater recharge pits and trenches, or the rooftop rain harvest system to collect the rainwater as and when it reaches the land.

The metal or concrete cover on the roads does not allow the storm water to seep down below the soil water zone. The civil engineers all over the country, especially in the arid and semi-arid zones, would do well if they realize that provision of trenches or pits on either side of the road at the time of laying and/or maintenance of the roads would yield positive effects in groundwater recharging. Likewise, strict monitoring by the civil engineers on the implementation of the Water Policy, which stipulates the provision of rooftop rainwater harvesting system in apartments, group housing and individual residences would greatly ease the stress on the urban water supply system. The responsibility of building bunds and *johads*, as well as regular desilting operations of the tanks, reservoirs and lakes lies entirely on the civil engineers of the public-works and irrigation departments. For innovating and improving the designs of appropriate water filter/treatment plants to suit the needs of the country's urban and rural communities, no professionals other than the civil engineers are better equipped.

It needs to be stressed that the role of the civil engineers of this country in augmenting the surface water, soil water and groundwater resources is most pivotal. With far-sightedness, fertile imagination and ethical execution of the public works undertaken, the civil engineers can organize and see through water augmentation movements in the country by which they can:

1. Develop the watersheds and groundwater basins by construction of percolation structures like trenches and pits, as also innumerable tanks and reservoirs to harvest the rainwater.
2. Build the integrated rooftop rainwater harvest-cum-groundwater recharge structures in the existing urban sprawls by minimal altering of the existing structure of the building and at affordable low costs, so that more urban population is encouraged to adopt water resource augmentation by rainwater harvesting and groundwater recharging in the local area around their residence.
3. Innovate low-cost techniques and invent cost-efficient building materials in the rooftop rainwater system so as to make the system affordable by the masses of the Indian population.
4. Function as vital links between the government and the semi-skilled and unskilled work force in organizing movements such as desilting of tanks, construction of bunds and *johads,* and digging of the groundwater recharge trenches and pits on either side of the roads and highways, to facilitate maximum spread of storm water on the land resulting in efficient groundwater recharging over vast areas.

The civil engineers in the public works, water supply, irrigation, and roads and building departments need to realize that they present themselves as the governments' representatives to the people and therefore, are expected to assist the effective implementation of the governments' Water Policy in letter and spirit.

A large number of mechanical engineers hold the key to the successful installation of wastewater treatment plants and seawater desalination plants. Building efficient wastewater treatment and water augmentation plants to suit the Indian conditions of climate and to meet the heavy demand of large quantities of water by the masses require highly efficient mechanical engineers and dedicated mechanics. Since the water augmentation machines have the capacity to function 24 hours a day for several days at a stretch, the responsibility of these engineers and mechanics extends to work under tough conditions in any part of the day or night, so that water augmentation from these mechanical installations takes place on a continuous basis. The aerators and sludge pumps of wastewater treatment plants, and various components such as stills, chambers, valves, condensers and the delicate pipe-lining work of the seawater desalination plants need continuous innovation to reduce the cost, improve the efficiency and enhance the output of water in terms of quantity. The mechanical engineers and mechanics need to bring about innovations in these machines and their parts in such a way that hassle-free installation of these machines takes place in large numbers, followed by their periodic maintenance.

Strong structural support to the mechanical installations such as wastewater treatment and seawater desalination plants can be given only with the aid of appropriately strong civil structures. Therefore, the mechanical engineers need to work in perfect harmony with the structural and civil engineers as a well-coordinated team. As such, every task in the water resource management is a teamwork involving a variety of professionals and they have to work with a spirit of camaraderie providing vital links to one another in achieving the mission of water resource augmentation.

If the government on its part wants to correct the situation of allowing the cost of drinking water to escalate to reach at par with that of milk, perhaps the one and the only way to do so is to generate large quantities of potable grade water so that the drinking water is available liberally in every part of the country. Such potable grade water needs to be totally free from micro-organisms that cause water-borne diseases. One way of achieving this mission, besides strengthening the distribution system of protected water supply only for drinking and cooking purposes, is by flooding the Indian consumer markets with low-priced ultraviolet ray (UV)-based water treatment units in such a way that the common man can afford to buy it with ease, as a one-time investment.

The advantages of UV-based water treatment machines should be popularized through electronic media to the masses of population. Display of advertisements on UV-based water treatment machine to popularize them among the masses of population must be made mandatory for the electronic media, which make huge profits through advertisements of other consumer goods. All the multiple channels of the electronic media should adopt an ethical practice for some time of the day by voluntarily sparing a few seconds intermittently to beam the advertisement highlighting the positive aspects of rooftop rainwater harvesting, groundwater recharging, water treatment machines including the UV-based treatment units. It is obvious that the government-controlled information and broadcasting agencies in the country has a major responsibility in this matter.

Besides, the government has to ensure that the UV-irradiated, pathogen-free protected water is made available to the Indian masses for drinking as a welfare measure at a subsidized cost or, preferably, free of cost for a certain period of time. This facility can be gradually phased out till a sizable population has installed the UV-based water treatment units in their homes. Subsidy and sales tax exemption on the UV-based water treatment machines would facilitate a large number of populations to buy them for domestic installation. Such an action would serve the purposes of:

1. Sparing the common man from the exorbitant cost for drinking water.
2. Eradication of recurrent incidence of water-borne seasonal diseases such as diarrhoea, intestinal worm infestation and gastro-enteritis.
3. Encouraging the masses of population to install water treatment units at home so that they do not depend on the government for their drinking water needs.

Manufacture of such a large number of UV-based water purifiers is the responsibility of a vast number of electronic engineers who would continuously strive to introduce innovations in their designs, internal components and functional efficiency to treat a wide range of water quantities, from small amounts to serve a family to vast quantities to serve a large community such as a whole village or the dwellers of urban apartments. The electronic engineers, thus, have a crucial role in the society in the next decade in improving the drinking water quality and its cost, and eradicate a host of water-borne ailments. They have a mammoth responsibility to manufacture a few millions of UV-based water disinfectant machines at low cost and make them popular in every household of the whole country.

When it comes to identifying groundwater basins, water-bearing geomorphologic structures and aquifers, there can be no other persons better equipped than the hydro-geologists, geoscientists and remote-sensing scientists. The priority of these scientists during the next decade would be to *systematically* (in contrast to local explorations as conducted at present) locate groundwater basins, structures and aquifers over vast terrains, cutting across State boundaries. It is also essential that these geoscientists compute the reserves of groundwater in the new groundwater basins located by them on the basis of empirical field data using well-established scientific methods like geophysical and pumping tests, so that their findings gain authenticity. The approach of these scientists in their pursuit towards water augmentation employing the scientific inputs in their respective fields is inexpensive and largely non-invasive, and the results come much faster than the water augmentation by wastewater treatment and desalination methods. Therefore, a stupendous responsibility lies on these scientists in their endeavours to water resource augmentation.

The practice of sustainable agriculture is one sure way of conserving large quantities of groundwater, and the modern irrigation methods such as drip and sprinkler irrigation would undoubtedly be effective ways to bring about desirable changes in sustainable agriculture. Agricultural practice using Precision Farming technology is an area which is capable of ushering in a revolution in India, not only in terms of water conservation but also advanced technologies, particularly Information Technology and space science, making deep inroads into conventional segments like agriculture. The onus of introducing Precision Farming in the Indian agriculture fully rests with the veterans of Information Technology, computer specialists (both hardware and software), space scientists and remote-sensing scientists. The agricultural tool manufacturers too play a vital role in manufacturing cost-effective agricultural implements linked to the Variable Rate Applicators in Precision Farming as also the water-sprinklers and -drippers.

Normally, the industrial and technological establishments demand logistics and infrastructure from the government authorities for setting up their missions in any part of the country. Contrarily, sustainable agriculture practices such as Precision Farming and sprinkler irrigation expect the industrial, scientific and technological establishments to create congenial logistics and infrastructure and assist the government in infusing the successful and proven results of the technological researches in conventional segments of human activities, as for example in agriculture. Though India has every right to boast that she harbours a sea of knowledge workers distributed evenly all over the country, when it comes to transfer of the results of researches in science and technology, we have the dubious distinction of a pathetic track record too, as many of the original research findings in science and technology have remained on paper as fantasies without much of their application to meet the needs of the society. The reason for this is many. Whatever the reason, this trend should end. The responsibility of blending the results of researches in science, industry and technology rests squarely on the researchers themselves, and if they fail in this mission, they have to blame themselves for their failure in their mission of getting their research results highlighted. The researchers often enjoy the government's funding for basic and advanced researches and therefore they are morally bound to spare at least a part of their earnings for highlighting the results of their research. In the current era of .com and websites, ownership of self-supported websites to highlight and patent the research findings of individual scientists, industrialists, entrepreneurs and technologists has become much easier than ever before.

Sensor technology in our satellite missions need to be improved further in a way that very high resolution space data comparable to the Ikonos imagery are generated in a routine manner as also made available to those who are in *genuine need* of them for practicing sustainable agriculture and Precision Farming technology. Since the satellite data of high precision, and spectral, spatial and temporal resolutions are pre-requisites for Precision Farming, the onus of making the space data of appropriate resolutions available on sale to its practitioners is on all the concerned government organizations.

A lot many computer professionals like software developers and specialists in handling Geographic Information System (GIS) software face the responsibility of extending support to the Precision Farming scientist and the landowners who have lent, or would lend in future, their land holding for the conduct of the high precision technique. Same is the case with those who have the exposure to handle the Global Positioning System (GPS) Receivers and accurately interpret the data collected from those gadgets. Who else is better equipped to develop algorithms to interpret, classify and compute the standing crops, normalized difference vegetation index (NDVI), pest and weed attacks, and the water- and nutrient-contents in the fields than the computer professionals? The responsibility of launching the second Green Revolution of sustainable nature with simultaneous conservation of water resources in agricultural activity and catapulting the Indian agriculture into Information Technology depends heavily on the efficiency of our computer professionals and their willingness to take up their profession in rural setup in India in close association with the farmers and landowners.

In the same breath, it needs to be emphasized that agricultural tool manufacture in the country needs to undergo a revolution of sorts, as it has to be made essentially GIS- and GPS-linked so that variable rate application (as described in the context of Precision Farming) of water and nutrients to the needy fields becomes possible. So much so, even the drip and sprinkler irrigation needs to be GIS-linked so that the water that is sprinkled is optimal for the crop need, thus aiding effective water conservation.

Successful implementation of sustainable agriculture practices as well as technological missions in the lines of Precision Farming cannot be achieved without the active support of the local top-level administrative machinery such as the office of the Collector, for it is the rank and file of these offices who will be responsible to:

1. Influence the farmers and landowners to participate in these joint missions.
2. Organize co-operative movement in land holdings and bring several pieces of inter-connected lands under the scope of Precision Farming to ensure increased agricultural productivity and Win-Win situation for all concerned, *i.e.*, the farmers, government and the scientist-technologist team that conducts the technological mission.
3. Work out equitable profit distribution arising out of the technological mission to the fullest satisfaction of all concerned, as also enhance revenue to the government.
4. Conserve the national wealth like water in the water-intensive human activities such as agriculture.
5. Create and archive a wealth of knowledge-based data for Information Technology to facilitate the right type of decision in the agricultural practice in future.

The term *anti-incumbency* has gained currency in recent times in India to refer to the overall inability of the government to rise to the increasing expectations of the masses of population. A critical analysis of the phenomenon called as the 'anti-incumbency' would reveal that this term is used with the same connotation as 'misgovernance'. The issue of water unavailability for day-to-day use would ever remain in the minds of the general public as the foremost among the anti-incumbency factors. Undoubtedly, the near-drought years 2002 and 2003 have been extremely difficult for the governments because providing adequate water for the nationwide agricultural activities and the day-to-day needs of the entire population of the country had indeed become thorny issues on account of weak monsoon resulting in deficient rainfall and depleted groundwater resource. Furthermore, the disputes over water sharing between the riparian States have remained unresolved even in 2004. The population in major parts of the country had faced extreme hardships due to inadequate supply of water for their daily needs during those two summers. Those in power before the elections had cited that as one of the major reasons for the electoral losses in the 2004 elections.

River interlinking

In the recent past, around the year 2002, the central government had taken a route to manage the country's water resources through the inter-linking the peninsular Indian rivers, with an idea of transferring surplus water from one river to the water-deficient areas. The idea of river interlinking and the concept of 'National Water Grid' are not new, as these aspects are discussed time and again, ever since Rao (1975) had mooted these ideas. There have been enormous changes in the scenario relating to the inter-State relations, politics, social imbalances, climatic conditions and acquisition of technological capabilities since 1975 when K. L. Rao first highlighted this idea. At the outset, though this approach to the water resource management sounds attractive, it is bound to prove more tortuous and complex, yielding uncertain results, for the following reasons:

1. Complex meteorological phenomena like the earlier discussed Climate Change, Global Warming, El Nino and atmospheric disturbances have been causing unpredictable navigation of atmospheric wind systems over the Indian sub-continent. These phenomena have been cumulatively responsible for:

   (*a*) Marginally altering the biome characteristics.

   (*b*) Recurrence of weak or failing monsoons.

   (*c*) Absence or deficiency of rainfall.

   (*d*) Dry and arid types of climates.

   (*e*) Creating drought-like atmospheric conditions over thousands of hectares of arable land.

   The very availability of surplus water in peninsular Indian rivers has, thus, become questionable or highly uncertain due to these meteorological factors.

2. The link canals would intersect several *faults* (geological weaker planes) and seismically active zones. The civil engineering remedial measures to prevent possible damage to the link canals on account of the seismicity or movements of land along these faults are extremely challenging and therefore the cost thereupon is bound to escalate to unimaginable limits.

3. The matter of sharing the water from the link canals both *enroute* and at the destinations of the canals is most likely to lead to disputes

in the lines of the current river water disputes between the riparian States, and therefore, there would be difficulties in achieving the purpose for which the project is conceived.

4. Heavy compensation to be paid to the owners of the land through which the link canal would pass through is likely to be a dampener for the project. Further, attempts by the government for the acquisition of the lands to route the link canals and the quantum of compensation money would in all likelihood lead to a spate of court cases. Looking at the track record of similar cases, it would seem that the settlement of these cases might take several years or decades.

5. Illegal tapping of water from the link canals by some of the agriculturists and industry-owners on a large scale would be a drain on the State exchequer and lead to revenue loss. Further, such illegal tapping would hamper water supply to the water-deficient terrain that is identified as the major beneficiary area under the river link project.

6. The government and its engineering personnel who execute the construction of the link canals are most likely face stiff resistance for going ahead with building the canal from the self-styled 'human activists' or 'environmentalists' who might disrupt the canal-related work under the pretext of inadequate monetary compensation to the lands, loss of agricultural lands, loss of forests, disturbance to the wild life and so on.

In a nutshell, the river linking projects are essentially water resource *management* projects rather than water resource augmentation projects. Until we build up a large resource of water in the country by *augmentation* techniques, the water resource management through river interlinking would be untenable. Therefore, water resource augmentation in the whole country should be our top priority rather than the management of water resource, which is currently becoming scarcer day-by-day. Until the time we *augment* our water resources, the exercises pertaining to its management to the various segments of human life become somewhat futile and meaningless. Under the complex situation relating to the water sharing between the States, if those in the authority depend to augment and replenish our water resources only on the natural phenomenon of monsoon rainfall on which we have absolutely no control, they choose to play a risky game on the aspirations of a billion-strong population of the country. Though the river-linking projects may sound attractive, under complex conditions of human and social relations prevailing in our country, in fact they are unimaginative adventures performed at heavy costs in terms money and human efforts.

Wastewater treatment plants and seawater desalination plants, in contrast to river interlinking, are bound to *augment* the water resources available for day-to-day human activities. With steady annual investments in terms of funds and human efforts over a period of about fifteen years, an unimaginable quantity of water that would be sufficient to carry out our day-to-day human activities, as also agricultural and industrial activities, can be accreted to our water resources. To add strength to this, let us take a look at the approach suggested by Abdul Kalam (1998) to visualize India in the club of Developed Nations in the year 2020. One of the main routes to attain that status, as suggested by Abdul Kalam is through the *'advanced technologies'*.

Besides, it is the advanced technologies that are developed and employed to enrich the economic conditions in the country that attract the foreign direct investments. It is well known that the foreign direct investors consider favourable infrastructures and logistic supports (such as the uninterrupted supply of water and power provided to the industrial units into which they would invest) as congenial conditions to make their investments. Therefore, the current priority of the governments should be to encourage, nurture and proliferate water augmentation technologies such as wastewater treatment plants for wastewater recycling, and seawater desalination plants along the seacoasts to generate good quality water. The responsibility for this squarely lies on the governments, both in the States and the Centre.

The executives of the government should realize that the face of the governments is visible to the foreign direct investors through the faces of the country's engineers, scientists and technologists. Additionally, the government's efficiency is seen through the efficiency of the group of people who build technological missions to augment the water and power resources, which together constitute the infrastructure of successful and economically strong nations. Therefore, if the executives of the governments sincerely want successful implementation of the missions relating to water augmentation technologies to raise the quality of the infrastructure standards and thereby attract the foreign direct investments in the country, they should be serious in formulating concrete plans involving the technically qualified people of proven capability in all the abovementioned engineering branches and draw sustainable policies on water resource augmentation with a vision for the year 2020.

Our Water Policy should envisage steady annual increments in the generation of recycled and desalinated seawaters commensurate to the funds and human resources pumped into these missions. These annual increments should be in such a way that by the year 2020, at least fifty percent of the water resources available for various human activities in our country would come from wastewater treatment and seawater desalination plants. The onus for organizing this is entirely on the governments, both in the Centre and in the States.

Before winding up a discussion on the Water Policy relating to the potable grade water and the water for the human, agricultural and industrial activities, a brief summary on the approach to its formulation *to suit the current times* will be very much pertinent. Such a Water Policy should address taking into account the following two major aspects:

1. *Water Resource Augmentation.*
2. *Water Resource Management.*

Those who are at the helm of affairs dealing with the matters relating to the water resources in the country should clearly differentiate between the two and initiate step-by-step protocol in each item of work proposed, as per the merit of each proposal as to whether it addresses water augmentation or water management. In the same breath, it needs to be emphasized at this juncture that our efforts in current times, say up to 2020, should be finely focused to water resource *augmentation* (rather than management) to build a huge resource of water in the entire country. This is possible by steadily launching aggressive measures to:

(*i*) Install several seawater desalination plants on either (east and west) coast, preferably more number of them on the east coast considering the severity of water scarcity in the cities and towns located on the eastern part of the country (for example, Madurai, Chennai, Nellore, Ongole and Visakhapatnam).

(*ii*) Install sufficient number of wastewater treatment plants (both municipal and industrial) evenly distributed all over the country in such a way that:

1. The treated water is distributed in sufficient quantities for all the human activ ities where the water quality is not a concern.
2. The treated clean water can be led into our heritage water bodies like rivers and lakes, to restore their pride, purity and sanctity, which are currently under serious threat.

(*iii*) Install several filtration-cum-treatment plants to treat the hard water.

(*iv*) Arrange for the manufacture of innumerable ultraviolet ray-based water treatment machines and install some of them to facilitate easy availability of pathogen-free drinking water to the common man.

(*v*) Revive the practice of building tanks, ponds and reservoirs to collect surface runoff rainwater.

(*vi*) Insist upon the construction of the integrated rooftop rainwater harvest-cum-groundwater recharge structures in the residences.

(*vii*) Provide for trenches and pits on either side of the roads and national highways to facilitate maximum water spreading during monsoon, thereby charging the depleted groundwater basins periodically.

The *management* component of the water policy has to necessarily address the following aspects concerning:

(*i*) Recognition of desirable water qualities in various human activities and stipulation of adequate quantities of the desirable quality water in concerned activities.

(*ii*) Time to time budgeting of the available water resources to the human activities, in tune with the quality desirable in the human activities.

(*iii*) Implementation of sustainable agricultural practices and promotion of water conservation methods, by sponsoring and promoting technologies such as Precision Farming.

(*iv*) Promote drip and sprinkler irrigation methods wherever feasible and encourage the farmers to adopt those irrigation methods to pave way for water conservation measures.

(*v*) Organize regular programmes for desilting and deepening of the tanks, ponds and reservoirs, in close association with the civil engineers responsible for building those water-bearing structures.

(*vi*) Create a database on the available groundwater basins, aquifers and geomorphologic structures that are favourable for holding the groundwater.

(*vii*) Regularly interact with those dealing with water augmentation methods to take stock of the quantity of replenishment and depletion of the water resources, so as to plan uninterrupted supply of water to various human activities.

(*viii*) Treatment of optimum quantities of raw water to upgrade it into potable grade so as to conserve the raw water for prolonged times thereby keeping check on the misuse of good quality water.

(*ix*) Monitor the supply of recycled or treated wastewater to the type of industries or human activities in which it can find its rightful use, and collect the revenue from this exercise.

The responsibility of both augmenting and managing the water resources rests to a major extent on all the concerned technologists under the overall supervision of the government through its appropriate agencies or departments dealing with the matters of water. The presence of the governments is felt by, and visible to, the common man through the government departments providing basic facilities, such as water and electricity. It is the responsibility of the governments to closely monitor the functions these departments, as also ensure that the rank and file of these departments are above suspicion in the matter of their character ethics, from the *general public's viewpoint*. Otherwise, public antipathy in the form of 'anti-incumbency' would result, and the executives would have to vacate the corridors of government as has been happening in recent times.

The author of this book had the opportunity to interact with a large number of youth, mostly men in their twenties, living all over the country, during the trying periods of Kargil war and when there was a build up of war clouds consequent upon the December 2001 attack on our Parliament. A flash of patriotic spirit appeared on the faces of the these men when they were asked the hypothetical question as to whether they would go to the scene of war and fight the country's enemy. Most (if not all) of them offered themselves to go to the battlefield and do *whatever they can* to fight the enemy.

The country faces threats from external enemies only at times. But permanently, within the country, we come across several faceless and formless but veritable enemies. We can only *feel* the presence of these enemies. Most treacherous among these enemies are: *Hunger, Poverty* and *Disease*. In fact, sadly, a majority of the same members of our youth brigade who display unquestionable spirit of patriotism are themselves affected by these inimical factors. The patriotic feelings in our youth are unquestionable and they can perform any nation-building activity, if asked. In the same breath it has to be accepted that the vast potential of our youth in nation-building activities has remained largely untapped. There had been no cognizable movement in the country to channelize the power of the youth. Under a strong leadership whose character ethics is beyond doubt, the youth of the country are capable organizing themselves and get involved in national movements. One such potential national movement is the Water Resource Augmentation movement.

One can come to a systematic logical conclusion that being strongly agriculture-dependent, in our country, these three enemies—hunger, poverty and disease—which affect millions of people are, some way or other, related to the availability of plenty of good quality water for oral consumption and for agriculture. Our youth who show spontaneity in fighting the temporary external enemy, have to be prepared to fight these permanent enemies within the country. Potable grade water in plenty is the common weapon with which we can fight all the three enemies. The task of building up a vast resource of water, which can eradicate them, has to be taken up ungrudgingly by the youth of our country. That would offer some of the unemployed youth the chance of self-employment. They have a significant role in the watershed development by organizing themselves in small groups to build bunds and *johads* to harness the surface runoff. They have to assume the responsibility of designing and building filter plants in the lines of Iraq-type filters (described earlier) using locally available materials and generate potable grade water from raw water in sufficient quantities to cater to the needs of the communities in which they live.

The biggest responsibility of both augmenting and managing the water resource in our country rests, however, on *every citizen* of this country. It is all the more important that those citizens who plan to own a property, be it an independent house or one of the group houses or a flat in a multi-storied apartment complex, have to spend a little more money to provide for the rooftop rainwater harvest cum-groundwater recharge system. The munificence of the general public was on display all over the country several times in the past, at the times of threat to our country's defense from external source, and natural calamities such as earthquakes and killer cyclones. Liberal contributions flowed from millions of common people all over the country into the relief funds, often in unmanageable proportions, at trying times. Strangely, when it comes to allocating some funds for fighting the water shortage, many citizens resent the expenditure involved. Besides, they shy away from the responsibility of contributing their mite to a developmental cause of building up a rich water resource in the country by way of, for instance, rooftop rainwater harvest or groundwater recharge structures. This tendency of the citizens to show indifference and disinterest to the government-sponsored developmental programmes, such as mass movements for rainwater harvest and groundwater recharge, has to undergo a total change towards positive side.

If desired results in public missions and community development programmes are to be achieved in letter and spirit, the citizens, the sponsoring authorities and the human force that executes those missions and programmes are expected to belong to the category of *ethical persons*, a concept originally propounded by James and Jongeward (1971). The concept of being ethical persons has tremendous relevance and significance even in the current times, not only in our country but in the whole world as well. There is an urgent necessity that the citizens being ethical persons in the matters relating to the water resource augmentation and management. In current times, the *ethical persons* in the context of water augmentation and water management programmes are those who *preserve the health and dignity of the persons and of the human race*. In order to achieve this, the ethical persons of current times conduct themselves in the following ways:

1. Promptly pay the amounts due to the government and to the community of which he is a part. These due amounts include the water bills and other cess that are charged by the government, in order to meet the expenses on account of the community development and providing amenities to them.
2. Assist the government in achieving the missions such as water augmentation and whole-heartedly participate in those missions, abiding by various clauses stipulated to accomplish those missions.
3. Do not amass or waste the water or any other natural substance obtained from the Earth.
4. Install rooftop rainwater harvest and groundwater recharge systems within their habitats to augment the water resources in their immediate environs.
5. Promote all kinds of water resource augmentation techniques, popularize them in the community amidst which they live, and influence others to adopt these techniques.
6. Preserve the heritage, sanctity and cleanliness of the water bodies like rivers, lakes and reservoirs.
7. Protect, enhance and look after the well-being of the people, as also protect, enhance and look after the well-being of the natural *biomes*, as well as the *inanimate* and *animate* natural world of which they are also a part.
8. Work for an environment in which people can become winners by enhancing the available natural resources and putting them to use optimally.

In a democratic setup, the citizens can evolve into ethical persons only when they conduct their lives and transact with his fellow beings based on strong *character ethics*. In fact, in the case of the nationwide water resource augmentation and management movements, the character ethics of all concerned persons involved in planning and execution of the projects need to be of very high order and beyond doubt. Even a temporary lapse in observance of character ethics by them not only affects the availability of water to the individual citizens, but also successively in turn as a cascading effect, to the small community in which they live, then to the people of a large region, and finally to the entire country. Additionally, such a lapse would result in the very fabric of moral values on which the human life rests to give way to develop a decadent society.

The water resource augmentation and management exercises have to be continued in our country with a clear mission—'*Make India an agriculturally and industrially developed nation by 2020, after eradicating hunger, poverty and disease, and by transforming the whole nation "richly watered" using water resource augmentation technologies*'. Till then, we—the civil, mechanical and electronic engineers, scientists in the branches of space, agriculture, soil, remote-sensing and geosciences, information technologists, administrators, executives and citizens—resolve to work relentlessly, to realize our mission. In this joint endeavour, we shall mutually support and remind:

"Arise! Alert! Stop not till the Goal is reached!"

REFERENCES

Abdul Kalam, A.P.J. with Rajan, Y.S., 1998. India 2020 A Vision for the new millennium. Penguin Books.

Agarwal, Anil and Narain, Sunita. 1991. Global Warming in an unequal world. Centre for Science and Environment, New Delhi.

Agarwal, Anil., Narain, Sunita and Khurana, Indira (Editors). 2001. Making Water Everybody's Business. Practice and Policy of Water Harvesting. Centre for Science and Environment, New Delhi.

Agarwal, Anil. 2001. Water Harvesting in a New Age. (*In*: Making Water Everybody's Business. Practice and Policy of Water Harvesting. *Eds*: Anil Agarwal, Sunita Narain and Indira Khurana.) Centre for Science and Environment, New Delhi. pp. 1-7.

Ahern, M.J. and McMichael, A.J. 2002. Global Environmental Changes and Human Health. *In:* Issues in Environmental Science and Technology, No.17—Global Environmental Change (Eds: R.E. Hester and R.M. Harrison. Royal Society of Chemistry, UK, pp. 139-160.

Aly, Narmine H. and El-Fiqi, Adel K. 2003. Mechanical vapor compression desalination systems—a case study. Desalination, volume-158, page 143-150. Elsevier.

Ansari, M.H., Kelkar, P.S., Hasan, M.S. and Paramasivam, R. 1998. Drinking water supply through RO desalination plants in Ramanathapuram district of Tamil Nadu. International Specialized Conference on Water Quality and its Management, New Delhi. Oxford & IBH Publishing Co. Pvt. Ltd.

Antala, Shamjibhai. 2001. The Awakening. (*In*: Making Water Everybody's Business. Practice and Policy of Water Harvesting. *Eds*: Anil Agarwal, Sunita Narain and Indira Khurana.) Centre for Science and Environment, New Delhi. pp. 75-77.

Brown, Paul. 1996. Global Warming—Can Civilization Survive? Blandford, London.

Bureau of Indian Standards (BIS) 2003, Indian Standards Drinking Water–Specification (First Revision). Publication No. IS: 10500: 1991 Edition 2.1 (1993-01).

Buttner, Hannah. 2001. Learning the Value of Raindrop. (*In*: Making Water Everybody's Business. Practice and Policy of Water Harvesting. *Eds*: Anil Agarwal, Sunita Narain and Indira Khurana.) Centre for Science and Environment, New Delhi. pp. 228-232.

Cole, H. A. 1988. Understanding Nuclear Power—A technical guide to the industry and its processes. Gower Technical Press (England & USA).

Devasia, Leelamma. 2002. Rural Development with Special Emphasis on Women, Water and Environment. (*In:* Sustainable Water and Soil Management. *Eds:* S. Kunst, T. Kruse and A. Burmester). Springer. pp. 29-56.

Dhruva Narayana, V.V., 2002. Soil and water conservation research in India. Directorate of Information and Publications of Agriculture, Indian Council of Agricultural Research, New Delhi.

Dormuth, K.W., Gillespie, P.A. and Whitaker, S.H. 1995. Disposal of Nuclear Fuel Waste. (*In:* Wastewater Treatment and Disposal—Issues in Environmental Science and Technology. *Eds:* R.E. Hester and R.M. Harrison). Royal Society of Chemistry, Cambridge, pp. 115-130.

Dunglena. 2001. Rainwater Retains Relevance. (*In*: Making Water Everybody's Business. Practice and Policy of Water Harvesting. *Eds*: Anil Agarwal, Sunita Narain and Indira Khurana.) Centre for Science and Environment, New Delhi. pp. 207-210.

Dwivedi, R.S., Wani, S.P., Bhattacharya, A. and Navalgund, R.R., 2002. Space Technology inputs for Precision Farming. ISPRS Technical Commission VII Symposium on Resource and Environmental Monitoring (Hyderabad, December 2002), National Remote Sensing Agency. Abstract in page 19.

Friedman, Thomas. 2000. The Lexus and the Olive Tree. Anchor Books, New York. pp. 89-91.

Gangadhar, V. 2001. Riches of Rayalaseema. (*In*: Making Water Everybody's Business. Practice and Policy of Water Harvesting. *Eds*: Anil Agarwal, Sunita Narain and Indira Khurana.) Centre for Science and Environment, New Delhi. pp. 33-42.

Garbrecht, Gunther. 1976. Ways of improving the efficiency of water utilization in irrigation systems. (*In*: Applied Sciences and Development—Volume-8). Institute for Scientific Co-operation, Tubingen, FRG.

Heywood, I., Cornelius, S. and Carver, S. 2002. An Introduction to Geographical Information Systems. Pearson Education Limited, Essex, England.

Holmes, Arthur. 1945. Principles of Physical Geology. Thomas Nelson and Sons Ltd., Edinburgh.

James, Muriel and Jongeward, Dorothy. 1971. Born to Win. Signet. p. 301-302.

Jeyakumar, R. 2001. Quenching Chennai's Thirst. (*In*: Making Water Everybody's Business. Practice and Policy of Water Harvesting. *Eds*: Anil Agarwal, Sunita Narain and Indira Khurana.) Centre for Science and Environment, New Delhi. pp. 201-206.

Johnston, Cliff. S. 1992. Managing Risk from Chemicals in Liquid Effluents. (*In*: Risk Management of Chemicals. *Ed*. M. L. Richardson). The Royal Society of Chemistry, Cambridge.

Kazmi, Absar Ahmad. and Muttamara, Samorn. 1998. Water Quality Modeling of the Yamuna River. International Specialized Conference on Water Quality and its Management, New Delhi. Oxford & IBH Publishing Co. Pvt. Ltd.

Kulkarni, Anil, V., Randhawa, S.S., Rathore, B.P., Bahuguna, I.M. and Sood, R.K. 2002. Snow and Glacier Melt Runoff Model to estimate Hydropower Potential. Journal of the Indian Society of Remote Sensing, 30(4), pp. 221-228.

Kunst, Sabine., Mennerich, Artur., and Wichern, Marc. 2002. Wastewater Treatment. (*In*: Sustainable Water and Soil Management. *Eds:* S. Kunst, T. Kruse and A. Burmester). Springer. pp 91-136.

Kunst, Sabine, and Pathak, Namrata. 2002. Alternative Technologies for Sanitation, Recycling and Reuse. (*In*: Sustainable Water and Soil Management. *Eds*: S. Kunst, T. Kruse and A. Burmester). Springer. pp 183-219.

Mamoria, C.B. 1979. (Revised by Anupam Goel. 1998.) Economic & Commercial Geography of India. Shiva Lal Agarwala & Co., Agra.

Mittal, B.K., Gupta, M.S. and Prasad, Anupam. 1998. Water Quality Scenario in Damodar River Basin. International Specialized Conference on Water Quality and its Management, New Delhi. Oxford & IBH Publishing Co. Pvt. Ltd.

Murty, J.V.S. 1994. Watershed Management in India. Wiley Eastern Ltd, New Delhi.

Namasivayam, C. and Thamaraiselvi, K. 2001. Removal of 4-chlorophenol by adsorption on to industrial solid "wastes" Fe(III)/Cr(III) hydroxide and mud from aqueous solution. (*In*: Emerging trends in environmental science *Editor*: C.S.P. Iyer). Asiatech Publishers, Inc., pp. 268-272.

Parry, Martin and Livermore, Matthew. 2002. Climate Change, Global Food Supply and Risk of Hunger. (*In*: Issues in Environmental Science and Technology, No.17—Global Environmental Change *Eds*: R.E. Hester and R.M. Harrison.) Royal Society of Chemistry, UK, pp. 109-137.

Patel, J.V. and Joshi, M.D. 2001. Hydrogeomorphological mapping and Landuse Classification for the part of Baran, Rajasthan—A Remote Sensing Approach. National Symposium on Advances in Remote Sensing Technology with special emphasis on High Resolution Imagery (Ahmedabad, December 2001). Indian Society of Remote Sensing. Abstract No. WR-21

Pathak, Namrata. 2002. Water Treatment and Rainwater Harvesting. (*In*: Sustainable Water and Soil Management. *Eds*: S. Kunst, T. Kruse and A. Burmester). Springer. pp. 57-90.

Rajankar, Manish and Dolke, Yogini. 2001. Decline of a Grand Tradition. (*In*: Making Water Everybody's Business. Practice and Policy of Water Harvesting. *Eds*: Anil Agarwal, Sunita Narain and Indira Khurana.) Centre for Science and Environment, New Delhi. pp. 30-32.

Rao, K. L. 1975. India's Water Wealth. Its Assessment, Uses and Projections. Orient Longman, New Delhi.

Ray, S.S., Singh, Ashvir., Panigrahy, S., Singh, J.P. and Parihar, J.S. 2001. Precision Farming using high resolution remote sensing data—an exploratory study for Indian context. National Symposium on Advances in Remote Sensing Technology with special emphasis on High Resolution Imagery (Ahmedabad, December 2001), Indian Society of Remote Sensing. Abstract No.HR-13.

Razdan, Tej. 2001. A Languishing Legacy. (*In*: Making Water Everybody's Business. Practice and Policy of Water Harvesting. *Eds*: Anil Agarwal, Sunita Narain and Indira Khurana.) Centre for Science and Environment, New Delhi. pp.217-220.

Research, Reference and Training Division, 2003. INDIA 2003. Publications Division, Ministry of Information and Broadcasting, Government of India, New Delhi.

Sawyer, Clair, N. and McCarty, Perry, L., 1978. Chemistry for Environmental Engineering (Third Edition). McGraw-Hill Book Company.

Smethurst, George. 1988. Basic Water Treatment for Application World-wide (Second Edition). Thomas Telford Ltd., U.K.

Smith, Vincent. A. 1958. The Oxford History of India (Fourth Edition; *Ed*: Percival Spear). Oxford University Press, Chennai.

Stern, Peter. 1989. Rain-water harvesting. (*In*: Community Water Development. *Ed*: Charles Kerr). Intermediate Technology Publication, London, UK. pp. 73-76.

Strahler, A. N. 1984. Elements of Physical Geography. John Wiley & Sons, USA.

Tangri, A.K., Ram Chandra and Yadav, S.K.S. 2001. Deciphering climatic variability in Uttaranchal Himalaya through temporal monitoring of the equilibrium line of Gangotri glacier, using multi-date satellite data. National Symposium on Advances in Remote Sensing Technology with special emphasis on High Resolution Imagery (Ahmedabad, December 2001), Indian Society of Remote Sensing. Abstract No. WR 47.

Tebbut, T.H.Y. 1992. Principles of Water Quality Control. Pergamon Press, Oxford, U.K.

Thornbury, William, D. 1969. Principles of Geomorphology. Wiley Eastern Ltd., New Delhi.

Todd, David, K. 1959. Ground Water Hydrology. John Wiley & Sons, USA.

Toffler, Alvin., 1990. Power Shift. Bantam Books, New York, USA.

Try, P.M and Price, G.J. 1995. Sewage and Industrial Effluents. (*In:* Wastewater Treatment and Disposal—Issues in Environmental Science and Technology. *Eds:* R.E. Hester and R.M. Harrison). Royal Society of Chemistry, Cambridge, pp. 17-41.

Wasteland Atlas of India, 2000. Government of India, Ministry of Rural Development, Department of Land Resources, in collaboration with National Remote Sensing Agency.

WHO. 1993. Guidelines for drinking water quality (second edition), Volume 1, Recommendations. World Health Organization, Geneva.

Williamson, Ken. Reverse Osmosis Water Treatment. Agri-Facts—Practical Information for Alberta's Agricultural Industry. (pdf file).

Zhu, Qiang., Liu, Changming., Mou, Haisheng., Wang, Huilin., Kung, Hsiang-te., Wanjung, Zhang., Zhijun, Liu., and Jiansheng, Cao. 2001. Weak Water becomes a Growing Point. (*In*: Making Water Everybody's Business. Practice and Policy of Water Harvesting. *Eds*: Anil Agarwal, Sunita Narain and Indira Khurana.) Centre for Science and Environment, New Delhi. pp. 170-182

WEB SITES REFERRED TO

1. **On Climate**

   http://www.blueplanetbiomes.org/world_biomes.htm

   http://beringclimate.noaa.gov/reports/no_03.pdf

   Rodionov, S., Overland, J. and Bond, N. 2003. Climate Overview-2003.

2. **On Drought and El Nino:**

   http://www.drought.unl.edu/whatis/predict.htm

   U.S.Department of Commerce / NOAA / PMEL /TAO / El Nino theme page.

   NOAA El Nino Page and NOAA La Nina Page.

3. **On Electro-Dialysis**

   http://www.lenntech.com/electrodialysis.htm

   http://www.tifac.org.in/news/memb.htm (News and Views. Newsletter; Staff Publication: Membrane Separation Processes—Technology and Business Opportunities—by G. Srikanth, page 6-7).

   http://www.unep.or.jp/ietc/Publications/TechPublications/TechPub-8a/electro.asp (United Nations Environment Programme: Division of Technology, Industry, and Economics: Internatiocnal Environmental Technology Centre: Sourcebook of Alternative Technologies for Freshwater Augmentation in Africa: Water quality improvement technologies: Electrodialysis).

   http://www.usace.army.mil/publications/armytm/tm5-813-8/c-7.pdf

4. **On Eutrophication:**

   http:/www.epa.gov/maia/html/eutroph/html (of US Environmental Protection Agency)

   http:/www.umanitoba.ca/institutes/fisheries/eutro.html

5. **On Global Positioning system, GPS:**

   www.ces.uga/ pubcd/b.1186.htm (article by Glen C. Rains and Daniel L. Thomas)

6. **On Humidity, Evaporation and Transpiration:**

   http://www.physicalgeography.net/fundamentals/8i.html (Physical Geography.net.— Fundamentals of Physical Geography).

   http://ww2010.atmos.uiucedu/(Gh)/guides/mtr/hyd/evap/rxml

   http://www.wildwildweather.com/humidity.htm

7. On Membrane Types, Properties and Modules:

http://www.tifac.org.in/news/memb.htm (News and Views. Newsletter; Staff Publication: Membrane Separation Processes—Technology and Business Opportunities—by G. Srikanth, page 1-3).

8. On Multiple Effect Evaporation:

http://cape.uwaterloo.ca/che100projects/sea/med.html

http://www.lapump.com/html/desalination.html

http://www.oas.org/usde/publications/Unit/oea59e/ch21.htm

9. On Multi-Stage Flash Distillation:

http://www.aquatech.com/aquachem/Process/multistageflash.htm

http://www.e-monitoring.com/bws/desalinationTechnology multiStageFlashDistillation.htm

http://www.serve.com/damien/home/solarweb/desal/msf.html

http://cape.uwaterloo.ca/che100projects/sea/msf.html

10. On Precision Farming:

http://www.gov.mb.ca/agriculture/facts/fbc08s00.html (An overview of precision farming technology: Is it for you?—Manitoba Agriculture and Food)

http://www.peisland.com/agrtour/pre.html (Sustainable Agriculture Tour—Precision Farming)

http://www.ces.uga.edu/pubcd/b1186.htm ('Precision Farming, An Introduction' by Glen C. Rains and Daniel L. Thomas)

11. On Reservoir Induced Seismicity:

www.worldbank.org/html/fpd/em/hydro/rie.stm

www.civ.hw.ac.uk.research/swmg/hydrogeology_pages/aderson.html

(Aderson Nascimento and others).

www.quakes.uq.edu.au/-quakes/gen_info/networks/ris.html (Russell Cuthbertson)

12. On Reverse Osmosis

http://agbiopubs.sdstate.edu/articles/FS877RO.pdf

(Reverse Osmosis—by Freg Bergsrud, Bruce Seelig and Russell Derickson)

http://www.tifac.org.in/news/memb.htm (News and Views. Newsletter; Staff Publication: Membrane Separation Processes—Technology and Business Opportunities—by G. Srikanth, page 3-6).

http://hrosystems.com/text/ro/html

http://www.visionengineer.com/env/reverse_osmosis.shtml

http://www.unu.edu/unupress/unupbooks/80858e/80858E2M.GIF

13. On Seawater Chemistry:

www.usask.ca/geology/classes/ geol206/geol206rr2.html

14. On Seawater Properties:

http://www.guilford.edu/original/academic/geology/Seawater.html

http://sam.ucsd.edu/sio210/lect_2/lecture_2.html

15. On Vapour Compression Distillation:

http://cape.uwaterloo.ca/che100projects/sea/vc.html

http://www.serve.com/damien/home/solarweb/desal/vc.html

16. On Water Desalination Technologies & Distillation Desalination:

http://www.unep.or.jp/ietc/Publications/TechPublications/TechPub-8f/B/Desalination1-1.asp

(United Nations Environment Programme—Division of Technology, Industry, and Economics—International Environmental Technology Centre—Newsletter and Technical Publications—<Sourcebook of Alternative Technologies for Freshwater Augmentation in West Asia>

http://www.coastal.ca.gov./desalrpt/dkeyfact.html (Seawater Desalination in California).

http://www.us-israel.org/jsource/History/desal.html (Jewish Virtual Library—Seawater Desalination Projects: The Challenge and the Options to meet the water shortage)

INDEX

E

F

G

H

I

J

K

L

M

T

U

V

W

Y

Z